Principles
of Electricity

Arthur Morley
OBE, DSc, Hon MIMechE.
Formerly Professor of Mechanical Engineering, University College,
Nottingham and formerly His Majesty's Inspector of Technical Schools

Edward Hughes
DSc(Eng), PhD, CEng, FIEE.
Fellow of Heriot-Watt College, Edinburgh and formerly Vice-Principal
and Head of the Engineering Department, Brighton College of
Technology

Revised by W. Bolton
BSc, M.InstP, C.Phys.
Formerly Head of Research, Development and Monitoring at the
Business and Technician Education Council

Longman
Scientific &
Technical

Longman Scientific & Technical
Longman Group UK Limited,
Longman House, Burnt Mill, Harlow
Essex CM20 2JE, England

and Associated Companies throughout the world

First published 1953
Ninth impression 1960
Second edition 1964
Fourth impression 1967
Third edition 1970
Seventh impression 1984
Fourth edition 1986
Second impression 1987

British Library Cataloguing in Publication Data

Morley, Arthur
    Principles of electricity.——4th ed.
    1. Electricity
    I. Title   II. Hughes, Edward, *1888–*
    III. Bolton, W. (*William*), *1933–*   IV. Morley,
    Arthur. Principles of electricity in SI units
    537   QC522

ISBN 0 582 41375 3

Produced by Longman Singapore Publishers (Pte) Ltd.
Printed in Singapore.

Contents

Preface to First Edition

Electrical theory has been burdened with several systems of units. In 1901, it was suggested by Prof. Giorgi, an Italian scientist, that the confusion caused by these different systems could be eliminated by the adoption of the metre, the kilogramme and the second as the units of length, mass and time respectively and the adoption of one of the practical units, such as the ampere, as a fourth fundamental unit. This metre–kilogramme–second (M.K.S.) system in its rationalized form was adopted unanimously by the International Electrotechnical Commission in 1950; and in April, 1952, the Council of the Institution of Electrical Engineers recommended that this system 'should be employed by authors in papers submitted to the Institution and that all students of electrical engineering should become conversant with its use'. The rationalized M.K.S. system has, therefore, been adopted in this volume and, apart from the conversion table given on p. 383, no reference has been made to the C.G.S. electromagnetic and electrostatic systems.

The symbols and nomenclature are in accordance with the recommendations of the British Standards Institution and the Institution of Electrical Engineers; and for the convenience of students, the symbols and abbreviations used in this book have been tabulated on pp. xi–xiv.

The text contains 83 worked examples and 435 problems. Most of the latter have been taken from examination papers; and for permission to publish these questions, we are grateful to the Institution of Electrical Engineers, the City and Guilds of London Institute, the East Midland Educational Union, the Northern Counties Technical Examinations Council, the Union of Educational Institutions and the Union of Lancashire and Cheshire Institutes. The greatest care has been taken to eliminate errors in the text and in the answers, but should any mistakes be found, we shall be very grateful to have them brought to our notice.

We wish to express our thanks to Dr F. T. Chapman, C.B.E., M.I.E.E. and Mr E. F. Piper, A.M.I.E.E., for reading through the manuscript and making a number of valuable suggestions.

E. H.
A. M.

▬▬▬ Preface to Second Edition

Recent changes in the National Certificate scheme have made it necessary to revise the contents of *Principles of Electricity*. In the new Two-year Senior Course for the Ordinary National Certificate in Engineering, Electrical Engineering Science is a compulsory subject in the first year (O.1) whilst Electrical Engineering A and B are two optional subjects in the second year (O.2). This new edition has been planned to cover the whole of the syllabuses in Electrical Engineering Science (O.1) of the five examining unions and practically the whole of their Electrical Engineering A (O.2) syllabuses.

Much of the text has been rewritten, new sections dealing with electrolysis, thermionics, semiconductors and rectifiers have been added and the whole of the book has been re-set in clearer type.

The term *phasor* has been used instead of *vector* to represent the magnitude and phase of a sinusoidal alternating current or voltage. This change is in line with continental and American practice.

E. H.

▬▬▬ Preface to Third Edition

In 1960 the General Conference of Weights and Measures recommended that the International System of Units should be universally adopted. This system is an extension and refinement of the traditional metric system. It embodies features which make it logically superior to any other system as well as being more convenient in practice: it is rational, coherent and comprehensive.

In this revised edition, references to British units have been deleted, but a Conversion Table has been added, giving the relationships between these units and the corresponding SI units. The changes also include the substitution of *magnetic field strength* and *electric field strength* for *magnetizing force* and *electric force* respectively, and the use of the terms *hertz, tesla and siemens*. The *joule* is used as the unit of energy, whether mechanical, electrical or thermal, but the *kilowatt hour* is retained when the latter is the more convenient unit.

E. H.

▬▬▬ Preface to Fourth Edition

In this revised edition the opportunity has been taken to considerably increase the length of the book, and to revise some of the chapters, in order to give a complete coverage of all the electrical principles likely to be met by students taking the BTEC National Certificates and National Diplomas in Electrical Engineering, Electronics, Telecommunications and related subjects. New chapters have been added on transformers, circuit theorems, d.c. transients and three-phase supply. The electronics have been completely rewritten to reflect the change from valves to semiconductors and the chapter on electrical measurements extended to give a coverage of a greater range of measurements.

The opportunity has also been taken to revise the graphical symbols used in the circuit diagrams and line them up with BS 3939.

W. B.

Symbols, Abbreviations and Definitions

Based upon British Standard 3763: 1970 and *The International System of Units*, 1970 (HMSO).

Notes on the Use of Symbols and Abbreviations

1. A unit symbol is the same for the singular and the plural: for example, 10 kg, 5 V.
2. In a compound unit-symbol, the product of two units is indicated by a space occurring between the units; for example N m.
3. In a unit symbol involving a multiple or sub-multiple, there should be no space between the prefix and the unit; for example kA not k A.
4. A solidus can be used to indicate division of units, for example m/s. An alternative way of representing such units is to use negative indices; for example m/s could be written as $m\ s^{-1}$.
5. Only only one multiplying prefix should be applied to a given unit: for example, picofarad (pF), not micromicrofarad ($\mu\mu F$).
6. Subscripts can be added to quantity symbols to distinguish between different aspects of that quantity: for example N_s for the number of secondary turns on a transformer and N_p for the number of primary turns.

Quantity and Unit Symbols

Quantity	Quantity symbol	Unit	Unit symbol
Angular velocity	ω	radian/second	rad/s
Area	A	square metre	m^2
Capacitance	C	farad	F
		microfarad	μF
Charge or Quantity of electricity	Q	coulomb	C
Conductance	G	siemens	S
Current:			
Steady or r.m.s.			
value	I	ampere	A
		milliampere	mA
		microampere	μA
Instantaneous value	i		
Maximum value	I_m		
Difference of potential:			
Steady or r.m.s.		volt	V
value	V	millivolt	mV
		kilovolt	kV
Instantaneous value	v		

Quantity and Unit Symbols—(continued)

Quantity	Quantity symbol	Unit	Unit symbol
Maximum value	V_m		
Electric field strength	**E**	volt/metre	V/m
Electric flux	Ψ	coulomb	C
Electric flux density	**D**	coulomb/square metre	C/m^2
Electromotive force:			
Steady or r.m.s. value	E	volt	V
Instantaneous value	e		
Maximum value	E_m		
Energy	W	joule	J
		kilojoule	kJ
		watt hour	W h
		kilowatt hour	kW h
		electron volt	eV
Force	F	newton	N
		kilonewton	kN
Frequency	f	hertz	Hz
		kilohertz	kHz
		megahertz	MHz
Heat, quantity of	Q	joule	J
Impedance	Z	ohm	Ω
Inductance, self	L	henry (plural, henrys)	H
Inductance, mutual	M	henry (plural, henrys)	H
Length	l	metre	m
		kilometre	km
Magnetic field strength	H	ampere/metre	A/m
Magnetic flux	Φ	weber	Wb
		milliweber	mWb
Magnetic flux density	B	tesla	T
Magnetomotive force	F	ampere	A
Mass	m	kilogram	kg
Permeability of free space or Magnetic constant	μ_0	henry/metre	H/m
Permeability, relative	μ_r	—	
Permeability, absolute	μ	henry/metre	H/m
Permittivity of free space or Electric constant	ε_0	farad/metre	F/m
Permittivity, relative	ε_r	—	
Permittivity, absolute	ε	farad/metre	F/m
Power	P	watt	W
		kilowatt	kW

Quantity and Unit Symbols—(*continued*)

Quantity	Quantity symbol	Unit	Unit symbol
Pressure	p	newton/square metre	N/m^2
		pascal ($= 1\ N/m^2$)	Pa
Reactance	X	ohm	Ω
Reluctance	S	ampere/weber	A/Wb
Resistance	R	ohm	Ω
		microhm	$\mu\Omega$
		megohm	$M\Omega$
Resistivity	ρ	ohm metre	$\Omega\ m$
		microhm metre	$\mu\Omega\ m$
Specific heat capacity	c	joule/kilogram kelvin	J/kg K
Temperature, thermodynamic	T	kelvin	K
Temperature, Celsius	t, θ	degree Celsius	$°C$
Temperature interval	t	kelvin, degree Celsius	K, $°C$
Time	t	second	s
		minute	min
		hour	h
Torque	T	newton metre	N m
Voltampere	—	voltampere	VA
		kilovoltampere	kVA
Volume	V	cubic metre	m^3
		litre ($= 0.001\ m^3$)	l
Weight	W	newton	N

Abbreviations for Words

Term	Abbreviation
Alternating current } (used adjectivally	a.c.
Direct current } as in 'a.c. motor')	d.c.
Electromotive force	e.m.f.
Magnetomotive force	m.m.f.
Phase (used adjectivally as in '3-ph supply')	ph
Potential difference	p.d.
Power factor	p.f.
Root-mean-square	r.m.s.

Electrical Machines

Term	Symbol
Number of armature conductors	Z
— parallel paths	c
— pairs of poles	p
— turns	N

Greek Letters used as Symbols in this book

Letter	Capital	Small
Alpha	—	α (angle, temperature coefficient of resistance)
Epsilon	—	ε (permittivity)
Eta	—	η (efficiency)
Theta	—	θ (angle)
Mu	—	μ (micro, permeability)
Pi	—	π (circumference/diameter)
Rho	—	ρ (resistivity)
Phi	Φ (magnetic flux)	ϕ (phase difference)
Psi	Ψ (electric flux)	—
Omega	Ω (ohm)	ω (angular velocity)

Prefixes and Symbols for Multiples and Sub-multiples

Multiplying factor	Prefix	Symbol
10^{12}	tera	T
10^{9}	giga*	G
10^{6}	mega	M
10^{3}	kilo	k
10^{2}†	hecto	h
10†	deca	da
10^{-1}†	deci	d
10^{-2}†	centi	c
10^{-3}	milli	m
10^{-6}	micro	μ
10^{-9}	nano	n
10^{-12}	pico	p

* *Giga* is derived from a Greek word meaning giant and dictionaries state that its pronunciation should be the same as 'giga' in gigantic.
† Powers which are multiples of 3 are generally preferred, but because of the large value that may result in the case of some derived units, such as volume, these smaller multiples are permissible, but their use should be limited as far as possible.

Definitions of electric and magnetic units

The *ampere* (A) is that constant *current* which, if maintained in two straight parallel conductors of infinite length, of negligible circular cross-section, and placed 1 m apart in vacuum, would produce between these conductors a force equal to 2×10^{-7} N per metre of length.

The *coulomb* (C) is the *quantity of electricity* transported in 1 s by 1 A.

The *volt* (V) is the *difference of electrical potential* between two points of a conductor carrying a constant current of 1 A, when the power dissipated between these points is equal to 1 W.

The *ohm* (Ω) is the *resistance* between two points of a conductor when a constant difference of potential of 1 V, applied between these points, produces in this conductor a current of 1 A, the conductor not being a source of any electromotive force.

The *henry* (H) is the inductance of a closed circuit in which an e.m.f. of 1 V is produced when the electric current in the circuit varies uniformly at the rate of 1 A/s.

(*Note:* this also applies to the e.m.f. in one circuit produced by a varying current in a second circuit, i.e. mutual inductance.)

The *farad* (F) is the *capacitance* of a capacitor between the plates of which there appears a difference of potential of 1 V when it is charged by 1 C of electricity.

The *weber* (Wb) is the *magnetic flux* which, linking a circuit of one turn, produces in it an e.m.f. of 1 V when it is reduced to zero at a uniform rate in 1 s.

The *tesla* (T) is the *magnetic flux density* equal to 1 Wb/m^2.

Definitions of other derived SI units

The *newton* (N) is the *force* which, when applied to a mass of 1 kg, gives it an acceleration of 1 m/s^2.

The *pascal* (Pa) is the *stress* or *pressure* equal to 1 N/m^2.

The *joule* (J) is the *work done* when a force of 1 N is exerted through a distance of 1 m in the direction of the force.

The *watt* (W) is the *power* equal to 1 J/s.

The *hertz* (Hz) is the unit of *frequency*, namely the number of cycles per second.

CHAPTER 1

Units

1.1 The International System of Units (SI)

The International System of Units, known as SI in every language, derives all the units used in the various technologies from *seven* base units. These are given in Table 1.1.

Table 1.1

Quantity	Unit	Symbol
length	metre	m
mass	kilogram	kg
time	second	s
electric current	ampere	A
temperature	kelvin	K
luminous intensity	candela	cd
amount of substance	mole	mol

The candela and the mole are not dealt with in this book, and will therefore not be referred to again.

The SI* is a *coherent* system of units, i.e. the product or quotient of any two unit quantities in the system is the unit of the resultant quantity. For example, unit area (= 1 square metre) results when unit length (= 1 metre) is multiplied by unit length, or

1 square metre = 1 metre × 1 metre.

Further examples illustrating the coherent nature of SI are:

(a) unit velocity (= 1 m/s) results when unit length (= 1 m) is divided by unit time (= 1 s);
(b) unit force (= 1 newton) results when unit mass (= 1 kg) is multiplied by unit acceleration (= 1 m/s^2).

---

* It is incorrect to speak of 'SI system'.

In practice, it is often convenient to use a multiple or submultiple of the SI unit, but the choice should, in general, be confined to powers of ten* which are a multiple of ± 3, thereby effecting a considerable reduction in the number of multiples and submultiples,

e.g.   1 kilometre $= 10^3$ metres,
       1 millimetre $= 10^{-3}$ metre,
       1 micrometre $= 10^{-6}$ metre.

For *communication* purposes it is usually convenient to express a quantity in terms of a multiple or submultiple of the SI unit, thereby avoiding the use of very large or very small numbers. For example, it is convenient to express the calorific of coal as, say, 30 megajoules per kilogram rather than 30 000 000 joules per kilogram. On the other hand, for *calculation* purposes and especially in equations, it is advisable to express the quantities in terms of the SI units and to insert the unit symbol in brackets† after the numerical value. For example, if a wire has a length of 10 km and a cross-sectional area of 4 mm², and if the resistivity of the material is 0.02 $\mu\Omega$ m, then:

$$\text{length} = 10[\text{km}] \times 1000[\text{m/km}] = 10^4 \text{ m},$$

$$\text{cross-sectional area} = 4[\text{mm}^2] \times 10^{-6}[\text{m}^2/\text{mm}^2] = 4 \times 10^{-6} \text{ m}^2$$

and $$\text{resistivity} = 0.02[\mu\Omega\,\text{m}] \times 10^{-6}[\Omega/\mu\Omega] = 0.02 \times 10^{-6}\,\Omega\,\text{m}.$$

Hence, from expression (4.4) given on page 49,

$$\text{resistance} = \frac{0.02 \times 10^{-6}[\Omega\,\text{m}] \times 10^4[\text{m}]}{4 \times 10^{-6}[\text{m}^2]} = 50\,\Omega.$$

Only one prefix should be used in forming the multiple or submultiple of a given unit. Thus one thousand kilograms should be referred to as 1 megagram (1 Mg) rather than 1 kilokilogram, and picofarad (pF) should be used rather than micromicrofarad ($\mu\mu$F). The prefix should be attached to a unit in the numerator rather than the denominator. For example, stress and pressure should be expressed as, say, 5 MN/m² rather than 5 N/mm².

---

* It is important that students should be familiar with the use of indices and with the multiplication and division of numbers expressed in the form 10^n. For example, 2 000 000 can more conveniently be written as 2×10^6, and an error of a nought is far more likely to be made in the extended than in the shorter form.

When multiplying, say, 10^6 by 10^3, we add the indices; and when dividing 10^6 by 10^3, we subtract the indices: thus,

$$10^6 \times 10^3 = 10^9 \qquad \text{and} \qquad 10^6 \div 10^3 = 10^3.$$

In general,

$$10^a \times 10^b = 10^{(a+b)} \qquad \text{and} \qquad 10^a \div 10^b = 10^{(a-b)}.$$

† Brackets, [ ], are more satisfactory than parentheses, ( ), as they indicate more definitely that, say, 10[m] means 10 metres and not 10 × m.

1.2 SI base units

(a) Metre

The metre is defined as the length equal to 1 650 763.73 wavelengths of the orange line in the spectrum of an internationally-specified krypton discharge lamp. This definition reproduces the metre with an accuracy of one part in a hundred million (10^8). The metre was formerly defined as the distance between two lines on a certain platinum-iridium bar at 0°C. This bar is kept at the International Bureau of Weights and Measures at Sèvres, near Paris, and is still used as a reference standard. Its exact length is periodically determined in terms of the wavelengths of radiation from the krypton lamp.

(b) Kilogram

The kilogram is defined as the mass of a platinum-iridium cylinder kept at Sèvres. The mass of another body can be compared with that of the standard cylinder with an accuracy of one part in a hundred million (10^8) by means of a specially-constructed balance.

(c) Second

The second is defined as the interval occupied by 9 192 631 770 cycles of radiation corresponding to the transition of the caesium-133 atom. This definition enables the fantastic precision of one part in ten thousand million (10^{10}) to be achieved. The second is approximately 1/86 400 of the mean solar day.

1 minute [min] = 60 seconds [s]
and
1 hour [h] = 3600 seconds.

The minute and the hour are not decimal multiples of the second and are therefore non-SI units, but they are so firmly established in practice that their use is likely to continue indefinitely.

(d) Ampere

The definition of the ampere is given on page 20.

(e) Kelvin

The kelvin, named after Lord Kelvin (1824–1907), is defined as 1/273.16 of the thermodynamic temperature of the triple point of water. On this scale, the temperature at which ice melts under standard atmospheric pressure is 273.15 K.

A temperature change of 1 kelvin is precisely the same as the temperature change of 1 degree Celsius which is the unit of the Celsius (or centigrade) scale. The temperature of melting ice can be stated as 0°C or 273.15 K and that of boiling water under standard atmospheric pressure can be stated as 100°C or 373.15 K. In general, if t be the temperature of a body in degrees Celsius and T be the same temperature in kelvins, then:

$$t = T - 273.15 \qquad [1.1]$$

The symbols °C and K can be used to denote temperature interval* and also the number of temperature units above or below a specified datum. For example, a temperature rise from, say, 30 degrees Celsius to 50 degrees Celsius can be expressed as 20°C or 20 K. It will be noted that the convention is to use the term 'degree Celsius' but not 'degree Kelvin'— merely 'kelvin'.

The degree of precision in the determination of the base units of length, mass and time is really beyond human comprehension and is possible only with the elaborate apparatus available at national laboratories such as the National Physical Laboratory in this country. No student should be expected to memorize the figures given above.

For normal practical purposes, sub-standards of *length* and *mass* are used. The exact values of these sub-standards are periodically checked against those of the standards kept at the national laboratories. Sub-standards of *time*, such as clocks and watches, can easily be checked against time signals transmitted from observatories or from broadcasting sources such as the B.B.C.

1.3 Unit of force

The SI unit of force is termed the *newton* (symbol, N) to commemorate the great English scientist, Sir Isaac Newton (1642–1727). The newton is defined as *the force which, when applied to a mass of 1 kg, gives it an acceleration of 1 m/s²*. Hence the force F, in newtons, required to give a mass m, in kilograms, an acceleration a, in metres/second², is:

$$F = ma \qquad [1.2]$$

If a body is allowed to fall freely (i.e. if the air resistance is negligible), the acceleration at sea level in the vicinity of London is almost exactly 9.81 m/s², whereas at the equator the acceleration is about 9.780 m/s² and that at each pole is about 9.832 m/s². From the definition of the newton, it follows that the gravitation force at sea level on a mass of 1 kg is almost exactly 9.81 N in London, 9.78 N at the equator and 9.832 N at each pole.

---

* The abbreviation 'deg', used to express a temperature interval, is now regarded as obsolescent by the International General Conference of Weights and Measures.

The gravitational force on a body is termed the *weight* of that body and its direction is always towards the centre of the earth. The weight of a given object varies from place to place on the earth's surface and is smaller the further the object is from the centre of the earth. The *mass* of the object, on the other hand, is constant irrespective of its position in space.

For most practical purposes we can say that:

the weight of a body $\simeq 9.81\ m$ newtons   [1.3]

where m represents the mass of the body, in kilograms. The sign '$\simeq$' will be used in this relationship wherever it appears in this book, to indicate that the '9.81' may be an approximate value and that the precise value depends upon the latitude and height of the point at which the relationship is used. Thus, if a body having a mass of 1 kg rests on a table, as shown in fig. 1.1, the downward force exerted on the table, i.e. the weight of the body, is practically 9.81 N.

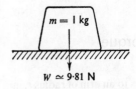

Fig. 1.1 Mass and weight

Example 1.1 *A force of 50 N is applied to a mass of 200 kg. Calculate the acceleration.*

Substituting for F and m in the expression $F = ma$, we have:

$50\ [\text{N}] = 200\ [\text{kg}] \times a$

$\therefore\quad a = 0.25\ \text{m/s}^2.$

Example 1.2 *A mass of 3 Mg is to be given an acceleration of $2\ \text{m/s}^2$. Calculate the force required, in kilonewtons.*

$\text{Mass} = 3\ \text{Mg} = 3000\ \text{kg}.$

$\therefore\quad \text{force} = 3000\ [\text{kg}] \times 2\ [\text{m/s}^2] = 6000\ \text{N}$

$\qquad\qquad = 6\ \text{kN}.$

Example 1.3 *A steel block having a mass of 80 kg rests on a table, the area of contact with the table being $2000\ mm^2$. Calculate (a) the downward force* *on the table and* (b) *the average pressure on the table, in kilonewtons per square metre.*

(a)   Downward force on table = weight of block

$$\simeq 80\,[\text{kg}] \times 9.81\,[\text{N/kg}]^*$$

$$= 784.8\,\text{N}.$$

(b)   Area of contact = 2000 mm²

$$= \frac{2000\,[\text{mm}^2]}{1\,000\,000\,[\text{mm}^2/\text{m}^2]} = 0.002\,\text{m}^2.$$

∴ average pressure on table $= \dfrac{\text{downward force}}{\text{area of contact}}$

$$= 784.8\,[\text{N}]/0.002\,[\text{m}^2]$$

$$= 392\,400\,\text{N/m}^2 = 392.4\,\text{kN/m}^2.$$

1.4 Unit of turning moment or torque

If a force F, in newtons, is applied at right angles to an arm of radius r, in metres, pivoted at a point O as shown in fig. 1.2,

turning moment or torque about point O
$$= T = Fr \text{ newton metres.} \qquad [1.4]$$

Fig. 1.2

1.5 Unit of work or energy

The SI unit of work is the *joule* (symbol, J), named after the English physicist, James P. Joule (1818–89), famous for his experiments on the relationship between mechanical and thermal energies. The joule is defined as *the work done when a force of 1 newton is exerted through a*

---

* This expression can alternatively be stated thus:

    weight of block $\simeq 80\,[\text{kg}] \times 9.81\,[\text{m/s}^2] = 784.8\,\text{N}.$

When the body under consideration is *stationary*, [N/kg] is better since it emphasizes the relationship between mass and weight, namely that a mass of 1 kg has a weight of approximately 9.81 N.

distance of 1 metre in the direction of the force. Hence, if a force F, in newtons, is exerted through a distance s, in metres, in the direction of the force,

$$\text{work done} = Fs \text{ joules} \tag{1.5}$$

$$1 \text{ kilojoule (kJ)} = 10^3 \text{ J}$$

and

$$1 \text{ megajoule (MJ)} = 10^6 \text{ J}.$$

If a force F is maintained tangentially at a radius r from axis O, as in fig. 1.2, the distance through which the point of application of the force travels in 1 revolution in $2\pi r$ metres,

$\therefore$ work done, in joules, in 1 revolution

$$= F \times 2\pi r = 2\pi \times Fr$$

$$= 2\pi \times \text{torque in newton metres.} \tag{1.6}$$

If a body of mass m, in kilograms, initially at rest, is given a uniform acceleration a, in metres/second^2, by a force F, in newtons, over a period t, in seconds,

$$\text{final speed} = v = at \text{ metres/second.}$$

Distance travelled = average speed × time

i.e. $$s = \tfrac{1}{2}vt \text{ metres}$$

$\therefore$ *kinetic energy* acquired by body $= F \times s = ma \times s$

$$= m \times \frac{z}{t} \times \frac{vt}{2}$$

$$= \tfrac{1}{2}mv^2 \text{ joules} \tag{1.7}$$

If a body having a mass m, in kilograms, is lifted vertically through a height h, in metres, and if g is the gravitational acceleration in metres/second^2 in that region,

$$\text{force required} = \text{weight of body}$$

$$= mg \text{ newtons}$$

$\therefore$ *potential energy* acquired by body

$$= \text{work done in lifting the body}$$

$$= \text{weight of body} \times \text{height}$$

$$= mgh \text{ joules} \tag{1.8}$$

$$\simeq 9.81mh \text{ joules.} \tag{1.9}$$

Example 1.4 *The work done in moving a body through a distance of 30 m is 600 J. Assuming the force to act in the direction of motion, calculate the* *value of the force.*

Since work done $= Fs$

$$\therefore \quad 600\,[\text{J}] = F \times 30\,[\text{m}]$$

$$\therefore \quad F = 20\,\text{N}.$$

Example 1.5   *A force of 70 N is exerted tangentially at the circumference of a pulley having a radius of 200 mm. Calculate (a) the torque and (b) the work done when the pulley is rotated through 20 revolutions.*

(a)   Radius of pulley $= 200\,\text{mm} = 0.2\,\text{m}$.

$$\therefore \quad \text{torque} = 70\,[\text{N}] \times 0.2\,[\text{m}] = 14\,\text{N m}.$$

(b)   From expression (1.6),

work done in 1 revolution $= 2\pi \times T$

$$= 2\pi \times 14\,[\text{N m}] = 88\,\text{J},$$

$\therefore$ work done in 20 revolutions $= 88 \times 20 = 1760\,\text{J}$

$$= 1.76\,\text{kJ}.$$

Example 1.6   *If a body having a mass of 20 kg is travelling at 10 km/h, what is the value of the kinetic energy possessed by the body?*

$10\,\text{km/h} = 10\,000\,\text{m/h}$

$$= \frac{10\,000\,[\text{m/h}]}{3600\,[\text{s/h}]} = 2.78\,\text{m/s}.$$

From expression [1.7] we have:

kinetic energy $= \frac{1}{2}mv^2$

$$= \frac{1}{2} \times 20\,[\text{kg}] \times (2.78)^2\,[\text{m/s}]^2$$

$$= 77.3\,\text{J}.$$

Example 1.7   *A body having a mass of 30 kg is supported 50 m above the earth's surface. Calculate its potential energy.*

Weight of body $\simeq 30\,[\text{kg}] \times 9.81\,[\text{N/kg}] = 294.3\,\text{N}$,

$\therefore$ work done in lifting the body through a height of 50 m

$$= 294.3\,[\text{N}] \times 50\,[\text{m}] = 14\,715\,\text{J}$$

$$= 14.715\,\text{kJ}.$$

1.6 Unit of power

Power is defined as *the rate of doing work*, and the SI unit of power is the *watt* (symbol, W), named after the famous Scottish engineer James Watt (1736–1819). *The watt is equal to 1 joule per second.*

In practice, the watt is often found to be inconveniently small; consequently the kilowatt (kW) is frequently used, the kilowatt being 1000 watts. For still larger powers, the megawatt (MW) is used, where

$$1 \text{ MW} = 1000 \text{ kW} = 1\,000\,000 \text{ W}.$$

When we are dealing with a large amount of work (or energy), it is often convenient to express the latter in *kilowatt hours* [kW h].

$$
\begin{aligned}
1 \text{ kW h} &= 1000 \text{ watt hours} \\
&= 1000 \times 3600 \text{ watt seconds or joules} \\
&= 3\,600\,000 \text{ J} = 3.6 \text{ MJ}. \qquad \text{[1.10]}
\end{aligned}
$$

If T be the torque or turning moment, in newton metres, exerted on a shaft, and if n be the speed of the shaft in revolutions per second, then from expression [1.6],

work done/second

$$
\begin{aligned}
&= \text{work done/revolution} \times \text{no. of revolutions/second} \\
&= 2\pi \times T \times n \text{ joules/second}
\end{aligned}
$$

i.e. power $= 2\pi n T$ watts   [1.11]

$\qquad\qquad = \omega T$ watts   [1.12]

where $\omega =$ angular velocity in radians/second

$\qquad\qquad = 2\pi n.$

Example 1.8 *A force of 60 N is applied to a body to move it at a uniform velocity through a distance of 20 m in 8 s in the direction of the force. Calculate the value of the power.*

Work done $= 60 \text{ N}[\text{N}] \times 20 \text{ [m]} = 1200 \text{ J}$,

$\therefore$ power $=$ work done per second

$\qquad\qquad = 1200 \text{ [J]}/8 \text{ [s]} = 150 \text{ W}.$

Example 1.9 *Calculate the power required to lift a mass of 300 kg at a constant speed through a vertical distance of 200 m in 4 min.*

Force required to lift load $\simeq 300 \text{ [kg]} \times 9.81 \text{ [N/kg]} = 2943 \text{ N}.$

Work done in 4 min $= 2943 \text{ [N]} \times 200 \text{ [m]} = 588\,600 \text{ J}.$

$$\therefore \quad \text{power} = \frac{588\,600 \text{ [J]}}{(4 \times 60) \text{ [s]}} = 2453 \text{ W}$$

$$= 2.453 \text{ kW}.$$

Example 1.10 *A motor vehicle exerts a steady pull of 800 N on a trailer when hauling it at 75 km/h. Calculate (a) the work done in 20 min (i) in megajoules, (ii) in kilowatt hours, and (b) the power required.*

(a) (i)

$$\text{Distance travelled} = 75\,[\text{km/h}] \times (20/60)\,[\text{h}] = 25\,\text{km}$$

$$= 25\,000\,\text{m},$$

$$\therefore \quad \text{work done} = 800\,[\text{N}] \times 25\,000\,[\text{m}] = 20\,000\,000\,\text{J}$$

$$= 20\,\text{MJ}.$$

(ii)

$$\text{Since } 1\,\text{kW h} = 3.6\,\text{MJ}$$

$$\text{work done} = \frac{20\,[\text{MJ}]}{3.6\,[\text{MJ/kW h}]} = 5.56\,\text{kW h}.$$

(b)

$$\text{Power} = \frac{\text{work done in joules}}{\text{time in seconds}}$$

$$= \frac{20\,000\,000\,[\text{J}]}{(20 \times 60)\,[\text{s}]} = 16\,670\,\text{W}$$

$$= 16.67\,\text{kW}.$$

Alternatively,

$$\text{power} = \frac{\text{work done in kilowatt hours}}{\text{time in hours}}$$

$$= \frac{5.56\,[\text{kW h}]}{(20/60)\,[\text{h}]} = 16.68\,\text{kW}.$$

Example 1.11 *An electric motor is developing 8 kW at a speed of 1200 rev/min. Calculate (a) the work done in 45 min (i) in kilowatt hours, (ii) in megajoules, and (b) the torque in newton metres.*

(a) (i)

$$\text{Work done} = 8\,[\text{kW}] \times (45/60)\,[\text{h}] = 6\,\text{kW h}.$$

(ii)

$$\text{Since } 1\,\text{kW h} = 3.6\,\text{MJ},$$

$$\therefore \quad \text{work done} = 6\,[\text{kW h}] \times 3.6\,[\text{MJ/kW h}] = 21.6\,\text{MJ}.$$

(b)

$$\text{Speed} = 1200\,\text{rev/min}$$

$$= 1200/60 = 20\,\text{rev/s},$$

$$\text{and power} = 8\,\text{kW}$$

$$= 8 \times 1000 = 8000\,\text{W}.$$

From expression [1.11],

$$\text{power} = 2\pi T \times n$$

$$\therefore \quad 8000\,[\text{W}] = 2\pi T \times 20\,[\text{rev/s}]$$

$$\therefore \quad T = 63.7\,\text{N m}.$$

Example 1.12 *The tensions on the two sides of a belt passing round a pulley are 2200 N and 460 N respectively. The effective diameter of the pulley is 400 mm and the speed is 700 rev/min. Calculate the power transmitted.*

The effective driving effort of a belt is the difference between the tensions on the tight and the slack sides;

$$\therefore \quad \text{net driving force} = 2200 - 460 = 1740 \text{ N.}$$

Effective radius of pulley $= 200 \text{ mm} = 0.2 \text{ m,}$

$$\therefore \quad \text{torque} = 1740 \, [\text{N}] \times 0.2 \, [\text{m}] = 348 \text{ N m.}$$

$$\text{Speed} = 700 \text{ rev/min}$$

$$= 700/60 = 11.67 \text{ rev/s,}$$

$$\therefore \quad \text{power} = 2\pi n T$$

$$= 2\pi \times 11.67 \, [\text{rev/s}] \times 348 \, [\text{N m}]$$

$$= 25\,500 \text{ W} = 25.5 \text{ kW.}$$

1.7 Thermal energy

The SI unit of thermal energy is the *joule*—exactly as for mechanical and electrical energies. The thermal unit 'calorie' is obsolete; and the term 'mechanical equivalent of heat' no longer has any significance. In other words, mechanical, electrical and thermal energies are interchangeable and are all expressed in joules.

Specific heat capacity

Different substances absorb different amounts of heat to raise the temperature of a given mass of the substance by one degree. The heat required to raise the temperature of 1 kg of a substance by 1°C is termed the *specific heat capacity* of that substance. Hence, if c represents the specific heat capacity of a substance in joules per kilogram kelvin, the heat required to raise the temperature of m kilograms of the substance by t degrees

$$= mct \text{ joules} \qquad\qquad [1.13]$$

Table 1.2 gives the approximate values of the specific heat capacity of some well-known substances for a temperature range between 0°C and 100°C:

Table 1.2

Substance	Specific heat capacity
	J/kg K
Water	4190
Ice (0°C to −20°C)	2100
Copper	390
Iron	500
Aluminium	950
Brass	370
Dry air at standard atmospheric pressure	1015

Example 1.13   *Calculate the heat required to raise the temperature of 6 kg of water from 10°C to 25°C.*

Increase of temperature $= 25 - 10 = 15°C = 15$ K.

Since specific heat capacity of water is 4190 J/kg K

heat required $= 6$ [kg] $\times 4190$ [J/kg K] $\times 15$ [K]

$\qquad\qquad = 377\,000$ J $= 377$ kJ.

Example 1.14   *How many kilograms of copper can be raised from 15°C to 60°C by the absorption of 80 kJ of heat?*

Increase of temperature $= 60 - 15 = 45°C = 45$ K.

Quantity of heat $= 80$ kJ $= 80\,000$ J.

Assuming the specific heat capacity of copper to be 390 J/kg K,

$80\,000$ [J] $= m \times 390$ [J/kg K] $\times 45$ [K]

$\therefore \quad m = 4.56$ kg.

Example 1.15   *An electric heater is required to heat 15 litres of water from 15°C to the boiling point (100°C) in 40 min. Assuming the efficiency of the heater to be 80 per cent, calculate (a) the energy consumed in (i) megajoules, (ii) in kilowatt hours, and (b) the cost of the energy consumed if the charge is 2 p/kW h. Assume the specific heat capacity of water to be 4190 J/kg K and the mass of 1 litre of water to be 1 kg.*

(a)   Mass of 15 litres of water $= 15$ kg.

Increase of temperature $= 100 - 15 = 85°C = 85$ K.

$\therefore$ useful heat required $= 15$ [kg] $\times 4190$ [J/kg K] $\times 85$ [K]

$\qquad\qquad\qquad = 5\,340\,000$ J $= 5.34$ MJ.

Since the efficiency of the heater is 80 per cent,

$$0.8 = \frac{\text{useful thermal energy required}}{\text{total energy supplied}}$$

$\therefore$ total energy supplied $= 5.34\,[\text{MJ}]/0.8 = 6.67\,\text{MJ}.$

Since $1\,\text{kW h} = 3.6\,\text{MJ},$

$\therefore$ total energy supplied $= 6.67\,[\text{MJ}]/3.6\,[\text{MJ/kW h}]$

$\qquad\qquad\qquad\qquad = 1.85\,\text{kW h}.$

(b)   Cost of energy $= 1.85\,[\text{kW h}] \times 2\,[\text{p/kW h}]$

$\qquad\qquad\qquad\qquad = 3.7\,\text{p}.$

1.8 Summary of important formulae and relationships

$$t\,[\text{degrees Celsius}] = T\,[\text{kelvins}] - 273.15 \qquad [1.1]$$

$$F\,[\text{newtons}] = m\,[\text{kg}] \times z\,[\text{m/s}^2] \qquad [1.2]$$

$$\text{Weight of a body, in newtons} \simeq 9.81\,[\text{N/kg}] \times m\,[\text{kg}] \qquad [1.3]$$

$$T\,[\text{newton metres}] = F\,[\text{newtons}] \times r\,[\text{metres}] \qquad [1.4]$$

$$\text{Work done}\,[\text{joules}] = F\,[\text{newtons}] \times s\,[\text{metres}] \qquad [1.5]$$

$$\text{Potential energy}\,[\text{joules}] = m\,[\text{kg}] \times g\,[\text{m/s}^2] \times h\,[\text{m}] \qquad [1.8]$$

$$\simeq 9.81mh \qquad [1.9]$$

$$\text{Kinetic energy}\,[\text{joules}] = \tfrac{1}{2}m\,[\text{kg}] \times v^2\,[\text{m/s}]^2 \qquad [1.7]$$

$$1\,\text{kW h} = 3.6\,\text{MJ} \qquad [1.10]$$

$$\text{Power} = 2\pi n\,[\text{rev/s}] \times T\,[\text{N m}] \qquad [1.11]$$

$$= \omega\,[\text{rad/s}] \times T\,[\text{N m}] \qquad [1.12]$$

$$\text{Density of water} \simeq 1000\,\text{kg/m}^3.$$

Heat required to raise temperature of m kilograms of a substance having specific heat capacity c joules per kilogram kelvin through t degrees

$$= mct\,\text{joules} \qquad [1.13]$$

1.9 Examples

1. Express 60°C in kelvin and 320 K in degrees Celsius.
2. A 10-kg mass is suspended at the end of a cord. Calculate the approximate value of the force, in newtons, exerted by the cord.

3. A body is suspended from a spring balance. The reading on the balance is 4 N. Assuming the local gravitational acceleration to be 9.81 m/s^2, calculate the mass of the body in grams.

4. Calculate the force required to give a mass of 10 kg an acceleration of 5 m/s^2.

5. Calculate the acceleration when a force of 30 N acts on a mass of 50 kg.

6. A force of 30 N acts through a distance of 4 m in the direction of the force. What is the work done?

7. Calculate the work done, in kilojoules, in lifting a mass of 800 kg through a vertical height of 40 m.

8. The work done in lifting a body vertically through a distance of 120 m is 5 W h. Calculate the mass of the body.
   If the time taken is 2 min, what is the average value of the power?

9. If the work done in moving a body is 1500 J and the time taken is 12 s, what is the average value of the power?

10. Calculate the kinetic energy (i) in megajoules, (ii) in kilowatt hours, possessed by a body having a mass of 500 kg when travelling at 120 m/s.

11. A body having a mass of 50 kg is supported 40 m above ground. What is its potential energy? Assume $g = 9.81 \text{ m/s}^2$.

12. The output of an electric motor is 8 kW and is maintained constant for 6 h. Calculate the work done (a) in kilowatt hours and (b) in megajoules.

13. If the speed of the motor in Q. 12 is 500 rev/min, what is the value of the torque in newton metres?

14. A motor develops a torque of 3 kN m. At what speed must it run in order that it may develop 200 kW?

15. An engine running at 1800 rev/min develops a torque of 4 kN m. Calculate (a) the power developed and (b) the work done, in kilowatt hours, in 3 h.

16. Energy is transmitted from the shaft of an engine by a belt passing over a pulley having a diameter of 600 mm. The effective pull on the belt is 2.8 kN. Calculate (a) the torque and (b) the power transmitted when the speed is 240 rev/min.

17. The pull on the tight side of a belt is 720 N and that on the slack side is 80 N. The pulley has a diameter of 460 mm and is rotating at 400 rev/min. Calculate (a) the torque and (b) the power transmitted.

18. An electrically-driven pump lifts 80 m^3 of water per minute through a height of 12 m. Allowing an overall efficiency of 70 per cent for the motor and pump, calculate the input power, in kilowatts, to the motor.
    If the pump is in operation for an average of 2 hours per day for 30 days, calculate the energy consumed, in kilowatt hours, and the cost of the energy at 1.5 p/kW h. Assume 1 m^3 of water to have a mass of 1000 kg and $g = 9.81 \text{ m/s}^2$.

19. Calculate the quantity of heat, in megajoules, required to raise the temperature of 200 kg of water from 15°C to 90°C. Assume the specific heat capacity of water to be 4190 J/kg K.

20. 90 MJ of heat are absorbed by a body having a mass of 1 Mg to raise its temperature from 20°C to 200°C. Assuming no loss of heat, calculate the specific heat capacity of the body.

21. Calculate the heat required to raise the temperature of 30 kg of copper from 12°C to 70°C. Assume the specific heat capacity of copper to be 390 J/kg K.

22. Calculate the time taken for a heating element, rated at 2 kW, to raise the temperature of 1.5 litres of water by 70°C. The efficiency of the heating equipment is 80 per cent. Also calculate the energy absorbed (a) in kilowatt hours and (b) in kilojoules.

23. An electric furnace is being used to melt 10 kg of aluminium. The initial temperature of the aluminium is 20°C. Assume the melting point of aluminium to be 660°C, its specific heat capacity to be 950 J/kg K and its specific latent heat of fusion to be 387 kJ/kg. Calculate the power required to accomplish the conversion in 20 min, assuming the efficiency of the furnace to be 75 per cent. What is the cost of the energy consumed if the tariff is 2 p/kW h?

CHAPTER 2

Electric Current

2.1 Effects of an electric current

All the phenomena that an electric current can produce may be grouped under one or other of the following headings:

    (a) magnetic effect,
    (b) heating effect,
    (c) chemical effect.

Let us assume that we have available at the terminals of a double-pole switch S (fig. 2.1) a direct-current supply, namely, a supply that causes the electric current to flow in one direction only. This direct-current supply may be obtained either from a direct-current generator or from a secondary battery, the principle of action of which is referred to later in this section. To the output side of S, let us connect the following items in series, so that the same current passes through each of them:

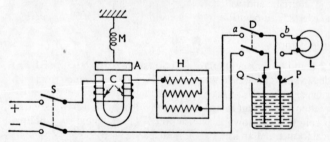

Fig. 2.1 Magnetic, heating and chemical effects of an electric current

    (1) coil C wound on an iron core bent into a U shape,
    (2) electric fire element H, of suitable size, and
    (3) double-pole change-over switch D, to which are also connected a flash-lamp bulb L and a pair of lead plates, P and Q, dipping into a dilute solution of sulphuric acid in water.

When S is closed and D is put over to side a, an electric current passes through C and H, from which it flows to P, through the liquid to Q, and then returns to the generator or battery.

If an iron plate or armature A be suspended by a spring M a little above the ends of the iron core, it will be attracted downwards immediately S is closed. In other words, the iron core becomes magnetized and mechanical work is done in extending M.

The element H indicates the presence of an electric current by the fact that it begins to give out heat, and its temperature may rise sufficiently for it to emit an appreciable amount of light. Hence electrical energy is being converted into heat and light energies.

After the current has been flowing for a few minutes, let us open S and switch D over to side *b*. The filament of L becomes incandescent, though its brightness gradually fades away. If the plates P and Q are withdrawn from the solution, P is found to have a faint chocolate-colour coating while Q appears unaltered. It follows that in this case electrical energy has been converted into chemical energy in changing the lead to an oxide of lead. This chemical action happens to be of a kind that is reversible; consequently, when the plates are connected to the lamp, chemical energy is converted back into electrical energy which is then converted into thermal and light energies. This reversible chemical action is the basis upon which secondary cells operate (see Chap. 19).

2.2 Measurement of an electric current

It is important that we should have some idea of the magnitude of the current in a circuit and we shall now consider how the magnetic and chemical effects* can be utilized for determining the value of the current.

Magnetic effect

In fig. 2.2, E and F represent the cross-section of two co-axial coils, E being suspended from a spring balance G about 2 cm above F which rests on a table or bench. The coils are connected so that they carry current in the *same* direction, the current being led into and out of coil E by flexible wires. Another coil, B, also shown in section in fig. 2.2, is wound on a hollow cylindrical former and an iron core C is suspended from a spring balance S.

The three coils, E, F and B, are connected in series with a variable resistor R, an ammeter A (i.e. an instrument calibrated to indicate the magnitude of the current directly in amperes) and a switch N across a direct-current supply.

---

* The heating effect of an electric current can be utilized for determining the value of the current by arranging for the heater carrying the current to raise the temperature of one of the two junctions of a thermocouple, as described in section 18.7. The thermo-e.m.f. is practically proportional to the difference of temperature for a limited range of temperature difference.

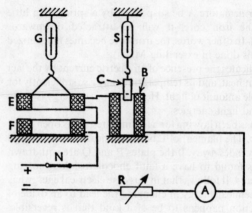

Fig. 2.2 Use of the magnetic effect to measure an electric current

With switch N closed, an electric current flows in the circuit. The magnetic effect produced by the current in coils E and F causes them to attract each other (the reason for this behaviour is given in section 5.13). Also, the iron core C is attracted towards coil B. The greater the current, the larger are the readings on balances G and S, and marks can be put on the two scales to register the various currents indicated by ammeter A.

These two methods are comparatively crude but are given at this stage merely to indicate the principles of methods applying the magnetic effect to the measurement of an electric current. The method using the two co-axial coils, E and F, illustrates the principle of operation of the balance used in National Laboratories for determining the value of the unit of electric current (see section 5.13).

The second method, namely the attraction of an iron core towards a coil carrying a current, illustrates the principle of action of the moving-iron instrument described in section 18.6(i).

Chemical effect

The glass vessel B in fig. 2.3 contains a solution of copper sulphate in water. Two copper plates, P and Q, are partially immersed in an *electrolyte* (an electrolyte being a liquid that can be decomposed electrically) and are connected in series with an ammeter A, a variable resistor R and a switch N to a direct-current supply.

Plate Q, namely the plate connected to the negative terminal of the supply, is washed, dried and weighed on a chemical balance before it is placed in the copper sulphate solution. Switch N is then closed and the current is maintained constant at a known value, for, say, 10 minutes. Switch N is then opened and plate Q is removed, washed, dried and again weighed. It is found that the mass of the plate has increased and the exact amount is determined from the difference between the balance readings.

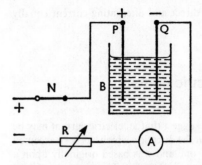

Fig. 2.3 Use of the chemical effect to measure an electric current

In this way we can measure accurately the amount of copper deposited by a *given current* in, say, 10, 20 and 30 minutes, and also the amount deposited in a *given time* by different values of current. It is found that the mass of copper deposited is directly proportional to the current and to the time; in other words, it is proportional to the *quantity* of electricity.

Owing to the accuracy and the ease with which the increase in the mass of a plate can be determined, this chemical effect of an electric current may be employed for measuring the quantity of electricity with a high degree of precision. At one time, this was the principle of the method used for defining the magnitude of the unit current.

2.3 Effect of reversing the current

Let us change over the supply connections to the circuits of figs. 2.2 and 2.3 so that the supply wire marked '−' will now go to coil F in fig. 2.2 and to plate P in fig. 2.3.

It is found that the effect of the reversed current is still to attract coil E and core C downwards in fig. 2.2. In fact, a given current, as indicated by ammeter A, produces exactly the same readings as before on the two scales.

In the case of fig. 2.3, it is found that the mass of plate Q decreases, so that the chemical action at this plate must now be taking place in the reverse direction. Further, it is found that the decrease in the mass of plate Q due to a given current for a given time is the same as the increase that took place before the connections were reversed.

Let us go a step further and connect the circuits of figs. 2.2 and 2.3 across an alternating-current supply. The electric current now reverses its direction many times every second. Coil E and core C are still found to be attracted downwards, but no change takes place in the mass of plate Q. The chemical method is therefore useless for measuring the value of an alternating current, whereas it is possible to construct instruments of the magnetic type that will measure direct and alternating current equally accurately.

2.4 Direction of an electric current

The experiments described above suggest that an electric current may be regarded as a flow of electricity, but the direction which has been accepted for this flow is a purely arbitrary one and was based originally upon a theory of Benjamin Franklin, long since discarded. This matter is referred to more fully in section 9.1. At this stage it will suffice to state that if a circuit such as the filament of a lamp be connected across a direct-current supply, the electric current is regarded as flowing from the positive terminal (usually painted red) through the circuit to the negative terminal (painted black). If this convention be applied to fig. 2.3, it will be seen that plate Q becomes heavier when current flows from P through the liquid (or electrolyte) to Q, and that it becomes lighter when current flows in the reverse direction.

It is therefore possible to utilize the chemical effect of a current for determining its direction in accordance with the convention universally accepted.

2.5 Units of electric current and of quantity of electricity

By an electric current is meant the rate at which electricity flows past a given point of an electric circuit, and the *unit of current* is termed the *ampere*, to commemorate a famous French scientist, André-Marie Ampère (1775–1836). It follows that the *unit quantity* of electricity, called the *coulomb* after another great French scientist, flows past a given point of a circuit when a current of 1 ampere is maintained for 1 second. Consequently, if a current of I ampere is maintained constant for t seconds, the corresponding quantity of electricity is represented by Q coulombs, where

$$Q \text{ [coulombs]} = I \text{ [amperes]} \times t \text{ [seconds]}. \qquad [2.1]$$

The symbols for 'ampere' and 'coulomb' are 'A' and 'C' respectively.

The value of the ampere, adopted internationally in 1948, is defined as the constant current which, if maintained in two straight parallel conductors of infinite length, of negligible circular cross-section and placed at a distance of 1 metre apart in a vacuum, will produce between them a force equal to 2×10^{-7} *newton per metre length* (section 5.12). The apparatus required to measure a current accurately in terms of this definition is very elaborate and expensive and is seldom available outside the principal national laboratories, such as the National Physical Laboratory in Great Britain. By means of very carefully-conducted experiments, however, it has been found that when a current of 1 ampere is passed between two copper plates immersed in a cooper sulphate solution, copper is deposited on the negative plate (i.e. plate Q in fig. 2.3) at the rate of 0.3294 mg/s. When the experiment is performed with a silver *voltameter* or *coulometer*, namely two plates of silver immersed in a silver nitrate solution, it is found that a current of 1 ampere causes silver to be deposited on the negative plate at the rate of 1.1182 mg/s.

Since 1 ampere is 1 coulomb per second, the unit quantity of electricity may also be termed an *ampere second*. For many purposes, such as for stating the quantity of electricity a primary or secondary battery is capable of supplying, the coulomb is inconveniently small, and a larger unit, called the *ampere hour* (A h), is preferable. Thus, if a battery gives 4 A for 10 hours, the quantity of electricity is 4 (amperes) × 10 (hours), namely, 40 A h.

2.6 Electrochemical equivalent

In section 2.2 it was found that the mass of copper deposited on the negative plate (or *cathode*) was proportional to the quantity of electricity flowing through the electrolyte. This relationship was discovered by Michael Faraday in 1832 when he enunciated two laws:

(1) the amount of chemical change produced by an electric current is proportional to the quantity of electricity, and

(2) the amounts of different substances liberated by a given quantity of electricity are proportional to their chemical equivalent mass, where

$$\text{chemical equivalent mass} = \frac{\text{relative atomic mass}}{\text{valency}}.$$

The relative atomic masses and valencies of some of the most common elements are given in table 2.1.

The mass of a substance liberated from an electrolyte by 1 coulomb is termed the *electrochemical equivalent* of that substance; thus, the electrochemical equivalents of copper and silver are respectively 0.3294 and 1.1182 milligrams/coulomb.

If z = electrochemical equivalent of a substance in
milligrams per coulomb,

and I = current in amperes for time t seconds,

mass of substance liberated = zIt milligrams   [2.2]

This section deals with the *facts* of electrolysis; the *mechanism* of electrolysis is discussed in section 19.2.

Table 2.1

Element	Relative atomic mass	Valency	Electrochemical equivalent
Aluminium	27.0	3	
Chlorine	35.5	1	
Chromium	52.0	3 or 6	
Copper (cuprous)	63.6	1	
— (cupric)	63.6	2	0.3294 mg/C
Gold	197.2	3	
Hydrogen	1.008	1	
Iron	55.8	2 or 3	
Lead	207.2	2	
Nickel	58.7	2	0.304 mg/C
Oxygen	16.0	2	
Potassium	39.1	1	
Silver	107.9	1	1.1182 mg/C
Sodium	23.0	1	
Tin	118.7	2 or 4	
Zinc	65.4	2	0.388 mg/C

Example 2.1 *If a current of 15 A is maintained constant for 20 min, calculate the quantity of electricity in (a) coulombs, (b) ampere hours.*

(a) Quantity of electricity (coulombs)

$$= \text{current (amperes)} \times \text{time (seconds)}$$
$$= 15 \times 20 \times 60 = 18\,000 \text{ coulombs.}$$

(b) Quantity of electricity (ampere hours)

$$= \text{current (amperes)} \times \text{time (hours)}$$
$$= 15 \times 20/60 = 5 \text{ A h.}$$

Example 2.2 *A steady current of 6.3 A is passed for 45 min through a solution of copper sulphate. Calculate the mass of copper deposited.*

Quantity of electricity (coulombs)

$$= \text{current (amperes)} \times \text{time (seconds)}$$
$$= 6.3 \times 45 \times 60$$
$$= 17\,010 \text{ coulombs.}$$

But 1 coulomb deposits 0.3294 mg of copper,

∴   mass of copper deposited $= 0.3294\,[\text{mg/C}] \times 17\,010\,[\text{C}]$

$$= 5600\,\text{mg} = 5.6\,\text{g}.$$

Example 2.3   *A current of 2 A is passed for 30 min between two platinum plates (or electrodes) immersed in a dilute solution of sulphuric acid in water. Calculate the mass of hydrogen and oxygen released.*

The effect of electrolysis in this case is to decompose water into its constituents, hydrogen and oxygen, the former being liberated at the negative plate (or cathode) and the latter at the positive plate (or anode).

Quantity of electricity $= 2 \times 30 \times 60 = 3600$ coulombs.

From the data given in the table on page 22,

chemical equivalent mass of hydrogen $= 1.008/1 = 1.008$.

Also, for silver, the electrochemical equivalent is $1.1182\,\text{mg/C}$ and the chemical equivalent mass is 107.9; hence, by Faraday's Second Law,

electrochemical equivalent of hydrogen $= 1.1182 \times \dfrac{1.008}{107.9}$

$$= 0.010\,45\,\text{mg/C},$$

∴   mass of hydrogen released $= 0.010\,45 \times 3600$

$$= 37.6\,\text{mg} = 0.0376\,\text{g}.$$

Similarly,

chemical equivalent mass of oxygen $= 16/2 = 8$,

∴   electrochemical equivalent of oxygen $= 1.1182 \times \dfrac{8}{107.9}$

$$= 0.0829\,\text{mg/C}$$

and mass of oxygen released $= 0.0829 \times 3600$

$$= 298\,\text{mg} = 0.298\,\text{g}.$$

2.7 Summary of important formulae

$Q\,[\text{coulombs}] = I\,[\text{amperes}] \times t\,[\text{seconds}]$ [2.1]

Mass of substance liberated from electrolyte, in milligrams

$= z\,[\text{mg/C}] \times I\,[\text{amperes}] \times t\,[\text{seconds}]$ [2.2]

$= z\,[\text{milligrams/coulomb}] \times Q\,[\text{coulombs}]$.

2.8 Examples

1. An ammeter is calibrated by being connected in series with a copper coulometer (or voltameter) through which a constant current is maintained for 15 min. The ammeter reading is 4 A. The initial and final masses of the cathode are 16.347 g and 17.518 g respectively. Calculate the error in the ammeter reading and state whether the ammeter is reading high or low.

2. A steady current was passed through a copper coulometer for half an hour. The mass of the cathode was found to increase by 1.65 g. Calculate the value of the current.

3. A metal plate having a surface of 12 000 mm^2 is to be copper-plated. If a current of 2 A is passed for 1 h, what thickness of copper is deposited? Should the plate be made the positive or the negative electrode? Density of copper is 8900 kg/m^3.

4. A solid metal cylinder, 30 mm diameter and 200 mm long, is to have copper deposited to a thickness of 0.1 mm over its *curved* surface. Find the time taken by a current of 40 A. Assume the electrochemical equivalent of copper to be 0.33 mg/C and the density of copper to be 8900 kg/m^3.

5. A steady current of 3 A is passed for 16 min through a silver coulometer. If 1 A deposits 0.001 118 g of silver per second, find the increase in the mass of the cathode. Also find the quantity of electricity: (a) in coulombs, (b) in ampere hours.

6. It is required to deposit a layer of nickel 0.2 mm thick on a surface area of 15 000 mm^2. Calculate the minimum time required if the maximum permissible current be 8 A. Assume the electrochemical equivalent of nickel to be 0.304 mg/C and the density of nickel to be 8800 kg/m^3. Also calculate the quantity of electricity required (a) in coulombs, (b) in ampere hours.

7. A metal plate having a surface area of 20 000 mm^2 is to be silver-plated. If a current of 5 A is maintained constant for a period of 1 h, what thickness of silver will be deposited on the plate? The electrochemical equivalent of silver is 1.118 mg/C and its density is 10 600 kg/m^3.

8. In an experiment to determine the electrochemical equivalent of copper, the following results were obtained:

   Initial mass of plate, 29.82 g; final mass of plate, 30.48 g; current, 2 A; time, 16 min 40 s.

   Calculate the experimental value of the electrochemical equivalent of copper. What precautions should be taken in carrying out this experiment?

   (U.E.I.)

9. A metal plate, having a total surface of 20 000 mm^2, is to be chromium-plated. If a current of 5 A is maintained for a period of 1 h, what thickness of chromium will be deposited on the plate? Assume the E.C.E. of chromium to be 0.0890 mg/C and the density to be 6600 kg/m^3.

10. A certain current is passed through two voltameters which are connected in series, one containing silver nitrate solution and the other a solution of copper sulphate. In 60 min it was found that 9 g of silver were deposited. Determine the mass of copper deposited in the same time. (E.C.E. for silver = 1.12 mg/C; E.C.E. for copper = 0.328 mg/C.)

(U.E.I.)

11. A steady current flowing through acidulated water liberates 1.248 litres of hydrogen, measured under standard temperature and pressure, per hour. Assuming the electrochemical equivalent of hydrogen to be 0.0104 mg/C and its density at standard temperature and pressure to be 90 g/m^3, calculate the value of the current.

12. From the data given in the table on p. 22, calculate the electrochemical equivalents of aluminium, gold and chlorine.

CHAPTER 3

Electric Circuit

3.1 Conductors and insulators

We have become so accustomed to the idea of an electric current being confined to a metallic circuit surrounded by a non-metallic substance, such as air or plastic, that we find it difficult to appreciate the significance of some remarkable experiments made in 1729 by an Englishman, Stephen Gray, when he discovered electrical conductors and insulators. He found that a brass wire allowed electricity to pass through it, whereas silk would not do so; and by means of a wire suspended by silk threads he transmitted electricity a distance of about 300 yards. Gray's discovery provided the key to the remarkable progress which electrical science made in the following one hundred years, to be crowned by Faraday's discovery of electromagnetic induction in 1831 (section 6.2).

Though it is usual to divide materials into two categories, conductors and insulators, it should be realized that these are only relative terms. No material is a perfect conductor and no material is a perfect insulator. In general, metals such as copper and silver are very good conductors, whereas non-metallic materials such as rubber and glass are good insulators.

During recent years, semiconducting materials, namely materials such as germanium and silicon whose electrical characteristics lie between those of metals and insulators, have attained considerable importance. These semiconductors, in a crystalline form and doped with suitable impurities, are used in rectifiers and transistors and are dealt with in Chapter 23.

The insulating property of oil and of fibrous materials, such as paper and cotton, is greatly affected by the amount of moisture they contain. The greater the moisture content, the poorer is the insulating property.

3.2 Heating effect of an electric current

When a block of stone is dragged along a horizontal surface, it is found that the friction between the two surfaces tends to resist the movement of one surface relative to the other and that the work done in overcoming friction is converted into heat. In an electric circuit we also find that the material of which the circuit is made tends to resist the passage of electricity through it, and that the electrical energy which has to be supplied to overcome this resistance is converted into heat. It follows that good conductors have very low electrical resistance, whereas good insulators have very high electrical resistance.

The laws relating to the heating effect of an electric current were discovered by James Prescott Joule whose name is perpetuated by the unit of energy, the *joule* (section 1.5). The general principle of the apparatus used by Joule is shown in fig. 3.1. A copper calorimeter B contains a known mass of water, and any loss of heat is minimized by a layer of cotton wool C between B and an outer container D. Stout copper wires pass through a wooden lid L and are attached at the lower ends to a helix R of known resistance. The temperature of the water is measured by a thermometer T. A constant current is passed through R for a known time and the rise of temperature of the water is noted. It is necessary to stir the water very thoroughly during the test to ensure uniform distribution of the heat.

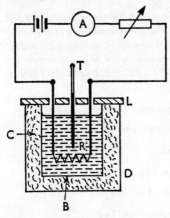

Fig. 3.1 Verification of Joule's laws

From his researches, Joule deduced that the heat generated in a wire is proportional to:

(a) the square of the current, e.g. if the current is doubled, the rate of heat generation is increased fourfold.
(b) the resistance of the wire; e.g. if the length of the wire is doubled, the rate of heat generation by a given current is also doubled.
(c) the time during which the current is flowing.

Consequently, if a current I flows through a circuit having resistance R for time t,

heat energy generated $\propto I^2Rt$.

The *unit of resistance* is that resistance in which a current of 1 ampere flowing for 1 second generates 1 joule of thermal energy. This unit of resistance is termed the *ohm* in commemoration of Georg Simon Ohm (1787–1854), the German physicist who enunciated Ohm's Law (section 3.6). Hence if a current I amperes flows through a circuit having resistance R ohms for t seconds,

$$\text{heat energy generated} = I^2 R r \text{ joules.} \qquad [3.1]$$

The ohm is represented by the abbreviation Ω (capital omega). Sometimes it is more convenient to express the resistance in millionths of an ohm, in which case the resistance is said to be so many *microhms* ($\mu\Omega$). On the other hand, when we are dealing with the resistance of insulating materials, the ohm is inconveniently small; consequently, another unit called the *megohm* ($M\Omega$) is used, one megohm being a million ohms.

Example 3.1　*A current of 5 A was maintained for 6 min through a 1.3-Ω resistor immersed in 0.44 litre of water. The water equivalent of the vessel and heater was 17 g. Calculate (a) the heat generated, in kilojoules, and (b) the temperature rise of the water. Neglect any loss of heat and assume the specific heat capacity of water to be 4190 J/kg K and the mass of 1 litre of water to be 1 kg.*

(a)　Heat generated in resistor $= I^2 R t$ joules

$$= 5^2 \times 1.3 \times 6 \times 60 = 11\ 700\,\text{J}$$

$$= 11.7\,\text{kJ.}$$

(b)　Since the mass of 1 litre of water is 1 kg,

$$\text{mass of water in vessel} = 0.44\,\text{kg.}$$

Water equivalent of vessel and heater $= 0.017\,\text{kg}$

$\therefore$　water equivalent of water, vessel and heater

$$= 0.44 + 0.017 = 0.457\,\text{kg.}$$

From expression [1.13],

$$\text{total heat required} = mct$$

$$= 0.457\,[\text{kg}] \times 4190\,[\text{J/kg K}] \times t$$

But heat generated in resistor

$$= \text{heat absorbed by water, vessel and heater}$$

i.e. $11\ 700\,[\text{J}] = 0.457\,[\text{kg}] \times 4190\,[\text{J/kg K}] \times t$

$\therefore$　temperature rise $= t = 6.11\,\text{K} = 6.11°\text{C.}$

3.3 Power in an electrical circuit

Power is the rate of doing work and is therefore expressed in joules per second or *watts* (see section 1.6). Hence, when a current I, in amperes, flows through a circuit having resistance R, in ohms, for time t, in seconds,

$$\text{power} = \frac{I^2 Rt}{t} \text{ joules/second}$$

$$= I^2 R \text{ watts.} \qquad [3.2]$$

Example 3.2 *The wire used in an electric heater has a resistance of 57 Ω. Calculate (a) the power, in kilowatts, when the heater is taking a current of 3.8 A, (b) the energy absorbed in 4 h (i) in kilowatt hours, (ii) in megajoules, and (c) the cost of the energy consumed if the charge is 1.5 p/kW h.*

(a) $\qquad\qquad\qquad$ Power $= 3.8^2 \times 57 = 823$ W
$\qquad\qquad\qquad\qquad\qquad = 0.823$ kW.

(b) (i) $\qquad$ Energy absorbed $= 0.823$ [kW] $\times 4$ [h] $= 3.292$ kW h

(ii) Since $\qquad\qquad 1$ kW h $= 3.6$ MJ

$\qquad \therefore$ energy absorbed $= 3.292$ [kW h] $\times 3.6$ [MJ/kW h]
$\qquad\qquad\qquad\qquad\qquad = 10.85$ MJ.

(c) $\qquad\qquad$ Cost of energy $= 3.292$ [kW h] $\times 1.5$ [p/kW h]
$\qquad\qquad\qquad\qquad\qquad = 4.94$ p.

3.4 Fall of electrical potential along a circuit

Suppose CD in fig. 3.2 to represent a long thin wire of uniform diameter made of an alloy such as Eureka (60 per cent copper and 40 per cent nickel) having a much higher resistance than the same length and diameter of copper. The wire is connected across the terminals of an accumulator. A

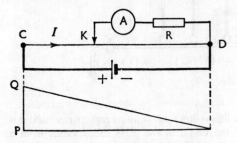

Fig. 3.2 Fall of electric potential

milliammeter A in series with a resistor R of, say, 1000 Ω is connected between terminal D and a contact A that can be moved along CD. The function of R is to limit the current through A, thereby protecting the latter from an excessive current and preventing an appreciable fraction of the current being diverted from length KD of the wire.

It is found that as the distance between K and D is increased, the current through A also increases, as indicated by the height of the graph shown in fig. 3.2, where PQ represents the ammeter reading with K at C. Since the direction of the current I in CD is assumed to be from C to D, C is said to be at a higher *potential* than D. In other words, when two points at different electrical potentials are connected together by a conductor, electricity is assumed to flow from the one at the higher potential to that at the lower potential. In fig. 3.2, the difference of potential between K and D is directly proportional to the deflection on A and therefore to the distance between D and the movable contact, the latter being at a higher potential than D.

3.5 Hydraulic analogy of fall of potential

A brass tube T (fig. 3.3) has a number of glass tubes attached to it. The tube is connected to a large jar A filled with a solution, such as methylene blue, which stands out clearly against a white background. At the other end of T, there is a tap C by which the flow of the liquid can be controlled. When C is shut, the level of the liquid in the glass tubes is the same as that in the jar, i.e. the whole pressure head of the liquid is available at C and there is no pressure drop in the pipe. Such a condition corresponds to a cell on open circuit, namely when there is no conducting path between the terminals. The whole of the electromotive force of the cell (section 3.7) then appears as a difference of potential between the terminals and there is no voltage drop in any of the connecting wires.

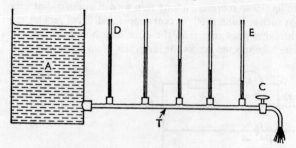

Fig. 3.3 Hydraulic analogy

As tape C is opened, the liquid flows out at an increasing rate, and it is found that the height of the liquid in the glass tubes varies from a minimum in E to a maximum in D; in fact, if a straight rod be placed opposite the heights of the liquid columns in D and E, it will also coincide with the heights in the intermediate tubes, showing that the difference of pressure between any two points along the tube is proportional to the distance between them. Also, it is found that the more rapidly the liquid is allowed to run out at C, the greater is the difference of pressure between two adjacent tubes; in other words, the greater is the fall of pressure in a given length of pipe. These effects are somewhat similar to the electrical relationships discussed in the next section, where it is shown that the difference of potential across a resistor is proportional to the current and to the resistance.

3.6 Ohm's Law

As long ago as 1827 Dr G. S. Ohm discovered that the current through a conductor, under constant conditions, was proportional to the difference of potential across the conductor. This fact can be demonstrated by connecting, as shown in fig. 3.4, a fixed resistor X, made of Eureka or other material whose resistance is not affected by temperature, in series with a variable resistor B and an ammeter A across the terminals of an accumulator. A voltmeter* is connected across X.

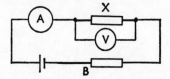

Fig. 3.4 Variation of p.d. with current

Different currents are obtained by varying B, and for each current the reading on V is noted. It is found that the ratio

$$\frac{\text{potential difference across X}}{\text{current through X}}$$

remains constant within the limits of experimental error, i.e. the current through a circuit having a constant† resistance is proportional to the difference of potential across that circuit.

---

* All moving coil voltmeters (see Chapter 18) consist of a milliammeter connected in series with a resistor having a high resistance, as shown in fig. 3.2. At this stage, however, a voltmeter may be regarded merely as an instrument that indicates the difference of electric potential between the two points across which it is connected.
† A resistor (i.e. a piece of apparatus used primarily because it possesses the property of resistance) is said to be *linear* if the current is proportional to the p.d. across its terminals. A material whose resistance varies with the applied voltage is referred to as a *non-linear* resistor, e.g. lamp filaments (section 4.6), metal and semiconductor rectifiers (sections 23.3 and 23.4).

Suppose X in fig. 3.5 to be a resistance box so constructed that the resistance of the wire between each pair of adjacent studs is 1 ohm and that by means of an arm C, the total resistance of X can be varied in steps of 1 Ω up to, say, 5 Ω. With C on stud 1, the resistance of B is adjusted to give a reading of, say, 0.3 A on ammeter A and the reading on voltmeter V is noted. Arm C is then moved to stud 2, B is readjusted to bring the current back to 0.3 A and the reading on V is again noted. The test is repeated with C on each of the other studs. It is found that the ratio

$$\frac{\text{potential difference across X}}{\text{resistance of X}}$$

remains constant, i.e. for a given current, the difference of potential between two points is directly proportional to the resistance of the circuit between those points.

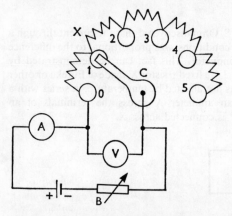

Fig. 3.5   Variation of p.d. with resistance

The *unit of potential difference* (or p.d.) is taken as the difference of potential across a 1-ohm resistor carrying a current of 1 ampere and is termed the *volt* (V) after Count Alessandro Volta, an Italian physicist who was the first to discover how to make an electric battery (section 19.1). It follows from the above experiments that if the current through a 1-Ω resistor is increased to, say, 3 A, the p.d. is 3 V, and that if the resistance is increased to, say, 4 Ω with the current maintained at 3 A, the p.d. across the resistance becomes 3 × 4, namely 12 V. Hence, if a circuit having a resistance of R ohms is carrying a current of I amperes, the p.d., V volts, across the circuit is given by:

$$V = IR, \qquad \text{or} \qquad I = V/R, \qquad \text{or} \qquad R = V/I.^* \qquad [3.3]$$

---

* It follows from this relationship that the *ohm* can be defined as *the resistance between two points of a circuit when a constant p.d. of 1 V, applied between these points, produces in the circuit a current of 1 A, assuming that there is no source of e.m.f. in the circuit between the two points.*

This relationship, known as *Ohm's Law*, is more complete and useful than that originally enunciated by Ohm. At this stage it is best to memorize Ohm's Law in one form only; and for this purpose the form $I = V/R$ is probably the most convenient.

From [3.2],

electrical power $= I^2 R$ watts

$$= I \times IR = IV\dagger \text{ watts} \qquad [3.4]$$

or $\qquad = \left(\dfrac{V}{R}\right)^2 \times R = \dfrac{V^2}{R} \text{ watts.} \qquad [3.5]$

Example 3.3 *An electric kettle takes 3 kW at 240 V. Calculate* (a) *the current and* (b) *the resistance of the heating element.*

(a) Since power $= IV = 3$ kW

$$= 3000 \text{ W,}$$

$\therefore \quad 3000 \, [\text{W}] = I \times 240 \, [\text{V}]$

so that $\quad I = 12.5 \, \text{A.}$

(b) From Ohm's Law,

$$I = V/R$$

$\therefore \quad 12.5 \, [\text{A}] = 240 \, [\text{V}]/R$

so that $\quad R = 19.2 \, \Omega.$

Example 3.4 *The kettle in Ex. 3.3 is required to raise the temperature of 1.7 l of water from 8°C to the boiling point, the input power remaining constant at 3 kW. If the efficiency of the kettle is 85 per cent, calculate the time taken. Assume the mass of 1 l of water to be 1 kg and the specific heat capacity of water to be 4190 J/kg K.*

Mass of water in kettle $= 1.7$ kg.

Rise of temperature $= 100 - 8 = 92°C = 92$ K.

$\therefore \quad$ useful heat required $= mct$

$$= 1.7 \, [\text{kg}] \times 4190 \, [\text{J/kg K}] \times 92 \, [\text{K}]$$

$$= 655\,000 \, \text{J.}$$

Since $\qquad$ efficiency of kettle $= 85$ per cent

$\therefore \quad$ energy supplied to kettle $= 655\,000/0.85$

$$= 771\,000 \, \text{J.}$$

---

† This relationship gives us an alternative definition of the *volt*, namely *the difference of potential between two points of a circuit carrying a constant current of 1 A when the power dissipated between these points is 1 W.*

Hence $$\text{time required} = \frac{\text{energy, in joules}}{\text{power, in watts}}$$

$$= \frac{771\,000\,[\text{J}]}{3000\,[\text{W}]}$$

$$= 257\,\text{s} = 4.28\,\text{min.}$$

Example 3.5 *An electric furnace is required to raise the temperature of 2 kg of brass from 12°C to 500°C in 15 min. The supply voltage is 230 V and the efficiency of the furnace is 80 per cent. Assuming the specific heat capacity of brass to be 370 J/kg K, calculate (a) the energy absorbed in kilowatt hours, (b) the power supplied to the furnace, (c) the current and (d) the resistance of the heating element.*

(a)   Rise of temperature $= 500 - 12 = 448°\text{C} = 488\,\text{K}$.

From expression [1.13],

$$\text{useful energy} = mct$$

$$= 2\,[\text{kg}] \times 370\,[\text{J/kg K}] \times 488\,[\text{K}]$$

$$= 361\,000\,\text{J}.$$

Since $$1\,\text{kW h} = 3\,600\,000\,\text{J},$$

$$\text{useful energy} = \frac{361\,000\,[\text{J}]}{3\,600\,000\,[\text{J/kW h}]}$$

$$= 0.1002\,\text{kW h}.$$

Allowing 80 per cent for the furnace efficiency,

$$\text{energy absorbed} = 0.1002/0.8 = 0.125\,\text{kW h}.$$

(b)   Input power, in kilowatts $= \dfrac{\text{energy, in kilowatt hours}}{\text{time, in hours}}$

$$= \frac{0.125\,[\text{kW h}]}{(15/60)\,[\text{h}]} = 0.5\,\text{kW}.$$

(c)   From expression (3.4),

$$(0.5 \times 1000)\,[\text{W}] = I \times 230\,[\text{V}]$$

$$\therefore \quad I = 105.8\,\Omega.$$

3.7 Electromotive force

It was shown in section 2.1 that when a current is passed for several minutes between two lead plates immersed in a dilute solution of sulphuric acid in water, a chocolate-colour coating is formed on the positive plate and that an electric current is obtained when the plates are then connected to a separate circuit; i.e. the combination of plates and acid became a voltaic cell capable of converting chemical energy into electrical energy (section 2.1). Such an arrangement is a source* of an *electromotive force* (e.m.f.); in other words, an electromotive force represents something in the cell which impels electricity through a conductor connected across the terminals of that cell. Electromotive force is represented by the symbol E, whereas difference of potential between two points is represented by V.

Consideration of theories accounting for the presence of an e.m.f. between the plates of an accumulator is outside the scope of this book. As far as we are concerned, the fact has to be accepted that when plates of different materials, such as lead and lead peroxide (as used in a lead-acid cell) or zinc and carbon (as used in the Leclanché cell), are placed in suitable solutions, an e.m.f. exists between the plates; and if a resistor is connected across them, an electric current flows through it.

Suppose E volts to be the e.m.f. of cell B in fig. 3.6 and I amperes to be the current when a circuit having a resistance R ohms is connected across the terminals. If the *internal resistance of the cell is negligible*,† the terminal voltage of the cell is the same as its e.m.f., namely E volts; hence

$$E = IR \qquad \text{or} \qquad I = E/R. \qquad\qquad [3.6]$$

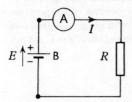

Fig. 3.6 E.M.F. of a cell

The e.m.f. of a cell can be measured by connecting a voltmeter across the terminals of the cell when the latter is on open circuit, i.e. when there is no other circuit connected across the terminals of the cell.

3.8 The standard cell

The most convenient reference standard on e.m.f. is the cadmium cell constructed in accordance with a specification drawn up by the International Electrotechnical Commission—a committee of experts from

---

* There are sources of e.m.f. other than voltaic cells, e.g. magnetic flux cutting a conductor and junctions of dissimilar metals at different temperatures (sections 6.1 and 18.7).

† The effect of the internal resistance of a cell is considered in section 4.3.

different countries. This standard cadmium cell is shown in section in fig. 3.7. The positive electrode is mercury and the negative electrode is an amalgam of mercury and cadmium. The electrolyte is a saturated solution of cadmium sulphate in water slightly acidulated with sulphuric acid. The mercurous sulphate acts as a depolarizer (section 19.4).

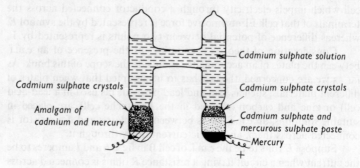

Cadmium sulphate solution

Cadmium sulphate crystals

Cadmium sulphate crystals

Amalgam of cadmium and mercury

Cadmium sulphate and mercurous sulphate paste

Mercury

Fig. 3.7   The cadmium standard cell

The e.m.f. of this cell is exactly 1.018 59 V at 20°C. Consequently the volt can be taken as 1/1.018 59 of the e.m.f. of the cadmium cell at 20°C. The e.m.f. of the cell falls by about 40 μV per degree rise of temperature over the temperature range 10°C to 25°C. Since the internal resistance is of the order of 1000 Ω, the cell is intended only as a standard of e.m.f. and not as a source of electrical energy.

3.9 Series connection of cells

Three accumulators were connected in series as in fig. 3.8, i.e. the negative terminal of the first cell was connected to the positive terminal of the second, and the negative of the second to the positive of the third. A

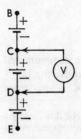

B

C

D

E

Fig. 3.8   Cells in series

voltmeter V was connected across various pairs of terminals in turn and the following results were obtained:

Terminals	E.M.F., volts
ED	2.16
DC	2.08
CB	2.1
EB	6.32

The sum of the e.m.f.s across ED, DC and CB = 6.34 volts, which is the same—within experimental error—as the total e.m.f. across EB. It is therefore evident that the e.m.f. of a number of cells connected in series is the sum of their individual e.m.f.s.

3.10 Parallel connection of cells

The positive ends of two similar accumulators were joined together to one end of a resistor R as shown in fig. 3.9. The negative ends were connected through milliammeters to the other end of R. When R was adjusted to 100 ohms, the readings on A, B and C were found to be 11.3, 8.5 and 19.8 milliamperes respectively. The milliammeters had relatively low resistance, so that the e.m.f. of the parallel cells is approximately $(19.8/1000) \times 100 = 1.98$ volts. These results indicate that when similar cells are in parallel, the e.m.f. is the same as that of one cell. On the other hand, the current is divided between the cells and the total current is the sum of the currents through the individual cells.

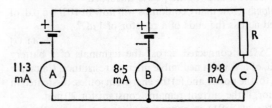

Fig. 3.9 Cells in parallel

3.11 Summary of important formulae

Electrical energy $= I^2Rt$ joules $\qquad\qquad$ [3.1]

$\qquad\qquad\quad = IVt$ joules

Electrical power $= I^2 R$ watts   [3.2]

$ = IV$ watts   [3.4]

$ = V^2/R$ watts   [3.5]

Ohm's Law:

$$I = VR/R, \qquad V = IR \qquad \text{or} \qquad R = V/I.$$ [3.3]

3.12 Examples

1. Calculate the current taken by a 100-W lamp connected across a 230-V supply. Also, find the corresponding resistance of the filament.

2. The field winding of an electric motor has a resistance of $157\,\Omega$. Calculate the power in kilowatts absorbed by the winding when the current is 2.8 A.

3. The heating element of an indirectly-heated thermionic valve takes a current of 0.4 A when the terminal voltage is 3.8 V. Calculate (a) the resistance of the element and (b) the power.

4. An electric fire takes 0.9 kW when connected across a 220-V supply. Calculate the resistance of the heating element.
    If the supply voltage is increased to 240 V, what is the percentage increase in the power? Assume the resistance to remain constant.

5. If a voltmeter has a resistance of $30\,\text{k}\Omega$, calculate the current and the power absorbed when it is connected across a 460-V supply.

6. The insulation resistance between a certain conductor and earth is $18\,\text{M}\Omega$. If the p.d. between the conductor and earth is 115 V, calculate (a) the leakage current and (b) the power absorbed.

7. Define the *coulomb*. How many coulombs will flow in 10 h if a p.d. of 1 mV is applied across the ends of a resistor of $1\,\mu\Omega$?
    (E.M.E.U.)

8. A resistor of $25\,\Omega$ is connected across the terminals of a battery having an e.m.f. of 6 V and a negligible internal resistance. Calculate (a) the current, (b) the power and (c) the energy, in joules, dissipated in the $25\text{-}\Omega$ resistor, if the current remains constant for 20 min.

9. When a resistor of $850\,\Omega$ is connected across a battery having negligible internal resistance, the current is 4.8 mA. Calculate (a) the e.m.f. of the battery and (b) the power.

10. A generator is supplying 80 lamps, each taking 60 W at 240 V. Calculate (a) the total current supplied by the generator, (b) the energy in kilowatt hours, consumed in 4 h and (c) the output power of the engine driving the generator, if the efficiency of the latter is 85 per cent.

11. An electric motor connected across a 440-V supply is developing 45 kW with an efficiency of 90 per cent. Calculate (a) the current, (b)

the input power and (c) the cost of running the motor at that load for 6 h if the charge for energy is 1.5 p/kW h.

12. State the name of and define the unit of power. Explain carefully the difference between *power* and *energy*.

    The energy absorbed in 10 min by a piece of electrical apparatus from a 240-V supply is 1.32×10^6 joules. Calculate (a) the current, (b) the quantity of electricity in coulombs taken in 1 min and (c) the energy in kilowatt hours absorbed in 96 h.

    (E.M.E.U.)

13. A 200-V Diesel-driven generator is required to supply the following: a lighting load comprising two hundred 100-W and five hundred 60-W lamps; a heating load of 35 kW; and miscellaneous loads totalling 25 A. Calculate the power output from the engine when the generator is supplying all the above loads at one time. Assume the efficiency of the generator to be 81 per cent.

    (N.C.T.E.C.)

14. A current of 5 A passes through a coil of 20-Ω resistance for 15 min. If the heat dissipated could be entirely used in heating water, what mass of water could have its temperature raised by 50°C? Assume the specific heat capacity of the water to be 4190 J/kg K.

    (U.L.C.I.)

15. The heating element of an electric kettle has a resistance of 40 Ω. Calculate the time required by a current of 6 A to raise the temperature of 1.8 litres of water from 14°C to the boiling point, if the efficiency of the kettle is 80 per cent. Also calculate: (a) the power and (b) the cost of the energy consumed at 1.4 p/kW h. Assume the specific heat capacity of water to be 4190 J/kg K.

16. The temperature of 100 litres of water is raised from 15°C to 65°C in 40 min in an electric heater having an efficiency of 90 per cent. If the supply voltage is 240 V, determine (a) the current taken, (b) the resistance of the heating element and (c) the power rating in kilowatts.

17. An electric furnace is required to raise the temperature of 3 kg of iron from 16°C to 750°C in 20 min. The supply voltage is 240 V. The efficiency of the furnace is 76 per cent and specific heat capacity of the iron is 480 J/kg K. Calculate (a) the current, (b) the resistance of the heating element, (c) the power and (d) the energy absorbed in kilowatt hours.

CHAPTER 4

Electrical Resistance

4.1 Resistors in series

Two resistors (i.e. a wire or other form of material used simply because of its resistance) R_1 and R_2 were connected in series across a battery, as shown in fig. 4.1, the current being indicated by an ammeter A. Also,

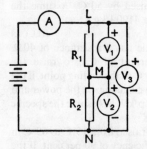

Fig. 4.1 Resistors in series

voltmeters V_1, V_2 and V_3 were connected to measure the p.d.s across R_1, R_2 and the whole circuit respectively. The following readings were obtained:

A	V_1	V_2	V_3
0.92 ampere	2.2 volts	3.6 volts	5.8 volts

By applying Ohm's Law, we find that $R_1 = 2.2/0.92 = 2.39\,\Omega$, $R_2 = 3.6/0.92 = 3.92\,\Omega$ and the resistance of the whole circuit $= 5.8/0.92 = 6.31\,\Omega$. But the sum of R_1 and $R_2 = 2.39 + 3.92 = 6.31\,\Omega$. Hence, it is seen that the total resistance of a circuit is the sum of the resistances connected in series; in other words, if R_1, R_2 and R_3 be in series, the total resistance R is given by

$$R = R_1 + R_2 + R_3 \qquad [4.1]$$

Also, if the voltmeters are of the moving-coil type (section 18.5), V_1 and V_2 indicate that L is at a higher potential than M and that M is at a higher potential than N; and V_3 indicates that L is at a higher potential than N by an amount equal to the sum of the p.d.s across LM and MN.

4.2 Resistors in parallel

Two resistors R_1 and R_2 were connected in parallel as in fig. 4.2, the currents being measured by ammeters A_1 and A_2. The total current was

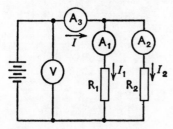

Fig. 4.2 Resistors in parallel

read on A_3. A voltmeter V read the p.d. across the circuits. It was found that the instrument readings were:

Voltmeter	A_1	A_2	A_3
5.9 volts	1.5 amperes	0.9 amperes	2.4 amperes

It will be seen that the total current is equal to the sum of the currents in the parallel circuits. Also, from Ohm's Law, it follows that $R_1 = 5.9/1.5 = 3.93\,\Omega$ and $R_2 = 5.9/0.9 = 6.55\,\Omega$.

The two resistors R_1 and R_2 in fig. 4.2 can be replaced by a single resistor R, as in fig. 4.3, the only condition being that the resistance of R must be such that the total current remains unaltered, i.e. in the above case:

$$2.4\,[\text{A}] = \frac{5.9\,[\text{V}]}{\text{resistance of R, in ohms}}$$

∴ resistance of $R = 5.9/2.4 = 2.46\,\Omega$.

Hence, $2.46\,\Omega$ may be said to be *equivalent* to $3.93\,\Omega$ and $6.55\,\Omega$ in parallel. It is evident that the equivalent resistance is less than either of the parallel resistances, but there does not seem to be any obvious connection between the values. For this problem it is more satisfactory to derive the relationship by considering the general case than by taking particular values.

Suppose I_1 and I_2 amperes to be the currents in parallel resistors R_1 and R_2 respectively when the p.d. is V volts (fig. 4.2). Then, by Ohm's Law, $I_1 = V/R_1$ and $I_2 = V/R_2$. If I is the total current indicated by A_3,

$$I = I_1 + I_2$$

$$= \frac{V}{R_1} + \frac{V}{R_2} = V\left(\frac{1}{R_1} + \frac{1}{R_2}\right).$$

If R in fig. 4.3 represents the resistance of a single or equivalent resistor through which a p.d. of V volts produces the same current I amperes, then $I = V/R$.

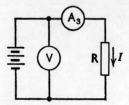

Fig. 4.3   Equivalent circuit of fig. 4.2

We have now derived two expressions for I; and by equating these expressions, we have

$$\frac{V}{R} = V\left(\frac{1}{R_1} + \frac{1}{R_2}\right)$$

$$\therefore \quad \frac{1}{R} = \frac{1}{R_1} + \frac{1}{R_2}. \qquad\qquad [4.2]$$

Let us apply this expression to the experimental results considered above:

$$\frac{1}{R_1} = \frac{1}{3.93} = 0.254 \quad \text{and} \quad \frac{1}{R_2} = \frac{1}{6.55} = 0.1527$$

$$\therefore \quad \frac{1}{R} = \frac{1}{R_1} + \frac{1}{R_2} = 0.254 + 0.1527 = 0.4067$$

and

$$R = \frac{1}{0.4067} = 2.46\,\Omega,$$

which is the same as the value previously derived.

The reciprocal of the resistance, that is, 1/resistance, is termed the *conductance*, the unit of conductance being 1 *siemens* (symbol, S). From expression [4.2], it follows that for resistors connected in parallel, the conductance of the equivalent resistor is the sum of the conductances of the parallel resistors.

The current in each of the parallel resistors R_1 and R_2 in fig. 4.2 can be expressed in terms of the total current thus:

$$V = I_1 R_1 = I_2 R_2 = (I - I_1) R_2$$

$$\therefore \quad (I - I_1) = I_1 R_1 / R_2$$

so that

$$I = I_1 \left(1 + \frac{R_1}{R_2} \right)$$

$$\therefore \quad I_1 = I \cdot \frac{R_2}{R_1 + R_2}. \qquad\qquad [4.3]$$

Similarly,

$$I_2 = I \cdot \frac{R_1}{R_1 + R_2}.$$

Example 4.1 *Three coils* A, B *and* C *have resistances 8, 12 and 15 Ω respectively. Find the equivalent resistance when they are connected* (a) *in series,* (b) *in parallel.*

(a) With the coils in series,

total resistance $= 8 + 12 + 15 = 35\,\Omega$.

(b) If R be the equivalent resistance of the three coils in parallel, then

$$\frac{1}{R} = \frac{1}{8} + \frac{1}{12} + \frac{1}{15} = 0.125 + 0.0833 + 0.0667$$

$$= 0.275 \text{ siemens}$$

$$\therefore \quad R = 3.64\,\Omega.$$

Example 4.2 *If* B *and* C *of example 4.1 are connected in parallel and* A *connected in series as in fig. 4.4, across a 20-V supply, find* (a) *the resistance of the combined circuit,* (b) *the current in each coil.*

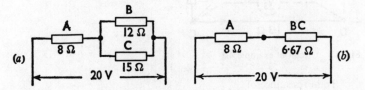

Fig. 4.4 Circuit of example 4.2

(a) Let R be the equivalent resistance of B and C, then

$$\therefore \quad \frac{1}{R} = \frac{1}{12} + \frac{1}{15} = 0.0833 + 0.0667 = 0.15 \text{ S},$$

$$\therefore \quad R = 6.67\,\Omega,$$

and total resistance $= 8 + 6.67 = 14.67\,\Omega$.

44   Principles of Electricity

(b) Total current $=20\,\text{V}/14.67\,\Omega=1.364\,\text{A}$, which is the current in A.

The p.d. across B and C in fig. 4.4(a) is the same as the p.d. across the equivalent resistance 6.67 in fig. 4.4(b),

$$\therefore\quad\text{total current}=\frac{\text{p.d. across B and C}}{\text{equivalent resistance of B and C}}$$

i.e.
$$1.364=\frac{\text{p.d. across B and C}}{6.67}$$

$\therefore$ p.d. across B and C $=1.364\times6.67=9.09\,\text{V}$.

Hence,
$$\text{current in B}=\frac{\text{p.d. across B}}{\text{resistance of B}}$$

$$=9.09/12=0.758\,\text{A}$$

and
$$\text{current in C}=1.364-0.758=0.606\,\text{A}.$$

Alternatively, using expression (4.3), we have:

$$\text{current in B}=1.364\times15/(12+15)=0.758\,\text{A}.$$

Example 4.3 *The resistance of the heating element of an electric iron is 180 Ω. It is connected to a 240-V supply by two conductors, each having a resistance of 1.3 Ω. Calculate: (a) the voltage across the heating element, (b) the voltage drop in the cable, (c) the power absorbed by the electric iron and (d) the power wasted in the cable.*

(a) The circuit is shown in fig. 4.5.

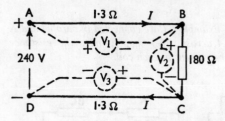

Fig. 4.5 Circuit of example 4.3

Total resistance of
$$\text{circuit}=1.3+180+1.3=182.6\,\Omega.$$

$$\therefore\quad\text{current}=\frac{\text{p.d. (volts) between A and D}}{\text{resistance (ohms) between A and D}}$$

$$=240/182.6=1.314\,\text{A}.$$

---

* When an arrow is drawn to represent the direction of the p.d. between two points the arrowhead should point towards the one which is at the higher potential, i.e. towards A in fig. 4.5.

But

$$\text{current also} = \frac{\text{p.d. between B and C}}{\text{resistance between B and C}}$$

$$\therefore \quad 1.314 = \frac{\text{p.d. between B and C}}{180}$$

$\therefore$ p.d. between B and C $= 1.314 \times 180 = 236.6$ V.

(b) $\text{Current also} = \dfrac{\text{p.d. between A and B}}{\text{resistance between A and B}}$

$$\therefore \quad 1.314 = \frac{\text{p.d. between A and B}}{1.3}$$

$\therefore$ p.d. between A and B $= 1.314 \times 1.3 = 1.7$ V.

Similarly, p.d. between C and D $= 1.314 \times 1.3$

$$= 1.7 \text{ V.}$$

Hence,

total voltage drop in cable $= 1.7 + 1.7 = 3.4$ V.

The existence of these potential differences can be demonstrated by connecting moving-coil voltmeters V_1, V_2 and V_3, as shown dotted in fig. 4.5. The readings on the instruments indicate that the potential of A is 1.7 V above that of B, the potential of B is 236.6 V above that of C and the potential of C is 1.7 V above that of D. Further, it is seen that the sum of the readings on V_1, V_2 and V_3 is equal to the total voltage between A and D.

(c) Since

power (watts) = current (amperes) $\times$ p.d. (volts)

$\therefore$ power of electric iron $= 1.314 \times 236.6 = 311$ W.

(d) Similarly,

power wasted in cable = current $\times$ voltage drop in cable

$$= 1.314 \times 3.4 = 4.47 \text{ W.}$$

4.3 Effect of the internal resistance of a cell

In section 3.7 it was pointed out that when a cell is supplying a current and the internal resistance is negligibly small, the terminal voltage is equal to the e.m.f. of the cell. In actual practice, however, the resistance of the electrolyte of the cell is seldom negligible and can easily be taken into account. Thus, in fig. 4.6, TT represent the terminals of a cell having an e.m.f. E and r represents the internal resistance of the cell. If a circuit of resistance R is connected across TT, then, from expression (4.1), the total resistance of the circuit is $R+r$, and the current I is given by:

$$I = \frac{E}{R+r}$$

and the terminal voltage $= V = IR$

$$= E - Ir.$$

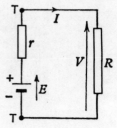

Fig. 4.6 Internal and external resistances

The internal resistance of a cell can be determined by connecting a very high resistance voltmeter across the terminals of the cell and noting the terminal voltage (a) with the cell on open circuit and (b) when the cell is supplying a known current I. The open-circuit reading gives the e.m.f., E, of the cell. If V be the terminal voltage when the current is I, then:

$$V = E - Ir$$

so that

$$r = (E - V)/I.$$

Example 4.4 *Two resistors, A and B, having resistances 10 Ω and 15 Ω respectively, are connected in parallel across a battery of four cells in series, as in fig. 4.7. Each cell has an e.m.f. of 2 V and an internal resistance of 0.2 Ω. Calculate: (a) the p.d. between battery terminals PQ, (b) the current through each resistor and (c) the total power dissipated in the resistors.*

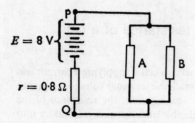

Fig. 4.7 Circuit diagram for example 4.4

(a)   Total e.m.f. of battery $= 2 \times 4 = 8$ V

and total internal resistance of battery $= 0.2 \times 4 = 0.8 \, \Omega$.

If R is the equivalent resistance of A and B:

$$\frac{1}{R} = \frac{1}{10} + \frac{1}{15} = 0.1 + 0.0667 = 0.1667 \, \text{S}$$

$\therefore \quad R = 6 \, \Omega$.

Hence, total resistance of circuit $= 6 + 0.8 = 6.8 \, \Omega$

and $\text{current} = \dfrac{\text{total e.m.f.}}{\text{total resistance}}$

$$= 8/6.8 = 1.177 \, \text{A}.$$

P.d. across A and B

$= $ total current $\times$ equivalent resistance of A and B

$= 1.177 \times 6 = 7.06$ V

$= $ terminal voltage of battery.

Alternatively, p.d. across the $0.8 \, \Omega$ in fig. 4.7

$$= 1.177 \times 0.8 = 0.94 \, \text{V}$$

$\therefore \quad$ terminal voltage of battery

$= $ battery e.m.f. $-$ voltage drop due to internal resistance

$$= 8 - 0.94 = 7.06 \, \text{V}.$$

(b)   Current in A $= \dfrac{\text{p.d. across A}}{\text{resistance of A}}$

$$= 7.06/10 = 0.706 \, \text{A}$$

and
      current in B $= 7.06/15 = 0.471$ A

or alternatively,

      current in B $= 1.177 - 0.706 = 0.471$ A.

(c)   Loss in resistors $=$ total current $\times$ p.d. across resistors

$$= 1.177 \times 7.06 = 8.3 \, \text{W}.$$

4.4 Comparison of the resistance of different materials

In fig. 4.8, CD represents 1 metre of Eureka wire having a diameter of about 0.5 mm, and DE and EF represent the same length and diameter of iron and copper wires respectively. The current is adjusted to about 0.5 A by means of R, and the differences of potential are measured by means of voltmeter V.

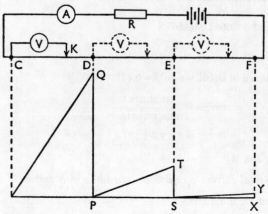

Fig. 4.8

One end of V is connected to terminal C and the other end to a sliding contact K. As the latter is moved from C to D, the p.d. increases uniformly from zero to PQ, as shown by the graph.

The test is repeated with one end of V connected to terminal D, and K is moved from D to E. The p.d. again increases uniformly from zero to ST. A repetition of the test on wire EF gives a p.d. increasing from zero to XY.

Since both the currents and the dimensions are the same for the three wires it follows that different materials having the same dimensions offer different resistances to the passage of an electric current. Thus, in the above experiment, ST is found to be about 7.5 times XY, while PO is about 30 times XY: in other words, an iron wire of given length and diameter has 7.5 times the resistance of a similar copper wire, while a Eureka wire of the same dimensions has 30 times the resistance of the copper wire.

4.5   Relationship between the resistance and the dimensions of a conductor

From the experiment described in the preceding section, it follows that for a uniform wire of a given material the value obtained by dividing the p.d. between any two points by the current, i.e. the resistance between those two points, is directly proportional to the distance between them.

Also, in section 4.2, it is explained that if two resistors, each having resistance R, are connected in parallel, the equivalent resistance R_e is given by:

$$\frac{1}{R_e} = \frac{1}{R} + \frac{1}{R} = \frac{2}{R}$$

$$\therefore \quad R_e = \tfrac{1}{2}R.$$

Hence, if two wires of the same material, having the same length and diameter, are connected in parallel, the resistance of the parallel wires is half that of one wire alone. But the effect of connecting two wires in parallel is exactly similar to doubling the area of the conductor. In just the same way the effect of connecting, say, five wires in parallel is the same as increasing the cross-sectional area of a wire five times, and the result is to reduce the resistance to a fifth of that of one wire. In general, we may therefore say that the resistance of a given length of a conductor is inversely proportional to its cross-sectional area.

Apart from the effect of temperature, referred to in section 4.6, the only other factor that influences the resistance is the nature of the material, as shown experimentally in section 4.4; hence we may now say that:

$$\left(\begin{array}{c}\text{resistance}\\ \text{of a wire}\end{array}\right) = \frac{\text{length of wire}}{\text{cross-sectional area}} \times \left(\begin{array}{c}\text{a constant for a}\\ \text{given material}\end{array}\right)$$

i.e. $R\,[\text{ohms}] = \dfrac{l\,[\text{metres}]}{a\,[\text{metres}^2]} \times \rho$ [4.4]

so that $\rho = Ra/l$ ohm metres

where ρ (Greek letter, pronounced 'rho') represents the constant.

If l is 1 m and a is 1 m² (for instance, if the resistance is being measured between the opposite faces of a metre cube of the material), the value of the resistance $= (1\,[\text{m}]/1\,[\text{m}^2]) \times \rho\,[\Omega\,\text{m}] = \rho$ ohm. Consequently the constant may be regarded as the resistance of a specimen, 1 m long and 1 m² in cross-sectional area, and is termed the *resistivity* of the material. For example, the resistivity of annealed copper at 20°C is 0.000 000 017 25 ohm metre. It is generally more convenient to use microhms rather than ohms in this connection, so that the above value then becomes 0.017 25 $\mu\Omega$ m.

Example 4.5 *Calculate the length of copper wire, 1.5 mm diameter, to have a resistance of 0.3 Ω, given that the resistivity of copper is 0.017 μΩ m.*

Cross-sectional area of wire $= (\pi/4) \times (1.5)^2 = 1.766\,\text{mm}^2$

$$= 1.766 \times 10^{-6}\,\text{m}^2$$

From expression [4.4], we have:

$$0.3\,[\Omega] = 0.017 \times 10^{-6}\,[\Omega\,\text{m}] \times \frac{\text{length}}{1.766 \times 10^{-6}\,[\text{m}^2]}$$

$\therefore$ length $= 31.2$ m.

4.6 Effect of temperature on resistance

Let us connect an incandescent lamp L in series with an ammeter A and a variable resistor R across a 240-V supply, as in fig. 4.9. A voltmeter V is connected across the lamp. Assuming the current through the voltmeter to

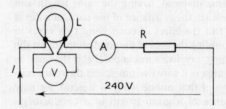

Fig. 4.9 Measurement of filament resistance at different voltages

be negligible compared with that through the lamp, we can obtain the resistance of the lamp by dividing the voltmeter reading by the corresponding ammeter reading. By varying the value of R, we can vary the current through the lamp and thus vary the filament temperature. In this way, we can determine the resistance of the lamp over a wide range of filament temperature.

The graph of fig. 4.10 shows the results obtained with a 100-W 240-V gas-filled lamp having a tungsten filament. It will be seen that as the temperature of the filament increases, so also does its resistance, and that the resistance at normal working temperature, i.e. with a terminal voltage of 240 V, is about ten times that of the lamp when cold.

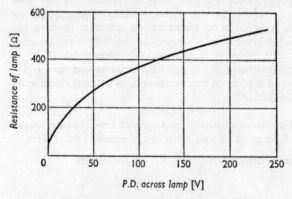

Fig. 4.10 Variation of filament resistance with voltage

The resistance of all pure metals, such as copper, iron, tungsten, etc., increases with increase of temperature. On the other hand, the resistance of carbon electrolytes and insulating materials such as rubber, paper, etc., decreases with increase of temperature. The resistance of certain alloys, such as manganin (copper, manganese and nickel), remains practically constant for a considerable variation of temperature.

4.7 Temperature coefficient of resistance

If the resistance of a coil of insulated copper wire is measured at various temperatures up to, say, 200°C, it is found to vary as shown in fig. 4.11, the resistance at 0°C being, for convenience, taken as 1 ohm. The resistance increases uniformly with the increase of temperature until it reaches 1.426 Ω at 100°C; i.e. the increase of resistance is 0.426 Ω for an increase of 100°C in the temperature, or 0.004 26 Ω/°C rise of temperature.

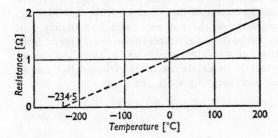

Fig. 4.11 Variation of resistance of copper with temperature

The ratio of the increase of resistance per degree Celsius rise of temperature to the resistance at some definite temperature, adopted as standard, is termed *the temperature coefficient of resistance* and is represented by the Greek letter α. From the figures given above, it follows that *if the standard temperature is assumed to be 0°C*, the temperature coefficient of resistance of annealed copper

$$= \frac{0.004\,26\,[\Omega/°C]}{1\,[\Omega]} = 0.004\,26/°C.$$

If the straight line of fig. 4.11 is extended backwards, the point of intersection with the horizontal axis is found to be −234.5°C. This means that for the range of temperature over which copper conductors are usually operated, the resistance varies as if it would be zero at −234.5°C. (Actually, the resistance/temperature relationship is not a straight line below about −50°C). Hence, over a range of 234.5°C, the variation of resistance in fig. 4.11 is $1\,\Omega$, so that the variation per degree Celsius is $1/234.5\,\Omega$, i.e.

temperature coefficient of resistance for standard annealed copper at $0°C$

$$= \frac{\text{change of resistance per degree change of temperature}}{\text{resistance at } 0°C}$$

$$= 1/234.5 = 0.004\,264/°C.$$

In general, if a material has a resistance R_0 at $0°C$ and a temperature coefficient of resistance α_0 at $0°C$, the increase of resistance for $1°C$ rise of temperature is $R_0\alpha_0$. If the temperature rises to t, the increase of resistance is $R_0\alpha_0 t$. Hence, if R_t be the resistance at t,

R_t = resistance at $0°C$ + increase of resistance

$$= R_0 + R_0\alpha_0 t = R_0(1 + \alpha_0 t) \qquad [4.5]$$

$$= R_0(1 + 0.004\,26t) \text{ for annealed copper.}$$

It is usually inconvenient and unnecessary to measure the resistance at $0°C$; for instance, in the case of the windings of electrical machines, it is often the practice to calculate the temperature rise after, say, three hours' operation at full load by measuring the resistance of the field windings before the commencement of the test and again immediately it is concluded. Suppose t_1 to be the initial temperature—usually taken as the temperature of the surrounding atmosphere—and t_2 to be the average temperature throughout the coils at the conclusion of the test. If R_1 and R_2 be the corresponding resistances (fig. 4.12), then:

$$R_1 = R_0(1 + \alpha_0 t_1) \qquad \text{and} \qquad R_2 = R_0(1 + \alpha_0 t_2)$$

$$\therefore \quad \frac{R_1}{R_2} = \frac{1 + \alpha_0 t_1}{1 + \alpha_0 t_2}$$

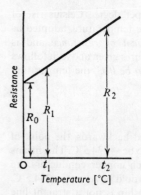

Fig. 4.12 Variation of resistance with temperature

Since $\alpha_0 = 1/234.5$ for standard annealed copper, then for this material,

$$\frac{R_1}{R_2} = \frac{1 + t_1/234.5}{1 + t_2/234.5} = \frac{234.5 + t_1}{234.5 + t_2} \qquad [4.6]$$

$$\therefore \quad t_2 = \frac{R_2}{R_1}(234.5 + t_1) - 234.5.$$

For a rough estimate of the increase of resistance for a given increase of temperature or of the increase of temperature for a given increase of resistance, it is often the practice to assume the standard temperature to be 20°C, i.e. to assume the temperature of the atmosphere surrounding the windings to be 20°C. This involves using a different value for the temperature coefficient of resistance; for instance, the temperature coefficient of resistance of annealed copper at 20°C is 0.003 92/°C. Hence, for a coil of copper wire having a resistance $R_2{}^0$ at 20°C, the resistance R_t at temperature t is given by:

$$R_t = R_{20}\{1 + 0.003\,92(t - 20)\}$$

In general,

$$R_t = R_{20}\{1 + \alpha_{20}(t - 20)\} \qquad [4.7]$$

Example 4.6 *The resistance of a coil of copper wire at the beginning of a heat test is 173 Ω, the temperature being 16°C. At the end of the test, the resistance is 212 Ω. Calculate the temperature rise of the coil. Assume the temperature coefficient of the resistance of copper to the 0.004 26/°C at 0°C.*

Substituting in expression [4.6], we have:

$$\frac{173}{212} = \frac{1 + (0.004\,26 \times 16)}{1 + (0.004\,26 \times t_2)}$$

$$\therefore \quad t_2 = 72.5°C,$$

so that temperature rise of coil $= 72.5 - 16 = 56.5°C$.

Example 4.7 *A certain length of aluminium wire has a resistance of 28.3 Ω at 20°C. What is its resistance at 60°C? The temperature coefficient of resistance of aluminium is 0.004 03/°C at 20°C.*

Substituting in expression [4.7], we have:

$$R_{60} = 28.3\{1 + 0.004\,03(60 - 20)\}$$

$$= 32.86\ \Omega.$$

4.8 Applications of the heating effect of an electric current

We have already seen in section 2.1 that when an electric current passes through a resistor, heat is generated and the temperature of the resistor is raised. The following applications are a few examples of the many ways in which this effect is utilized:

(a) Electric fires, cookers, etc.

The heating element is usually an alloy of nickel and chromium, since this material has a high resistivity and is capable of withstanding a high temperature without becoming oxidized when exposed to the air.

(b) Incandescent electric lamps

In the incandescent lamp, the filament must be capable of operating for long periods at a high temperature without appreciable deterioration, and for this duty there are only two materials that have proved satisfactory, namely carbon and tungsten. Owing to its relatively low efficiency, the carbon filament lamp, however, is obsolete.

The modern tungsten-filament lamp is gas-filled. If the gas were removed from the bulb, loss of heat from the filament by convection would be prevented; but the vacuum has the disadvantage that the filament vaporizes at a lower temperature than it does when gas is present. This effect is very similar to the variation of the boiling point of water with pressure.

The vaporization of the filament reduces the cross-sectional area of the filament, thereby increasing its resistance and reducing the temperature and the luminous intensity of the lamp. It also allows the tungsten to condense on the internal surface of the bulb, blackening the latter and reducing the luminous intensity still further. Consequently the highest temperature at which it is practicable to work the filament is limited to about 2100°C, with a vacuum corresponding to a pressure of about 0.0001 mm of mercury (about $13 \, \text{mN/m}^2$).

By the introduction of an inert gas (namely a gas which has no chemical action on the filament), such as nitrogen or argon, the temperature of the filament can be raised to about 2500°C before blackening takes place at an excessive rate. But if no other change were made except merely to introduce a gas, it would be found that the amount of heat lost by convection between the filament and the bulb would be so great that the power required to maintain the filament temperature at 2500°C would have increased more in proportion than the light given out by the lamp. Consequently the efficiency would be lower than that of the vacuum lamp. This difficulty is overcome by winding the filament as a very close helix (fig. 4.13), in fact so close that the gas is unable to pass between the turns and can merely glide over the outside of the helix. In other words, the surface with which the gas can come into contact is practically the same as that of a rod of diameter *d* and length *l* (fig. 4.13); and since this area is far less than the surface area of the filament itself, the loss of heat by convection is very considerably reduced.

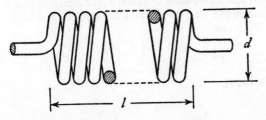

Fig. 4.13 A helical filament

In gas-filled lamps up to 100-watt size, the coiled filament is wound into a coarser helix, thereby reducing still further the effective filament surface exposed to the gas. This coiled-coil lamp has an efficiency of about 10 to 20 per cent higher than that of the corresponding lamp with the filament wound as a single helix.

(c) Fuses

A fuse is a wire or strip of metal inserted in a circuit for the purpose of interrupting or opening that circuit when the current exceeds a certain prearranged value. The wire is usually made of a metal, such as copper, tin, lead or an alloy which melts comparatively easily. The fusing current depends upon a large number of factors, such as the material, the diameter and length of the wire, the ventilation of the fuseholder and the duration of the current.

The fuse wire should be mounted on a fireproof holder, such as porcelain. Large fuses are often covered with asbestos sleeves to prevent any metal vapour being deposited on the porcelain when the fuse 'blows', since such a deposit may be sufficient to provide a conducting path between the fuse terminals.

(d) Temperature rise permissible in electrical machines

When a generator is supplying electrical power or an electric motor is supplying mechanical power, the machine is said to be *loaded*, and the power lost in the machine is converted into heat, thereby raising the temperature of the windings. The maximum temperature that is permissible depends upon the nature of the insulating materials employed; thus, materials such as paper and cotton become brittle if their temperature is allowed to exceed about 100°C, whereas materials, such as mica and fibrous glass, can withstand a much higher temperature without any injurious effect on their insulating and mechanical properties.

Since the temperature rise of an electrical machine, when loaded, is largely due to the I^2R losses in the windings, it follows that the greater the load, the greater are the losses and therefore the higher the temperature rise. The *full load* or *rated output* of a machine is the maximum output power obtainable from the machine under certain specified conditions, e.g. for a specified temperature rise after the machine has supplied that load continuously for several hours.

4.9 Kirchhoff's laws

A German physicist, Gustav Kirchhoff (1824–87), enunciated two laws which can be very useful when problems on the electric circuit have to be solved.

First Law *If several conductors meet at a point, the total current flowing towards that point is equal to the total current flowing away from it, i.e. the algebraic sum of the currents is zero.*

Thus, if five wires were joined together at J (fig. 4.14) and if the arrowheads represent the directions of the respective currents, then total current

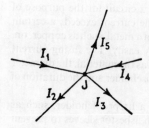

Fig. 4.14   Circuit to illustrate Kirchhoff's First Law

flowing towards $J = I_1 + I_4$, while total current flowing away from $J = I_2 + I_3 + I_5$. Since there is no accumulation of electric charge at J, it follows that:

$$I_1 + I_4 = I_2 + I_3 + I_5 \quad \text{or} \quad I_1 - I_2 - I_3 + I_4 - I_5 = 0$$

In general,

$$\sum I^* = 0. \tag{4.8}$$

---

* $\sum$ is the Greek letter *sigma* and is used to represent 'algebraic sum of'.

Second Law *In any closed circuit, the algebraic sum of the products of the current and the resistance of each part of the circuit is equal to the resultant e.m.f. in the circuit.*

Let us consider the simple case of a circuit consisting of a battery A (fig. 4.15) having an e.m.f. E_1 volts and another battery B having an e.m.f. E_2 volts, connected in opposition, and two resistances R_1 and R_2 in series as shown. The batteries are assumed to have negligible internal resistance.

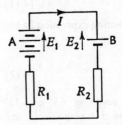

Fig. 4.15 Circuit to illustrate Kirchhoff's Second Law

If E_1 is greater than E_2, the resultant e.m.f. is $(E_1 - E_2)$ volts acting in a clockwise direction round the circuit and circulating a current I amperes. The sum of the products of the current and the resistances is $I(R_1 + R_2)$.

In such a case, the above law merely states that

$$I(R_1 + R_2) = E_1 - E_2$$

which is Ohm's Law.

Let us now consider the effect of adding a resistance R as shown in fig. 4.16. Suppose the current through A to be I_1 and that through B to be I_2, each current being assumed to be in the same direction as the corresponding e.m.f. Since both I_1 and I_2 are assumed to flow towards junction C, it follows from Kirchhoff's First Law that the current flowing from C towards R must be $I_1 + I_2$.

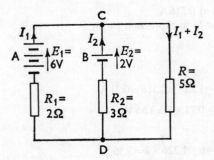

Fig. 4.16 Circuit to illustrate Kirchhoff's Laws

By applying Kirchhoff's Second Law to the circuit formed by A, R_1 and R, we have:

$$E_1 = I_1 R_1 + (I_1 + I_2)R. \tag{4.9}$$

Similarly for circuit B, R_2 and R, we have:

$$E_2 = I_2 R_2 + (I_1 + I_2)R. \tag{4.10}$$

Also, for circuit A, R_1, B and R_2,

$$E_1 - E_2 = I_1 R_1 - I_2 R_2. \tag{4.11}$$

Expression [4.11] can also be obtained by subtracting expression [4.10] from expression [4.9].

In general,

$$\sum E = \sum IR \tag{4.12}$$

Example 4.8 *Two batteries, A and B, having e.m.f.s of 6 V and 2 V respectively and internal resistances of 2 Ω and 3 Ω respectively, are connected in parallel across a 5-Ω resistor. Calculate (a) the current through each battery and (b) the terminal voltage.*

The batteries and resistances are as shown in fig. 4.16. For the circuit formed by A, R_1 and R,

$$6 = 2I_1 + 5(I_1 + I_2) = 7I_1 + 5I_2. \tag{4.13}$$

and for circuit B, R_2 and R,

$$2 = 3I_2 + 5(I_1 + I_2) = 5I_1 + 8I_2. \tag{4.14}$$

Multiplying expression [4.13] by 5 and expression [4.14] by 7,

$$35I_1 = 25I_2 = 30$$

and

$$35I_1 + 56I_2 = 14$$

$$\therefore \quad -31I_2 = 16$$

and

$$I_2 = -0.516 \, \text{A}.$$

i.e. battery B is being *charged* at 0.516 A.

Substituting this value for I_2 in expression [4.13],

$$7I_1 - 5 \times 0.516 = 6$$

$$\therefore \quad I_1 = 1.226 \, \text{A}.$$

Total current through $R = 1.226 - 0.516 = 0.71 \, \text{A}$

$$\therefore \quad \text{terminal voltage} = 0.71 \times 5 = 3.55 \, \text{V}.$$

Alternatively

$$\text{terminal voltage} = 6 - 1.226 \times 2 = 3.55 \, \text{V}$$

or

$$\text{terminal voltage} = 2 + 0.516 \times 3 = 3.55 \, \text{V}.$$

Example 4.9 *Two batteries, A and B, having e.m.f.s of 6 V and 4 V respectively and negligible internal resistance, are connected in series with two resistors, P and Q, having resistances of 20 Ω and 5 Ω respectively, as shown in fig. 4.17. Calculate the current and the p.d. across each resistor. Plot a graph showing the potentials of points D, E, F and G relative to the potential of point C.*

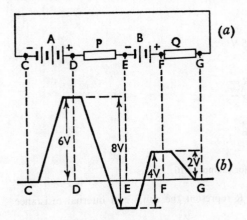

Fig. 4.17 Diagrams for example 4.9

Resultant e.m.f. = 6 + 4 = 10 V.

This e.m.f. acts in an anti-clockwise direction in the circuit shown in fig. 4.17(a).

Total resistance = 20 + 5 = 25 Ω.

∴ current = 10/25 = 0.4 A.

P.d. across P = 0.4 × 20 = 8 V

and p.d. across Q = 0.4 × 5 = 2 V.

The potentials of different points in the circuit relative to the potential of point C are represented by the graph in fig. 4.17(b); thus, between C and D there is a rise in potential of 6 V due to the e.m.f. of battery A. Owing to the voltage drop in resistor P, there is a fall of potential of 8 V between D and E, so that the voltage drop between C and E is 6 − 8 = −2 V, i.e. E is 2 V negative relative to C. Point F is 4 V positive relative to E and is therefore 2 V positive relative to C. The voltage drop in Q is 2 V, i.e. G is 2 V negative relative to F and is therefore at the same potential as C. It is assumed that the resistance of the conductor connecting G to C is negligible, so that G and C are at the same potential.

It will be seen that the resultant of the *IR* products is (0.4 × 20 + 0.4 × 5), namely 10 V, and that the resultant e.m.f. is (6 + 4), namely 10 V; in other words, between C and G there is a rise of potential of 10 V due to the e.m.f.s of A and B and a fall of potential of 10 V due to the *IR* drops in P and Q.

Example 4.10   *Three similar primary cells are connected in series to form a closed circuit as shown in fig. 4.18. Each cell has an e.m.f. of 1.5 V and an internal resistance of* 30 Ω. *Calculate the current and show that points* A, B *and* C *are at the same potential.**

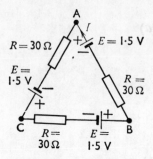

Fig. 4.18   Circuit diagram for example 4.10

In fig. 4.18, *E* and *R* represent the e.m.f. and internal resistance respectively of each cell.

$$\text{Total e.m.f.} = 1.5 \times 3 = 4.5\,\text{V},$$

$$\text{total resistance} = 30 \times 3 = 90\,\Omega$$

$$\therefore \quad \text{current} = 4.5/90 = 0.05\,\text{A}.$$

The voltage drop due to the internal resistance of each cell is 0.05 × 30, namely 1.5 V. Hence the e.m.f. of each cell is absorbed in sending the current through the internal resistance of that cell, so that there is no difference of potential between the two terminals of the cell. Consequently the three junctions A, B and C are at the same potential.†

4.10 Systems of distribution

At this stage we shall consider only direct-current (d.c.) systems. Distribution systems are usually supplied from d.c. generators or from rectifiers at

---

* It can be interesting to draw the circuit of fig. 4.18 on the blackboard and ask students which of points A and B is at the higher potential!
† This result has important practical applications, e.g. in connection with the non-existence of a third harmonic in the terminal voltage of a delta-connected 3-phase machine or transformer. The same principle can be applied to explain why there is no magnetic leakage in a toroid uniformly wound with a magnetizing winding—the ampere-turns per unit length are absorbed in sending the magnetic flux through the reluctance of that length, irrespective of how small that length may be. Hence, all points of the toroid are at the same magnetic potential.

a voltage that is maintained approximately constant on any particular system. In fact, one of the Regulations governing the distribution of electricity stipulates that the voltage at a consumer's premises must not vary by more than ± 6 per cent of the declared value. For instance, if a consumer is supplied at a nominal voltage of, say, 230 V, the actual voltage must not exceed 244 V and must not fall below 216 V. If lamps (or other apparatus) are connected in series across the supply mains as in fig. 4.19, it is evident that each lamp has to carry the same current; and if the five lamps of fig. 4.19 are exactly similar and the supply voltage is 230 V, the voltage across each lamp is 230/5, namely 46 V.

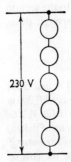

Fig. 4.19 Lamps in series

This system has the following disadvantages: (a) all the lamps must have the same current rating and have to be alight simultaneously; (b) if one lamp burns out, the other lamps are extinguished; and (c) if one lamp develops a short-circuit, the supply voltage is divided between the remaining four lamps and the excessive temperature will cause one of them to burn out very soon. Hence the system generally employed is that in which lamps, heaters, motors, etc., are connected in *parallel* across the supply mains.

Figure 4.20 gives the general arrangement of a d.c. two-wire system. Two d.c. generators DD are shown connected in parallel to bus-bars BB.

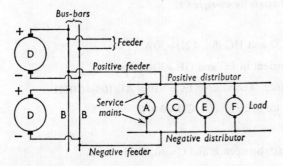

Fig. 4.20 A distribution system

The bus-bars are two copper bars that extend the whole length of the switchboard, 'bus' being an abbreviation of 'omnibus', a Latin word meaning 'for all'. Thus, all the generators at the generating station and all the cables connecting the station to various 'feeding' points are connected to the bus-bars. The cables radiating from the station are called *feeders*, whereas *distributors* are cables to which the *service mains* supplying the individual consumers are connected. A distributor is connected at one or more points to feeders, but no service mains are connected to the latter.

In practice the distributors are interconnected to form a network that is almost like that of a spider's web. This network is connected to feeders at the most suitable points. Such an arrangement has the advantage that if for some reason a feeder has to be disconnected, the current to the section normally supplied by that feeder can still be supplied through other feeders and distributors.

Example 4.11 *Two loads, A and B (fig. 4.21), taking 50 A and 30 A respectively, are connected to a two-wire distributor at distances of 200 m and 300 m respectively from the feeding point, the p.d. at which is 120 V. The resistance of the distributor is 0.01 Ω per 100 m of single conductor. Find: (a) the p.d. across each load, (b) the cost of the energy wasted in the distributor if the above loads are maintained constant for 10 hours. Assume the cost of energy to be 1.2 p/kW h.*

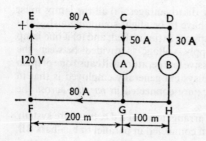

Fig. 4.21 Circuit diagram for example 4.11

(a) Current in CD and HG (fig. 4.21) = 30 A.

and   current in EC and GF = 30 + 50 = 80 A.

Resistance of conductor EC = 0.01 × 200/100 = 0.02 Ω

and   resistance of conductor CD = 0.01 Ω.

Hence,

p.d. between E and C = 80 × 0.02 = 1.6 V.

Similarly,   p.d. between G and F = 1.6 V.

But

p.d. between E and F = sum of the p.d.s in circuit ECGF

i.e. $120 = 1.6 + $ p.d. between C and G $+ 1.6$.

If the voltage drop in the service main be neglected,

p.d. across load A $= 120 - 3.2 = 116.8$ V.

Also

p.d. between C and D $= 30 \times 0.01 = 0.3$ V

and

p.d. between H and G $= 0.3$ V.

But

p.d. between C and G = sums of p.d.s in circuit CDHG

i.e. $116.8 = 0.3 + $ p.d. between D and H $+ 0.3$

so that

p.d. across load B $= 116.8 - 0.6 = 116.2$ V.

(b) Power wasted in conductors EC and FG

$= $ current $\times$ voltage drops in EC and FG

$= 80 \times 3.2 = 256$ W.

Power wasted in conductors CD and HG $= 30 \times 0.6 = 18$ W

∴ total power wasted in distributor $= 256 + 18 = 274$ W

$= 0.274$ kW

and

energy wasted in 10 hours $= 0.274 \times 10 = 2.74$ kW h

∴ cost of this energy $= 2.74 \times 1.2 = 3.288$ p.

Example 4.12 *A two-wire ring distributor (i.e. a distributor in which each conductor forms a complete circuit or loop, as in fig. 4.22) is 300 m long and is fed at 240 V at A. At point B, 150 m from A, there is a load of 120 A and at C, 100 m in the opposite direction, there is a load of 80 A. The resistance per 100 m of single conductor is 0.03 Ω. Find: (a) the current in each section, (b) the p.d.s at B and C.*

(a) Let x be the current, in amperes, from A to B in the positive conductor.

From Kirchhoff's First Law, it follows that the current from B to C in positive conductor

$= x - 120$ A

and current from C to A in positive conductor

$= x - 120 - 80 = x - 200$ A.

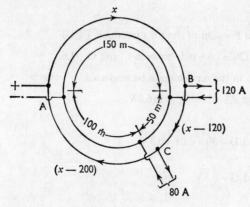

Fig. 4.22   Circuit diagram for example 4.12

Resistance of positive conductor between A and B

$$= 0.03 \times 150/100 = 0.045 \, \Omega$$

resistance of positive conductor between B and C

$$= 0.03 \times 50/100 = 0.015 \, \Omega$$

and resistance of positive conductor between C and A

$$= 0.03 \, \Omega.$$

Hence,

voltage drop in positive conductor between A and B

$$= x \times 0.045 \, \text{V},$$

voltage drop in positive conductor between B and C

$$= (x - 120) \times 0.015 \, \text{V}$$

and voltage drop in positive conductor between C and A

$$= (x - 200) \times 0.03 \, \text{V}.$$

Since there is no e.m.f. in the loop formed by the positive conductor, it follows from Kirchhoff's Second Law that:

$$x \times 0.045 + (x - 120) \times 0.015 + (x - 200) \times 0.03 = 0$$

∴ $x = 86.7 \, \text{A} = $ current in section AB.

Current in section BC $= 86.7 - 120 = -33.3 \, \text{A}$

$$= 33.3 \, \text{A from C to B in positive conductor}$$

and

current in section CA $= 86.7 - 200 = -113.3 \, \text{A}$

$$= 113.3 \, \text{A from A to C}$$

in positive conductor.

(b) Voltage drop in positive and negative conductors between A and B

$$= 86.7 \times 0.045 \times 2 = 7.8 \text{ V}$$

and voltage drop in positive and negative conductors between A and C

$$= 113.3 \times 0.03 \times 2 = 6.8 \text{ V}$$

$\therefore$ p.d. across load at B $= 240 - 7.8 = 232.2$ V

and

p.d. across load at C $= 240 - 6.8 = 233.2$ V.

The difference of 1 V between the p.d.s across the loads at B and C should agree with the voltage drop calculated from the current in section BC and the resistance of that section, namely $33.3 \times 0.015 \times 2 = 0.999$ V. The very slight discrepancy between the two values is due to the fact that the value of x was limited to three significant figures. This degree of accuracy is sufficient for most practical purposes.

4.11 Summary of important formulae

For resistors in series,

$$R = R_1 + R_2 + R_3 + \text{etc.} \qquad [4.1]$$

For resistors in parallel,

$$\frac{1}{R} = \frac{1}{R_1} + \frac{1}{R_2} + \frac{1}{R_3} + \text{etc.} \qquad [4.2]$$

For R_1 and R_2 in parallel,

$$I_1 = I \cdot \frac{R_2}{R_1 + R_2} \qquad [4.3]$$

$$R = \frac{\rho l}{a} \qquad [4.4]$$

$$R_t = R_0(1 + \alpha_0 t) \qquad [4.5]$$

or

$$R_t = R_{20}\{1 + \alpha_{20}(t - 20)\} \qquad [4.7]$$

For standard annealed copper,

$$\frac{R_1}{R_2} = \frac{234.5 + t_1}{234.5 + t_2} \qquad [4.6]$$

or

$$t_2 = \frac{R_2}{R_1}(234.5 + t_1) - 234.5$$

Kirchhoff's First Law. For several conductors meeting at a point,

algebraic sum of currents $= \sum I = 0.$ [4.8]

Kirchhoff's Second Law. For a closed network,

algebraic sum of e.m.f.s = algebraic sum of products of currents
and their respective resistances

i.e. $$\sum E = \sum IR.$$ [4.12]

4.12 Examples

1. The resistances of two coils A and B are 14 Ω and 18 Ω respectively. Calculate the combined resistance when they are (a) in series, (b) in parallel.

2. Two metal-filament lamps take 0.8 A and 0.9 A respectively when connected across a 110-V supply. Calculate the value of the current when they are connected in series across a 220-V supply, assuming the filament resistance to remain unaltered. Also find the voltage across each lamp.

3. Two circuits, A and B, are connected in parallel across a 50-V supply. Circuit A is found to take 120 W and the total current is 4.2 A. Calculate the resistances of A and B and the power absorbed by B.

4. The element of a 500-W electric iron was designed for use on a 200-V supply. Calculate the percentage increase of heat obtained if the iron is to be used on a 240-V supply. It can be assumed that the resistance of the element remains unaltered.

    What value of resistance is needed to be connected in series in order that the iron can be operated normally from this 240-V supply?

    If the iron is to be redesigned for the 240-V supply, calculate the resistance value of the new element if the power consumption is to remain at 500 W. (N.C.T.E.C.)

5. Three coils of resistance 8 Ω, 12 Ω and 24 Ω respectively are joined in (a) series, (b) parallel. What is their joint resistance for each grouping? What current would flow in each case if the group were connected to a 44-V supply? (U.L.C.I.)

6. Three resistors of 6 Ω, 8 Ω and 12 Ω are connected in parallel and thence to a source of supply. If a current of 10 A flows through the circuit, calculate the current flowing through each resistor. (U.E.I.)

7. State Ohm's Law both in words and in symbols.

    Two resistors of 10 Ω and 30 Ω respectively are connected in series across a 240-V supply. Calculate the value of the supply current when a resistor of 60 Ω is connected in parallel across (a) the 30-Ω resistor, (b) the whole-circuit. (N.C.T.E.C.)

8. Two resistors, R_1 and R_2, are connected in parallel to a certain supply. If the current taken from the supply is 5 A, calculate the value

of R_1 given that $R_2 = 6\,\Omega$ and the current through R_1 is 2 A.

Calculate also the total power absorbed by the two resistors when they are connected (a) as above, (b) in series, to the same supply.

(N.C.T.E.C.)

9. A circuit consists of two parallel resistors, having resistances of $20\,\Omega$ and $30\,\Omega$ respectively, connected in series with $15\,\Omega$. If the current through the 15-Ω resistor is 3 A, find (a) the currents through the 20-Ω and the 30-Ω resistors respectively, (b) the voltage across the whole circuit and (c) the total power.

10. A circuit consists of two resistors A and B, in parallel, connected in series with another resistor C. The resistances of A, B and C are $25\,\Omega$, $16\,\Omega$ and $5\,\Omega$ respectively. If the circuit is connected across a 30-V supply, calculate the current in each resistor.

11. Three resistors, $18\,\Omega$, $20\,\Omega$ and $30\,\Omega$ respectively, are connected in parallel; two more resistors, $3.6\,\Omega$ each, are connected in parallel with each other. The two groups of resistors are then connected in series to a 45-V battery. Draw a circuit diagram and calculate the total current, the current in each resistor, the power absorbed in each of the 3.6-Ω resistors and the p.d. across the 20-Ω resistor. (E.M.E.U.)

12. An accumulator has a terminal voltage of 1.9 V when supplying a current of 8 A. The terminal voltage rises to 2.03 V immediately the load is switched off. Calculate the internal resistance of the cell.

13. A battery has a terminal p.d. of 52 V on open circuit and 48 V when connected across a 10-Ω resistor. Find its internal resistance.

14. State Ohm's Law. What is meant by the internal voltage drop of a cell when current flows through it?

A cell has an e.m.f. of 1.45 V and an internal resistance of $0.5\,\Omega$. Calculate its internal voltage drop and the terminal potential difference when it is delivering a current of 0.5 A. (U.L.C.I.)

15. A voltmeter connected across an accumulator reads 2.06 V when the cell is on open circuit. The voltmeter reading immediately falls to 1.92 V when a 0.2-Ω resistor is connected across the terminals of the cell. Calculate (a) the current, (b) the internal resistance of the cell, (c) the total electrical power generated in the cell, (d) the power in the external resistor and (e) the power wasted due to the internal resistance of the cell.

16. A battery of ten primary cells has an open-circuit e.m.f. of 15 V. When a 30-Ω resistor is connected across the battery, the terminal p.d. is 12. V. Calculate (a) the current, (b) the internal resistance per cell and (c) the power in the external circuit.

17. (a) Show by means of diagrams what you understand by series and parallel grouping of cells.

(b) A 10-Ω resistor is connected across the terminals of a battery consisting of four cells in series. Each cell has an e.m.f. of 1.5 V and an internal resistance of $0.5\,\Omega$. Calculate (i) the current, (ii) the terminal voltage, (iii) the energy in joules absorbed in the external circuit if the current remains constant for 2 min. (E.M.E.U.)

18. Resistors having resistances of $4\,\Omega$ and $6\,\Omega$ in parallel are connected

to three 1.5-V cells in series, each cell having an internal resistance of 2.2 Ω. Calculate the current in the 4-Ω and the 6-Ω resistors.

Calculate the total current if a 2-V accumulator of negligible resistance is connected in circuit but with polarity opposing that of the three 1.5-V cells. (N.C.T.E.C.)

19. Two resistors, A and B, are connected in parallel across a battery of twenty cells in series. Each cell has an e.m.f. of 2.1 V and an internal resistance of 0.13 Ω. The resistances of A and B are 20 Ω and 28 Ω respectively. Calculate (a) the current from the battery, (b) the currents in A and B and (c) the terminal voltage of the battery.

20. Define the *coulomb*, the *ampere* and the *volt*.

A circuit consisting of three resistors in parallel of 4 Ω, 6 Ω and 12 Ω respectively is connected to a cell of internal resistance 1 Ω and e.m.f. 1.5 V. Assuming the e.m.f. of the cell to remain constant, calculate the number of coulombs delivered in 1 min. (U.L.C.I.)

21. Two cells are connected in series and the circuit is completed by means of two parallel coils of resistances 7 Ω and 5 Ω respectively, connected in parallel. One cell has an e.m.f. of 1.5 V and an internal resistance of 1.2 Ω and the other cell an e.m.f. of 1.1 V and an internal resistance of 0.7 Ω. Calculate (a) the current in each cell, (b) the current in each coil and (c) the terminal voltage of each cell. (E.M.E.U.)

22. What is the difference between a primary and a secondary cell? Six cells, each having an e.m.f. of 2 V and an internal resistance of 2 Ω, are connected in two groups of three in series, and the two groups connected in parallel to an external resistor of 30 Ω. Sketch the arrangement and calculate the current which will flow through the external resistor. (U.E.I.)

23. Derive expressions for the single equivalent resistances of three resistances r_1, r_2 and r_3 when they are (a) in series, (b) in parallel.

A coil of 20 Ω resistance is joined in parallel with a coil of x Ω resistance. This combination is then joined in series with a piece of apparatus A, and the whole circuit connected to 100-V mains. What must be the value of x so that A shall dissipate 600 W with 10 A passing through it? (U.L.C.I.)

24. Four resistors, each of 240 Ω, are connected in parallel, and the group then joined in series with a heater of 40 Ω resistance. The complete circuit is connected to a 200-V supply. Calculate the heat dissipated by the heater in joules per minute. (U.L.C.I.)

25. A circuit consists of three resistors of 3 Ω, 4 Ω and 6 Ω in parallel and a fourth resistor of 4 Ω in series. A battery of e.m.f. 12 V and internal resistance 6 Ω is connected across the circuit.

(a) Draw a diagram of the arrangement.
(b) What is the total external resistance?
(c) What is the total current in the circuit?
(d) What is the terminal voltage across the battery? (E.M.E.U.)

26. Calculate the resistance of 1 km of aluminium wire, given that the diameter of the wire is 6 mm and the resistivity of aluminium is 0.028 μm m.

27. Calculate the cross-sectional area of a copper conductor, 300 m long, such that it may carry 500 A with a voltage drop of 8 V. Assume the resistivity of copper to be $0.019 \, \mu\Omega$ m.

28. The resistance of 7.7 m of manganin wire, of uniform cross-section $1.3 \, mm^2$, is $2.6 \, \Omega$. Calculate the resistivity of manganin in microhm metre.

29. A coil has 10 000 turns of insulated copper wire, the mean length per turn being 150 mm and the area of cross-section $0.3 \, mm^2$. Calculate the resistance of the coil, assuming the resistivity of copper at working temperature to be $0.02 \, \mu\Omega$ m.

30. A copper rod, 6 mm diameter and 1 m long, has a resistance of $670 \, \mu\Omega$. If the rod is drawn out to a wire having a uniform cross-sectional area of $0.08 \, mm^2$, calculate its new resistance, assuming the temperature to remain unaltered.

31. Define the term *resistivity*. In a test on a 100-mm strip of copper, the resistance was found to be $171 \, \mu\Omega$. The average cross-section area was $9.92 \, mm^2$. Calculate the resistivity of the copper. (N.C.T.E.C.)

32. (a) Define *resistivity*.

    (b) A copper wire, 100 m long, has a diameter one-third the diameter of a length of manganin wire. The resistivities of manganin and copper are 0.44 and $0.018 \, \mu\Omega$ m. Calculate the length of the manganin wire if it has the same resistance as the copper wire.

    (c) Calculate the resistance which should be in parallel with a resistance of $4 \, \Omega$ to give an equivalent resistance of $3 \, \Omega$. (U.L.C.I.)

33. A coil of insulated copper wire has a resistance of $85 \, \Omega$ at $10°C$. What is its resistance at $80°C$? Assume the temperature coefficient of resistance of copper to be $0.0043/°C$ at $0°C$.

34. The coil of a relay takes 0.12 A when it is at the room temperature of $15°C$ and connected to a 60-V supply. After about 3 h, the current is 0.105 A, the voltage remaining unaltered at 60 V. Calculate the average temperature throughout the coil, assuming the wire to be copper having a temperature coefficient of resistance of $0.0043/°C$ at $0°C$.

35. A copper rod, 100 mm long and 2.5 mm diameter, has a resistance of $340 \, \mu\Omega$ at $15°C$. If the rod is drawn out into a wire of uniform diameter 0.5 mm, calculate its resistance at $60°C$. Assume the temperature coefficient of resistance to be $0.0043/°C$ at $0°C$.

36. A coil has a resistance of $100 \, \Omega$ at $20°C$. When a 200-V supply is connected to the coil, the current gradually falls and is ultimately steady at 1.7 A. Calculate the final value of the average temperature of the coil if the temperature coefficient of resistance at $20°C$ is $0.0039/°C$.

37. The copper field coils of a motor have a resistance of $200 \, \Omega$ at $80°C$. What is their resistance at $20°C$? The temperature coefficient of resistance of copper is $0.0039/°C$ at $20°C$.

38. Two batteries, A and B, are connected in parallel and a load of $10 \, \Omega$ is connected across their terminals. A has an e.m.f. of 12 V and an internal resistance of $2 \, \Omega$; B has an e.m.f. of 8 V and an internal

resistance of $1\,\Omega$. Use Kirchhoff's Laws to determine the values and direction of the currents flowing in each of the batteries and in the external resistor. Also determine the p.d. across the external resistor.

(U.E.I.)

39. A separately-excited d.c. generator and a battery are connected in parallel. An e.m.f. of 240 V is generated in the armature of the d.c. machine and the armature circuit has a resistance of $0.8\,\Omega$. The battery has an e.m.f. of 220 V and an internal resistance of $0.5\,\Omega$. A 20-Ω resistor is connected across the terminals of the combined circuits. Determine (a) the values and directions of the currents in the d.c. machine and in the battery and (b) the terminal voltage.

40. Two batteries, A and B, are joined in parallel. Connected across the battery terminals is a circuit consisting of a battery C in series with a 20-Ω resistor, the negative terminal of C being connected to the negative terminals of A and B. Battery A has an e.m.f. of 54 V and an internal resistance of $2\,\Omega$, and the corresponding values for battery B are 60 V and $1\,\Omega$. Battery C has an e.m.f. of 10 V and a negligible internal resistance. Determine the value and direction of the current in each battery.

41. A battery having an e.m.f. of 10 V and an internal resistance of $0.01\,\Omega$ is connected in parallel with a second battery of e.m.f. 10 V and internal resistance $0.008\,\Omega$. The two batteries in parallel are properly connected for charging from a d.c. supply of 20 V through a 0.9-Ω resistor. Calculate the current taken by each battery and the current from the supply.

(N.C.T.E.C.)

42. A motor is taking 80 A from a distribution point 200 m away. The p.d. at that point is 465 V, and the cable has a resistance of $0.03\,\Omega$ per 100 m of single conductor. Calculate (a) the p.d. across the motor terminals, (b) the power supplied to the motor and (c) the power wasted in the cable.

43. A cable, 0.4 km long, has two aluminium conductors, each having a cross-sectional area of $30\,\text{mm}^2$. Calculate the total voltage drop in the cable when the current is 15 A. Assume the resistivity of aluminium to be $0.033\,\mu\Omega\,\text{m}$ at working temperature.

44. A motor, the input to which is 20 kW at 220 V, is supplied by a two-core cable, 400 m long. If the power lost in the cable is not to exceed 5 per cent of the motor input, calculate the minimum permissible cross-sectional area of the cable conductor. What will be the voltage at the feeding point? Assume the resistivity of the conductor at working temperature to be $0.02\,\mu\Omega\,\text{m}$.

45. A cable having a resistance of $0.023\,\Omega$ per 100 m of single conductor has a load of 120 A at a point A. 70 m from the feeding end, and another load of 50 A at a point B, 40 m beyond. If the voltage at the feeding point is 242 V, calculate the voltages across the loads at A and B respectively. Also, find the total power wasted in the cable.

46. A two-wire distributor, ABC, is fed at A at 245 V. Loads of 80 A and 50 A are supplied at B and C, which are 120 m and 200 m respectively from A. Each of the aluminium conductors has a cross-sectional area

of 70 mm². Calculate the p.d. across each load point, assuming the resistivity of aluminium at working temperature to be 0.033 $\mu\Omega$ m.

47. A two-wire ring main, 3 km long, is formed of copper wires, each having a resistance of 0.003 Ω per 100 m. It is fed at 240 V. At a distance of 800 m in one direction from the feeding point there is a load of 80 A, and 1500 m in the other direction there is a load of 120 A. Calculate (a) the current in each section, (b) the voltage at each load point, and (c) the total power wasted in the ring main.

48. A two-wire distributor, 200 m long, is fed at each end at 250 V. At a distance of 50 m from one end there is a load of 120 A, and 80 m from the other end there is a load of 150 A. The resistance per 100 m of single conductor is 0.06 Ω. Calculate (a) the current in each section of the distributor, (b) the voltage at each load point and (c) the total power wasted in the distributor.

49. Calculate the ratio of the mass of an aluminium conductor of given length and resistance to that of a copper conductor of the same length and resistance. Assume the densities of aluminium and copper to be 2700 kg/m³ and 8900 kg/m³ respectively and their resistivities at working temperature to be 0.033 $\mu\Omega$ m and 0.02 $\mu\Omega$ m respectively.

CHAPTER 5

Electromagnetism

5.1 Magnetic field

Before dealing with the magnetic effect of an electric current, it is necessary to explain what is meant by a magnetic field.

If a permanent magnet be suspended so that it is free to swing in a horizontal plane, as in fig. 5.1, it is found that it always takes up a position such that a particular end points towards the earth's North Pole. That end

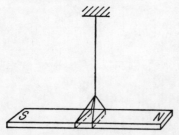

Fig. 5.1 A suspended permanent magnet

is, therefore, said to be the *north-seeking* end of the magnet; similarly, the other end is the *south-seeking* end. For short, these are referred to as the *north* (or N) and *south* (or S) *poles* respectively of the magnet. In the case of small compass-needles, the north pole is usually indicated by a small crosspiece, as shown in fig. 5.2.

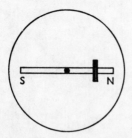

Fig. 5.2 Compass needle

If the N pole of another magnet is brought near the N pole of the suspended magnet, the latter is repelled; whereas attraction occurs if the S pole of the second magnet is brought near the N pole of the suspended magnet. In general we may therefore say that like poles repel each other whereas unlike poles attract each other.

Let us next place a permanent magnet on a table, cover it over with a sheet of smooth cardboard and sprinkle some iron filings uniformly over the sheet. Slight tapping of the latter causes the filings to set themselves in curved chains between the poles, as shown in fig. 5.3. The shape and density of these chains enables one to form a mental picture of the magnetic condition of the space or 'field' around a bar magnet and lead to the idea of *lines of magnetic flux*. Thus, any one of the chains shown in fig. 5.3 represents the direction of the magnetic field along the path occupied

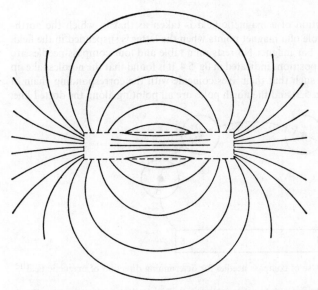

Fig. 5.3   Use of iron filings for determining distribution of magnetic field

by that chain. Also, the more intense the magnetic field, the closer together are the chains. Consequently, it has become the practice to refer to the density of a magnetic field as being so many lines of flux per unit area perpendicular to the direction of the flux. The basis upon which the numerical value of the number of lines of flux per square metre is calculated is dealt with in section 5.11.

It is necessary to emphasize at this stage that these lines of magnetic flux have no physical existence; they are purely imaginary and were introduced by Michael Faraday as a means of visualizing the distribution and density of a magnetic field. It is important, however, to realize that the magnetic flux permeates the *whole* of the space occupied by that flux.

As to the nature of the magnetic field, all we can say is that it appears to be some form of strain both in the space occupied by the magnet and in that around the magnet—somewhat analogous to the skin of a rubber balloon in that it tends to collapse, and in consequence to keep to the shape that involves the minimum of stress. Thus, if a balloon is dented inwards, say by pressing a finger against it, the skin reacts against the finger, tending to push the latter out of the way. In sections 5.9 and 6.6 it is shown that when a magnetic field is dented or distorted by an electric current, the magnetic field reacts in such a way that it tends to push away the conductor carrying the current.

5.2 Direction of magnetic field

The direction of a magnetic field is taken as that in which the north-seeking pole of a magnet points when the latter is suspended in the field. Thus, if a bar magnet NS rests on a table and four compass-needles are placed in positions indicated in fig. 5.4, it is found that the needles take up positions such that their axes coincide with the corresponding chain of filings (fig. 5.3) and the north poles are all pointing along the dotted line,

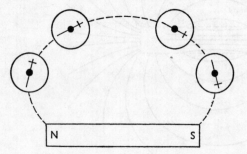

Fig. 5.4   Use of compass needles for determining direction of magnetic field

from the N pole of the bar magnet to its S pole. Hence the magnetic flux is assumed to pass through the magnet, emerge from the N pole and return to the S pole. This may be expressed in another way, thus: if a compass-needle is placed near one end of a magnetized iron rod and if its N pole is repelled from the rod, the direction of the magnetic flux in that region is outwards from the rod, and the adjacent surface of the rod has a north polarity.

5.3 Characteristics of lines of magnetic flux

In spite of the fact that lines of magnetic flux have no physical existence, they do form a very convenient and useful basis for explaining various magnetic effects and for calculating their magnitudes. For this purpose, lines of magnetic flux are assumed to have the following properties:

1. *The direction of a line of magnetic flux at any point in a non-magnetic medium, such as air, is that of the north-seeking pole of a compass-needle placed at that point*, as already described in section 5.2.

2. *Each line of magnetic flux forms a closed path*, as shown by the dotted lines in figs. 5.5 and 5.6. This means that a line of flux emerging from any point at the N-pole end of a magnet passes through the surrounding space back to the S-pole end and is then assumed to continue through the magnet to the point at which it emerged at the N-pole end.

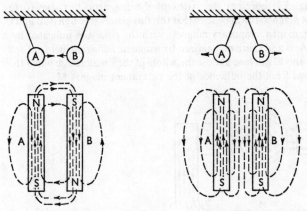

Fig. 5.5 Attraction between magnets   **Fig. 5.6** Repulsion between magnets

3. *Lines of magnetic flux never intersect*. This follows from the fact that if a compass needle is placed in a magnetic field, its north-seeking pole will point in one direction only, namely in the direction of the magnetic flux at that point.

4. *Lines of magnetic flux are like stretched elastic cords, always trying to shorten themselves*. This effect can be demonstrated by suspending two permanent magnets, A and B, parallel to each other, with their poles arranged as in fig. 5.5. The distribution of the resultant magnetic field is indicated by the dotted lines. The lines of magnetic flux passing between A and B behave as if they were in tension, trying to shorten themselves and thereby causing the magnets to be attracted towards each other. In other words, unlike poles attract each other.

5. *Lines of magnetic flux which are parallel and in the same direction repel one another*. This effect can be demonstrated by suspending the two permanent magnets, A and B, with their N poles pointing in the same direction, as in fig. 5.6. It will be seen that in the space between A and B the lines of flux are practically parallel and are in the same direction. These flux lines behave as if they exerted a lateral pressure on one another,

thereby causing magnets A and B to repel each other. Hence like poles repel each other.

5.4 Magnetic induction and magnetic screening

In fig. 5.7, N and S are the poles of a U-shaped permanent magnet M, A and B are soft-iron rectangular blocks attached to the magnet and C is a hollow cylinder of soft-iron placed midway between A and B. The dotted lines in fig. 5.7 represent the paths of the magnetic flux due to the permanent magnet. It will be seen that this flux passes through A, B and C, making them into temporary magnets with the polarities indicated by *n* and *s*, i.e. A, B and C are magnetized by *magnetic induction*. Being of soft-iron, A, B and C will lose almost the whole of their magnetism when they are removed from the influence of the permanent magnet M.

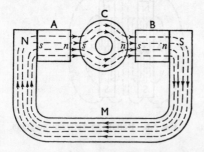

Fig. 5.7 Magnetic induction and screening

Figure 5.7 also shows that no* flux passes through the air space inside cylinder C. Consequently, a body placed in this space would be screened from the magnetic field around it. Magnetic screens are used to protect cathode-ray tubes (as used in television) and instruments such as moving-iron ammeters and voltmeters (section 18.6) from external magnetic fields.

The effect of magnetic induction can also be shown by suspending a soft-iron bar A by a spring B above the poles of a U-shaped permanent magnet M, as in fig. 5.8. Bar A is magnetized by induction as shown by the dotted lines, and is attracted towards M, thereby increasing the tension in B. This force of attraction may be explained as being due to the tension in the flux between A and the poles of M pulling A towards N and S, or merely as the attraction between N and *s* and between S and *n*; but these are simply two ways of stating the same thing.

---

* Actually, there must be some magnetic flux across the air space inside the soft-iron cylinder C, but the density of this flux is so low that, for most purposes, it can be assumed to be zero.

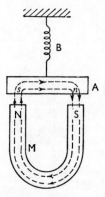

Fig. 5.8 Magnetic induction

5.5 Magnetic field due to an electric current

In section 2.1 it was demonstrated that one of the characteristics of an electric current is its ability to magnetize the space surrounding it. The discovery of this phenomenon by Oersted at Copenhagen, in 1820, was the first definite demonstration of a relationship between electricity and magnetism. Oersted found that if he placed a wire carrying an electric current above a magnetic needle (fig. 5.9) and in line with the normal

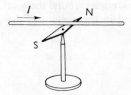

Fig. 5.9 Oersted's experiment

direction of the latter, the needle was deflected clockwise or counter-clockwise, depending upon the direction of the current. This phenomenon may be better understood from experiments made with the apparatus shown in fig. 5.10.* A stout copper wire W passes vertically through a hole in a sheet of cardboard or glass G placed horizontally, with a number of compass-needles arranged around W.

---

* It is usual to represent a current receding from the reader by a cross, as in fig. 5.10, and an approaching current by a dot, as on the left-hand conductor C in fig. 5.26. These conventions are based upon the cross being the back end elevation of the feathers of an arrow and the dot being the front view of the arrow point.

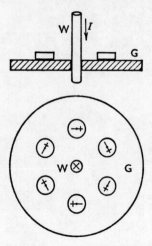

Fig. 5.10   Magnetic field of an electric current

With no current through the wire, the north-seeking poles of the needles all point towards the earth's north pole. Immediately a current of, say, 10 amperes is switched on, the needles are deflected, and after a number of oscillations they come to rest with their axes lying roughly on a circle having the conductor as its centre—as shown in fig. 5.10.

If the current is reversed, all the compass-needles reverse their direction.

Let us now remove the compass-needles and sprinkle plate G as uniformly as possible with iron filings. A current of, say, 20 or 30 amperes is then passed through wire W and the plate tapped gently. It is found that the filings tend to arrange themselves in concentric circles around the wire (fig. 5.11), this tendency being most pronounced in the vicinity of the conductor.

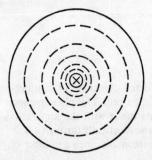

Fig. 5.11   Distribution of magnetic field in fig. 5.10

These experiments indicate that a magnetic state or field is produced around a wire carrying an electric current and that the intensity of this field decreases as the distance from the conductor increases.

5.6 Direction of the magnetic field due to an electric current

It was pointed out in section 5.2 that the direction of the magnetic field in any particular space is always taken as the direction of the N pole of a compass-needle placed in that field. Consequently, from an experiment such as that described in connection with fig. 5.10, we can determine the relationship between the direction of the magnetic field and that of the current. Thus, it is seen from fig. 5.10 that if we look along the conductor and if the current is flowing away from us, the magnetic field has a clockwise direction.

A convenient method of representing this relationship is to grip the conductor with the *right* hand, with the thumb outstretched parallel to the conductor and pointing in the direction of the current; the fingers then point in the direction of the magnetic field around the conductor.

Another way of representing the relationship between the direction of a current and that of its magnetic field is to place a cork-screw or a wood-screw (fig. 5.12) alongside the conductor carrying the current. In order that

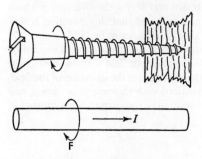

Fig. 5.12   Right-hand screw rule

the screw may travel in the same direction as the current, namely towards the right in fig. 5.12, it has to be turned clockwise when viewed from the left-hand side. Similarly, the direction of the magnetic field, viewed from the same side, is clockwise around the conductor, as indicated by curved arrow F.

5.7 Magnetic field of a solenoid

A solenoid consists of a number of turns of wire wound in the same direction, so that when the coil is carrying a current, all the turns are assisting one another in producing a magnetic field. Figure 5.13 shows a few turns of wire wound helically through holes in a horizontal board and

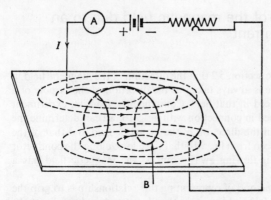

Fig. 5.13 Magnetic field of a solenoid

connected to a battery capable of supplying 15 to 20 amperes. Iron filings are evenly sprinkled over the board and the latter is gently tapped. The filings arrange themselves in concentric paths of the shape shown by the thin dotted lines. The direction of the magnetic field can be determined by placing a compass-needle on the board and noting the direction in which its north-seeking pole is pointing. It is found that this direction is that indicated by the arrowheads on the dotted lines in fig. 5.13.

The magnetic field may be intensified by inserting a rod of iron inside the solenoid, as shown in fig. 5.14. The magnetic field is again represented by the dotted lines and the arrowheads indicate the direction of the field. The iron core thus becomes magnetized with the polarities shown, and behaves like a permanent magnet so long as the current is maintained in the coil.

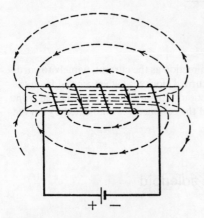

Fig. 5.14 Solenoid with an iron core

The direction of the magnetic field produced by a current in a solenoid may be deduced by applying either the grip or the screw rule.

Thus, if the solenoid is tripped with the *right* hand, with the fingers pointing in the direction of the current, then the thumb outstretched parallel to the axis of the solenoid points in the direction of the magnetic field *inside* the solenoid.

The screw rule can be expressed thus: If the axis of the screw is placed along that of the solenoid and if the screw is turned in the direction of the current, it travels in the direction of the magnetic field, namely towards the right in figs. 5.13 and 5.14.

5.8 Applications of the magnetic effect of a current

The following examples are given to indicate ways in which the magnetic effect of a current may be utilized, and are not intended to be exhaustive.

(a) Lifting magnet

A coil C, usually of insulated copper strip, is wound round a central core A forming part of an iron casting shaped as shown in fig. 5.15, where the upper half is a sectional elevation at YY and the lower half is a sectional plan at XX. Over the face of the electromagnet is a disc D of non-magnetic

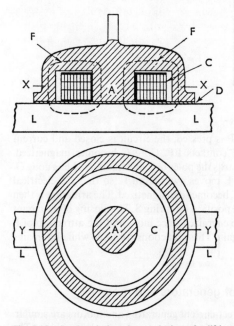

Fig. 5.15 Sectional elevation and plan of a lifting magnet

manganese steel which is capable of withstanding considerable impacts when the load L is picked up. The load must be of magnetic material and the dotted lines FF represent the paths of the magnetic flux. The lifting magnet is a large-scale application of the magnetic experiment described in section 2.1.

(b) Electric bell

Two coils AA (fig. 5.16) are wound on an iron core C and connected in series in such a manner that they help each other in setting up magnetic flux through C and an iron plate or armature B, as indicated by the dotted line. The armature is supported by a flat spring S fixed at the upper end. Attached to B is another flat spring D carrying a contact E which normally rests against an adjustable contact F.

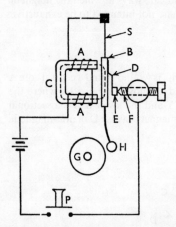

Fig. 5.16 Electric bell

When a push-button P is pressed, the circuit is closed and current flows through coils A and contacts EF. Core C becomes magnetized, armature B is attracted towards the poles of C and hammer H hits gong G. At the same time, contact E moves away from F, so that the electrical circuit is broken and core C becomes demagnetized. The armature is then brought back to its original position by spring S. The closing of the circuit at EF enables coils A to be energized once more and the attraction of B causes the gong to be hit again. These reactions continue while P remains closed.

(c) Magnetic circuits of generators and motors

The magnetic circuits of direct-current generators and motors are similar. Figure 5.17 shows the magnetic circuit of a four-pole machine. The

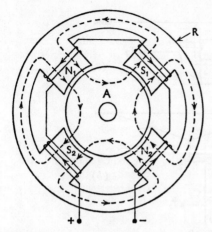

Fig. 5.17 Magnetic circuit of a 4-pole generator or motor

armature core A consists of iron laminations, assembled on the shaft. The fixed part of the machine is made of four iron cores attached to an iron ring R, called the *yoke*. It is usual to place a coil on each pole core and to connect them in series as shown. Since the poles must be alternately N and S, it is essential that the coils should be so connected that the current flows clockwise—when viewed from the armature end of the poles—round the cores which are to be south, and counterclockwise round those which are to be north.

It will be seen from fig. 5.17 that the magnetic flux which emerges from N_1 divides, half going towards S_1 and half towards S_2. Similarly, the flux which emerges from N_2 divides equally between S_1 and S_2.

(d) Telephone receiver

The simplest type of telephone receiver (fig. 5.18(a)) consists of a permanent magnet NS, two nickel-iron pole-pieces BB attached to the magnet, two coils CC and a thin cobalt-iron disc or diaphragm D arranged with a small clearance between it and the pole-faces of BB. Terminals T_1T_2 are connected through the telephone line to the transmitter, so that the alternating current through CC reverses its direction as many times per second as the vibrations of the sound waves at the transmitter. For instance, if the sound at the transmitter is the middle C of a piano tuned to the scientific scale, the vibrations of the air, and hence of the microphone diaphragm, are at the rate of 256 per second; and the current in the line between the transmitter and the receiver alternates at this rate.

When current is passing through CC from T_1 to T_2, the effect is to intensify the magnetism round the magnetic citcuit, and thus to increase the force attracting disc D towards the poles of BB. On the other hand,

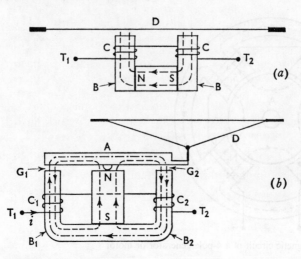

Fig. 5.18   Telephone receivers

when current is passing from T_2 to T_1, it tries to produce magnetism in opposition to that of the magnet, and the effect is to reduce the flux round the magnetic circuit and thus reduce the pull on D. Consequently the latter is made to vibrate at the same frequency as the alternating current. The movement of D sets up vibrations of the same frequency in the air around and thus produces a sound similar to that at the transmitter.

In the most modern type of telephone receiver—the Rocking Armature receiver—a permanent magnet NS has a nickel-iron U-plate $B_1 B_2$ attached to one end and a nickel-iron armature A pivoted on the other end as shown in fig. 5.18(b). A conical diaphragm D, fixed at its outer edge, has its apex attached to one end of A. Coils C_1 and C_2 are wound on limbs B_1 and B_2 respectively and connected via terminals $T_1 T_2$ to the telephone line.

When there is no current through $C_1 C_2$, the flux due to magnet NS crosses the airgap at pole N and then divides, so that roughly a half returns via limb B_1 and the other half via limb B_2, as indicated by the uniformly dotted lines.

If a current i flows from T_1 to T_2, the ampere-turns due to C_1 and C_2 can be regarded as producing a flux through limbs $B_1 B_2$ and armature A, as indicated by the chain-dotted line. This flux is superimposed on that due to the permanent magnet NS. From fig. 5.18(b) it will be seen that the effect is to reduce the flux in gap G_1 and increase that in gap G_2, thereby reducing the downward pull on the left-hand side of A and increasing that on the right-hand side. Consequently the armature is tilted clockwise and the apex of D is pulled downwards. The flux through magnet NS remains constant.

When the current through C_1 and C_2 flows in the reverse direction, i.e. from T_2 to T_1, the flux in gap G_1 is increased and that in gap G_2 is reduced. Hence armature A is tilted anti-clockwise and the apex of D is pushed upwards.

With an alternating current through coils $C_1 C_2$, armature A is rocked to and fro at the frequency of that current so that the diaphragm emits a sound having the same frequency as that of the sound waves at the transmitter.

The principle of action of a telephone bell is exactly similar to that of the receiver shown in fig. 5.18(b) except that armature A is fitted with a striker which can move between two gongs.

(e) Moving-iron ammeter and voltmeter

The construction and action of the moving-iron instrument are dealt with in section 18.6.

5.9 Force on a conductor carrying current across a magnetic field

In section 5.5, it was shown that a conductor carrying a current can produce a force on a magnet situated in the vicinity of the conductor. By Newton's Third Law of Motion, namely that to every force there must be an equal and opposite force, it follows that the magnet must exert an equal force on the conductor. One of the simplest methods of demonstrating this effect is to take a copper wire, about 2 mm diameter, and bend it into a rectangular loop as represented by BC in fig. 5.19. The two tapered ends of

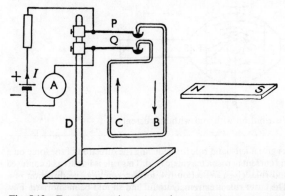

Fig. 5.19 Force on conductor carrying current across a magnetic field

the loop dip into mercury contained in cups, one directly above the other, the cups being attached to metal rods P and Q carried by a wooden upright rod D. A current of about 5 amperes is passed through the loop and the N pole of a permanent magnet NS is moved towards B. If the current in this wire is flowing downwards, as indicated by the arrow in fig. 5.19, it is found that the loop, when viewed from above, turns counter-clockwise, as shown in plan in fig. 5.20. If the magnet is reversed and again

Fig. 5.20   Direction of force on conductor in fig. 5.19

brought up to B, the loop turns clockwise.

If the magnet is placed on the other side of the loop, the latter turns clockwise when the N pole of the magnet is moved near to C, and counterclockwise when the magnet is reversed.

These effects can be explained* by the simple apparatus shown in elevation and plan in fig. 5.21. Two permanent magnets NS rest on a sheet

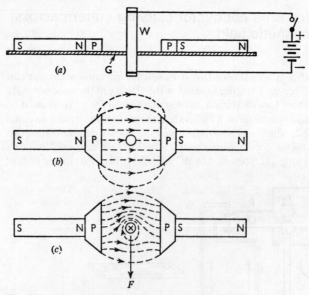

Fig. 5.21   Flux distribution with and without current

---

* Many textbooks give a *left*-hand rule for deducing the direction of the force on a conductor carrying current across a magnetic field. This rule is liable to be confused with the *right*-hand rule, given in section 6.4, for determining the direction of a generated e.m.f. The latter rule is extremely useful and should be memorized. Few students can memorize both rules correctly and it is suggested that the left-hand rule should be forgotten and that the direction of the force on a current-carrying conductor in a magnetic field should be deduced from first principles by drawing separately the magnetic field due to the current in the conductor and that due to the permanent magnet or electromagnet and thus derive the distribution of the resultant magnetic field as in fig. 5.21.

of paper or glass G, and soft-iron pole-pieces P are added to increase the area of the magnetic field in the gap between them. Midway between the pole-pieces is a wire W passing vertically downwards through G and connected through a switch to a 6-volt battery capable of giving a very large current for a short time.

With the switch open, iron filings are sprinkled over G and the latter is gently tapped. The filings in the space between PP take up the distribution shown in fig. 5.21(b). If the switch is closed momentarily, the filings rearrange themselves as in fig. 5.21(c). It will be seen that the magnetic flux has been so distorted that it partially surrounds the wire. This distorted field acts like a stretched elastic string bent out of the straight; the flux tries to return to the shortest path between PP, thereby exerting a force F urging the conductor out of the way.

It has already been shown in section 5.6 that a wire W carrying a current downwards in fig. 5.21(a) produces a magnetic field as shown in fig. 5.22. If this field is compared with that of fig. 5.21(b), it is seen that on the

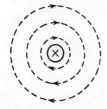

Fig. 5.22 Flux distribution due to current in a straight conductor

upper side the two fields are in the same direction, whereas on the lower side they are in opposition. Hence, the combined effect is to strengthen the magnetic field on the upper side and weaken it on the lower side, thus giving the distribution shown in fig. 5.21(c).

By combining diagrams similar to figs. 5.21(b) and 5.22, it is easy to understand that if either the current in W or the polarity of the magnets NS is reversed, the field is strengthened on the lower side and weakened on the upper side of diagrams corresponding to fig. 5.21(b), so that the direction of the force acting on W is the reverse of that shown in fig. 5.21(c).

On the other hand, if both the current through W and the polarity of the magnets are reversed, the *distribution* of the resultant magnetic field and therefore the direction of the force on W remain unaltered.

5.10 Applications of the mechanical force on a conductor carrying current across a magnetic field

The following are only a few typical examples of the mechanical force discussed in the preceding section.

(a) Moving-coil loudspeaker

Small loudspeakers are usually made with a central cylindrical magnet M of a steel alloy such as Alcomax (section 18.5), to which is attached a cylindrical soft-iron pole-piece P, as shown in fig. 5.23(a). The return path

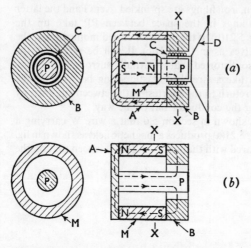

Fig. 5.23 Moving-coil loudspeakers

for the magnetic flux is provided by a soft-iron cylindrical pot A and a soft-iron ring B. In the uniform airgap between P and B is a coil C wound on a former carried by a conical diaphragm D, and current is led into and out of C by thin flexible wires. The dotted lines represent the paths of the magnetic flux due to the permanent magnet M.

The alternating current supplied by, say, a radio receiver passes through C, so that at one instant there is a force acting on the wires of C urging the coil and its diaphragm towards the right. At the next instant the current in C will have reversed and the diaphragm will be drawn to the left. Thus the air in the vicinity of D is set in vibration and the corresponding sound is produced.

Large loudspeakers are usually made with the permanent magnet in the form of a cylindrical ring M (fig. 5.23(b)), magnetized in the direction of its axis. The remainder of the magnetic circuit consists of a plate A, a solid cylindrical pole-piece P and a ring B, all of soft iron. This arrangement has the advantage of smaller internal leakage flux than would be the case with a larger centre-pole magnet surrounded by a soft-iron casing, as in fig. 5.23(a).

(b) The electric motor

Figure 5.24 shows a copper disc D supported by a spindle in such a way that its lower edge just dips into a mercury pool M. The portion of the disc

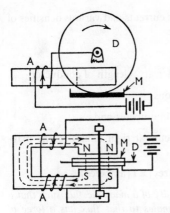

Fig. 5.24 Elementary form of electric motor

that is between the spindle and the mercury is situated in the field of an electromagnet excited by coils AA. One end of a battery is connected to the mercury pool and the other end to coils A and then, via the spindle, to disc D. The latter is therefore carrying current from its spindle to the mercury in a direction that is at right angles to the field of the electromagnet NS. By combining the magnetic field due to this current with that due to the electromagnet, it is found that the resultant field is distorted in such a direction as to exert a torque turning the disc clockwise.

This principle is applied in the rotating mercury meter used on d.c. supply systems for measuring the amount of electrical energy consumed. It is also the principle of action of the electric motor, though actual machines are much more complicated in construction and are referred to more fully in Chapter 16.

(c) Moving-coil ammeter and voltmeter

The construction and operation of the moving-coil instrument are dealt with in section 18.5.

5.11 Magnitude of the force on a conductor carrying current across a magnetic field

With the apparatus of fig. 5.19, it can be shown qualitatively that the force on a conductor carrying a current at right angles to a magnetic field is increased when (a) the current in the conductor is increased and (b) when the magnetic field is made stronger by bringing the magnet nearer to the conductor. With the aid of more elaborate apparatus, the force on the conductor can be measured for various currents and various densities of the magnetic flux, and it is found that:

force on conductor

∝ current × flux density × length of conductor

If F = force on conductor in newtons,

I = current through conductor in amperes and

l = length, in metres, of conductor at right angles to the magnetic flux,

F [newtons] ∝ flux density × l [metres] × I [amperes].

The unit of flux density is the density of a magnetic flux such that a conductor carrying 1 ampere at right angles to that flux has a force of 1 newton per metre acting upon it. This unit is termed a *tesla**(T).

Hence for a flux density B, in teslas,

force on conductor = BIl newtons   [5.1]

For a magnetic field having a cross-sectional area a, in square metres and a uniform flux density B, in teslas, the total flux, in *webers*† (Wb), is represented by the Greek *capital* letter Φ (phi), where

$$\Phi \, [\text{webers}] = B \, [\text{teslas}] \times a \, [\text{metres}^2]$$

or

$$B \, [\text{teslas}] = \frac{\Phi \, [\text{webers}]}{a \, [\text{metres}^2]} \qquad\qquad [5.2]$$

The *weber* may be defined either (i) as *that magnetic flux which, when cut at a uniform rate by a conductor in 1 second, generates an e.m.f. of 1 volt* (section 6.5) or (ii) as *that magnetic flux which, linking a circuit of one turn, induces in it an e.m.f. of 1 volt when the flux is reduced to zero at a uniform rate in 1 second* (section 6.6).

Example 5.1   *The pole core (fig. 5.17) of an electrical machine is circular in cross-section and has a diameter of 120 mm. If the total flux in the core is 16 mWb, calculate the flux density.*

Diameter of core = 120 mm = 0.12 m,

∴   cross-sectional area of core = $(\pi/4) \times (0.12)^2 = 0.011\,32\,\text{m}^2$.

---

* Nikola Tesla (1857–1943), a Yugoslav who emigrated to USA in 1884, was a very famous electrical inventor. In 1885 he patented 2-phase and 3-phase alternators and motors.
† Wilhelm Eduard Weber (1804–91), a German physicist, was the first to develop a system of absolute electrical and magnetic units.

Total magnetic flux $= 16\,\text{mWb} = 0.016\,\text{Wb}$

$\therefore$ flux density $= 0.016\,[\text{Wb}]/0.011\,32\,[\text{m}^2]$

$= 1.413\,\text{T}.$

Example 5.2 *A conductor carried a current of 800 A at right angles to a magnetic flux having a density of 0.5 tesla. Calculate the force on the conductor, in newtons per metre length.*

Substituting for B, l and I in expression (5.1), we have:

force per metre length $= 0.5\,[\text{T}] \times 1\,[\text{m}] \times 800\,[\text{A}]$

$= 400\,\text{N}.$

Example 5.3 *The coil of a moving-coil instrument (fig. 18.6) is wound with $42\frac{1}{2}$ turns. The mean width of the coil is 25 mm and the axial length of the magnetic field is 20 mm. If the flux density in the airgap is 0.2 T, calculate the torque, in newton metres, when the current is 15 mA.*

Since the coil has $42\frac{1}{2}$ turns, one side has 42 wires and the other side has 43 wires.

From expression [5.1],

force on the side having 42 wires

$= 0.2\,[\text{T}] \times 0.02\,[\text{m}] \times 0.015\,[\text{A}] \times 42 = 2520 \times 10^{-6}\,\text{N},$

$\therefore$ torque on that side of coil

$= 2520 \times 10^{-6}\,[\text{N}] \times 0.0125\,[\text{m}] = 31.5 \times 10^{-6}\,\text{N m}.$

Similarly,

torque on side of coil having 43 wires

$= 31.3 \times 10^{-6} \times 43/42 = 32.2 \times 10^{-6}\,\text{N m},$

$\therefore$ total torque on coil

$= (31.5 + 32.2) \times 10^{-6} = 63.7 \times 10^{-6}\,\text{N m}.$

5.12 Force between two long parallel conductors carrying electric current

It was shown in section 5.5 that a current-carrying conductor is surrounded by a magnetic field and in section 5.9 that a current-carrying conductor placed across a magnetic field has a force acting upon the conductor. It therefore follows that when two current-carrying conductors are parallel to each other, there is a force acting on each of the conductors. This effect can be very easily demonstrated by means of the

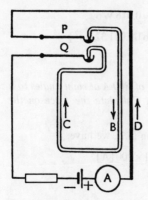

Fig. 5.25   Force between two parallel current-carrying conductors

apparatus referred to in section 5.9. Thus, in fig. 5.25, the rectangular loop
BC has its tapered ends dipping into mercury in cups supported one
directly above the other by rods P and Q, and a current of about 10 to
15 amperes is passed through the loop. Part of the electrical circuit
consists of a long straight rod D which can be placed alongside B, as
shown in plan in fig. 5.26. When the currents in D and B are in opposite

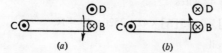

Fig. 5.26   Force between two parallel current-carrying conductors

directions, as in fig. 5.26(*a*), the two conductors repel each other and the
loop (viewed from above) is deflected clockwise. On the other hand, if rod
D is turned through 180° so that the currents in B and D are in the same
direction, as in fig. 5.26(*b*), the conductors attract each other and the loop
turns counterclockwise.

These effects are most easily explained by first drawing the magnetic
fields produced by each conductor and then combining these fields. Thus,
fig. 5.27(*a*) shows two conductors, A and B, each carrying current towards
the paper. The magnetic flux due to current in A alone is represented by
the uniformly dotted circles in fig. 5.27(*a*), and that due to B alone is
represented by the chain-dotted circles. It is evident that in the space
between A and B the two fluxes tend to neutralize each other, but in the
space outside A and B they assist each other. Hence the resultant
distribution is somewhat as shown in fig. 5.27(*b*). Since magnetic flux
behaves like a stretched elastic cord, the effect is to try to move conductors
A and B towards each other; in other words, there is a force of attraction
between A and B.

If the current in B is reversed, the magnetic fields due to A and B assist
each other in the space between the conductors and the resultant
distribution of the flux is shown in fig. 5.27(*c*). The lateral pressure between

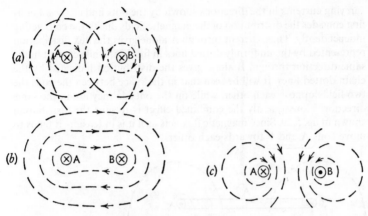

Fig. 5.27 Magnetic fields due to parallel current-carrying conductors

the lines of flux exerts a force on the conductors tending to push them apart (section 5.3(5)).

It is this force between parallel current-carrying conductors that forms the basis for the definition of the *ampere*, adopted internationally in 1948, namely, *the constant current which, if maintained in two straight parallel conductors of infinite length of negligible circular cross-section and placed at a distance of 1 metre apart in a vacuum, will produce between them a force equal to 2×10^{-7} newton per metre of length.*

5.13 Force between coils carrying electric current

We are now in a position to explain the effect observed in section 2.2, namely, that two co-axial coils, placed one above the other, attract or repel each other, depending upon the relative direction of their currents. Suppose A and B in fig. 5.28 to represent the cross-section of two coils

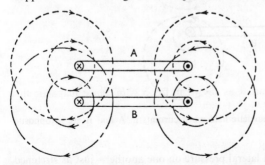

Fig. 5.28 Magnetic fields due to current in A and B separately

carrying currents in the directions shown by the dots and crosses. Let us first consider the distribution of the magnetic fields due to the coils acting independently. Thus, current through A alone gives the flux distribution represented by the uniformly dotted lines in fig. 5.28, while current in the same direction through B alone gives the distribution indicated by the chain-dotted lines. It will be seen that in the space between the coils the two fields oppose each other, while on the outside they are in the same direction. Consequently the combined effect is to give the distribution shown in fig. 5.29. Since magnetic flux acts as if it is in tension, it tends to move coils A and B towards each other.

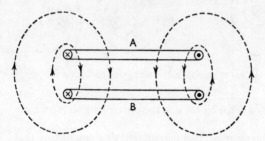

Fig. 5.29   Resultant magnetic field when currents in A and B are in the same direction

On the other hand, if the current through B is reversed, the direction of the arrowheads on the chain-dotted lines in fig. 5.28 is reversed. Consequently, the magnetic fields of A and B are in the same direction in the space between the coils and in opposition outside the coils, so that the resultant distribution becomes that shown in fig. 5.30. But lines of

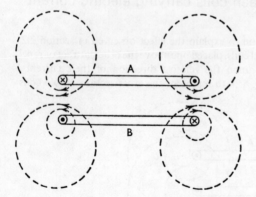

Fig. 5.30   Resultant magnetic field when currents in A and B are in opposite directions

magnetic flux exert a lateral pressure on one another—just as stretched rubber cords try to swell when allowed to contract in length, and thereby exert sideways pressure on other rubber cords that are alongside. This lateral pressure between the lines of flux makes them behave as if they were repelling one another. This force of repulsion is passed on to coils A and B, so that they try to move away from each other.

This attraction and repulsion between coils carrying an electric current has been applied in the current balance* used at the National Physical Laboratory for determining the value of an electric current in terms of the definition given in section 5.12.

5.14 Summary of important formulae

$$\text{Force on conductor } [\text{N}] = B[\text{T}] \times l[\text{m}] \times I[\text{A}] \qquad [5.1]$$

$$B[\text{teslas}] = \frac{\Phi[\text{webers}]}{a[\text{square metres}]} \qquad [5.2]$$

5.15 Examples

1. Give a diagram showing the direction of the magnetic field round (a) a straight wire, (b) one wire bent into the form of a circle, (c) a long solenoid, when each of these has an electric current passing through it. The direction of the current and of the magnetic field must be clearly indicated. (U.L.C.I.)

2. A straight conductor is carrying an electric current. What type of magnetic field is produced? What happens when such a current-carrying conductor is placed in a magnetic field? (U.E.I.)

3. Given a horseshoe-shaped piece of soft iron, some insulated copper wire, a battery, a switch, a variable resistor, an ammeter and a compass needle, state how you would make an electromagnet and test it for polarity. Give a circuit diagram and on it show the direction of current and the polarity of the magnet. (U.L.C.I.)

4. Describe, with diagrams, an experiment to show that a force is produced when a current-carrying conductor is in a magnetic field. What are the relative directions of the magnetic field, current and force? (N.C.T.E.C.)

5. Explain what is meant by a 'magnetic field' and the 'direction of a magnetic field'. A long, straight horizontal wire lies along the magnetic meridian and an observer holds a compass needle above and close to the wire. Describe, with sketches, the effect on the needle

---

* Further particulars, together with an illustration of this balance, can be found in the *Journal of the Institution of Electrical Engineers*, December 1951.

when a current is passed through the wire (a) flowing from north to south and (b) flowing from south to north, giving reasons for the effect produced.   (E.M.E.U.)

6. Describe an experiment for plotting the magnetic field around a bar magnet by the use of a small compass needle.

Draw diagrams to show the type of field to be expected with (a) one bar magnet alone, (b) two similar bar magnets placed in line with similar poles facing and about 10 cm apart.   (N.C.T.E.C.)

7. (a) Describe with the aid of a diagram the magnetic field associated with the current flowing in a long straight conductor. Mark clearly the directions of the current and the magnetic field.

(b) Draw a second diagram showing the field associated with two parallel conductors, each carrying a steady current flowing in the same direction. Indicate the direction of the force on each conductor.

(c) From a consideration of (b), give a definition of the unit of current.   (N.C.T.E.C.)

8. The flux in the pole of an electric motor is 0.013 Wb. If the pole has a circular cross-section and a diameter of 120 mm, calculate the value of the flux density.

9. If the flux density inside a solenoid is 0.08 T, and the cross-sectional area of the solenoid is 2000 mm^2, calculate the value of the total flux in microwebers.

10. A straight conductor is carrying a current of 2500 A at right angles to a magnetic field having a uniform density of 0.12 T. Calculate the force on the conductor in newtons per metre.

11. A conductor 0.3 m long, is carrying a current of 60 A at right angles to a magnetic field of uniform density. The force on the wire is 8 N. Calculate the density of the magnetic field.

12. The coil of a moving-coil loudspeaker (fig. 5.23) has a mean diameter of 15 mm and is wound with 80 turns. It is situated in a radial magnetic field of 0.5 T. Calculate the force, in millinewtons, on the coil when the current is 12 mA.

13. A straight horizontal wire carries a steady current of 150 A and is situated in a uniform magnetic field of 0.6 T acting vertically downwards. Determine the magnitude of the force per metre length of conductor and the direction in which it acts.

Explain how the principle of the force on a current-carrying conductor situated in a magnetic field is utilized in order to obtain rotation of the armature of a d.c. motor.   (N.C.T.E.C.)

14. The coil of a moving-coil instrument is wound with $36\frac{1}{2}$ turns. The mean width of the coil is 25 mm and the effective axial length of the magnetic field is also 25 mm. The flux density in the gap is 0.13 T. Calculate the torque when the current is 25 mA.

15. The coil of a moving-coil instrument is wound with $50\frac{1}{2}$ turns on a rectangular former. The axial length of the pole shoes is 23 mm and the mean width of the coil is 17 mm. If the flux density in the gap is 0.18 T, calculate the current to give a torque of 30×10^{-6} N m.

16. The coil of a moving-coil instrument is wound with $40\frac{1}{2}$ turns on a

former having an effective length of 25 mm and an effective breadth of 20 mm. The flux density in the gap is 0.16 T. Calculate the torque in newton metres when the current is 15 mA.

17. The armature of a certain electric motor has 900 conductors and the current per conductor is 24 A. The flux density in the airgap under the poles is 0.6 T. The armature core is 160 mm long and has a diameter of 250 mm. Assume that the core is smooth (i.e. there are no slots and the winding is on the cylindrical surface of the core) and also assume that only two-thirds of the conductors are simultaneously in the magnetic field. Calculate (a) the torque in newton metres and (b) the power developed, in kilowatts, if the speed is 700 rev/min.

(*Note*. In the case of slotted cores, the flux density in the slots is very low, so that there is very little torque on the conductors; nearly all the torque is exerted on the teeth.)

CHAPTER 6
▆▆▆ Electromagnetic Induction

6.1 Induced e.m.f.

It was mentioned in section 5.5 that the magnetic effect of an electric current was discovered by Oersted in 1820. The knowledge of this connection between electricity and magnetism caused many scientists of the time, particularly Michael Faraday in England, to try to discover a method of obtaining an electric current from a magnetic field. Failure after failure dogged Faraday's efforts until on 29 August 1831 he made the great discovery of *electromagnetic induction* with which his name will be for ever associated.

As far as we are concerned, it will be more convenient to approach this matter experimentally in a different sequence from that followed by Faraday. Let us take a coil C (fig. 6.1), wound with a large number of turns, and connect it to a galvanometer G, namely a very sensitive moving-coil ammeter. If a permanent magnet NS is moved up to and along the axis of C, as shown, the moving coil of G is deflected, thereby indicating that there must be an electromotive force induced or generated in coil C. Immediately the movement of NS ceases, the moving coil of G returns to its original position. This effect proves that e.m.f. is induced only while NS is moving relative to C.

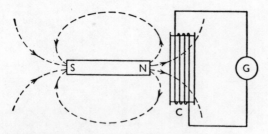

Fig. 6.1 Electromagnetic induction

Let us now move NS away from C. The galvanometer deflection is found to be in the reverse direction, showing that the direction of the induced e.m.f. depends upon the direction in which NS is moved relative to coil C.

If, next, we hold the magnet stationary but move the coil towards the magnet and then away from it, the deflection of the galvanometer is found to follow exactly the same sequence as it did when the magnet was moved and the coil held stationary. This result shows that the generation of an e.m.f. in C depends only upon the *relative* movement of the magnet and the coil.

If the permanent magnet is turned through 180 degrees so that its S pole is pointing towards the coil, it is found that a repetition of the movements described above is accompanied by galvanometer deflections similar to those previously obtained, except that their directions are reversed. Thus, the direction of the e.m.f. induced by bringing the S pole up to the coil is the same as that previously obtained when the N pole was moved away from the coil.

The arrowheads on the dotted lines in fig. 6.1 represent the direction of the magnetic field in their respective regions. It will be seen that as the magnet is moved towards the coil, the magnetic flux of NS also moves across the wires forming the coil; that is, the magnetic flux is said to *cut* the coil. Similarly, when the coil is moved towards the magnet, the magnetic flux is said *to be cut* by the coil. It is this relative movement of the magnetic flux and the coil that causes an e.m.f. to be induced (or generated) in the latter. Alternatively* we can say that the induced e.m.f. is due to a change in the value of the magnetic flux passing through the coil. The above experiments also show that the direction of the induced e.m.f. depends both upon the direction of the magnetic flux and upon that in which the coil moves relative to the magnetic flux.

Let us next bring the magnet up to the coil at different speeds. It is found that the greater the speed, the greater is the deflection of the galvanometer and, therefore, the greater must be the e.m.f. induced in the coil.

6.2 Induced e.m.f. (*continued*)

Let us now replace magnet NS of fig. 6.1 by a coil A (fig. 6.2) connected through a switch S to a cell. At the instant when S is closed, there is a momentary deflection on G; and when S is opened, G is deflected

---

* It is immaterial whether we consider the e.m.f. as being due to change of flux linked with a coil or due to the coil cutting or being cut by magnetic flux; the result is exactly the same. The fact of the matter is that we do not know what is really happening; but we can calculate the effect by imagining the magnetic field in the form of lines of flux, some of which expand from nothing when the field is increased or collapse to nothing when the field is reduced. In so doing, the flux may be regarded as cutting the turns of the coil, or alternatively, the effect may be regarded as being due to a change in the value of the flux passing through the coil.

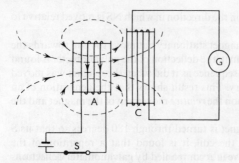

Fig. 6.2 Electromagnetic induction

momentarily in the reverse direction. On the other hand, if S is kept closed and coil A moved towards C, the galvanometer is deflected in the same direction as when S was closed with A stationary. The withdrawal of A causes a deflection in the reverse direction. Deflection of G continues only while there is relative movement between the two coils, i.e. while the magnetic flux passing through coil C is changing.

The dotted lines in fig. 6.2 represent the distribution of the magnetic flux due to current in coil A. When S is opened, the current falls to zero. Consequently, the magnetic flux of A must also disappear; in other words, the magnetic flux is said to *collapse* towards A, and in so doing, the flux that passed through (or was linked with) coil C cuts the latter and induces an e.m.f. in it.

Similarly, when S is closed, the current through A causes a magnetic field to come into existence; and in this process the magnetic flux may be regarded as spreading outwards from coil A, and some of this flux will extend sufficiently to cut soil C and thereby induce an e.m.f. in it. It will be seen that as far as the e.m.f. induced in C is concerned, both the closing of S in fig. 6.2 and the moving of A towards C, with S closed, have the same effect as moving the magnet towards C in fig. 6.1.

The effects observed with the apparatus of fig. 6.2 may be accentuated by placing an iron core inside the coils, thereby increasing the magnetic flux linked with C due to a given current in A. In fact, we may go still further and wind the two coils A and C on an iron ring R, as in fig. 6.3. When S is closed, the current in A sets up magnetic flux through R, as

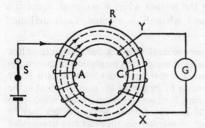

Fig. 6.3 Electromagnetic induction

indicated by the dotted circles. This flux, in becoming linked with coil C, induces in the latter an e.m.f. which circulates a current causing G to be deflected momentarily. So long as S remains closed, there is no further change of magnetic flux and therefore no e.m.f. induced in C. When S is opened, the magnetic flux decreases and an e.m.f. is induced in C in the reverse direction.

It was by means of apparatus similar to that shown in fig. 6.3 that Faraday discovered electromagnetic induction, namely that a change in the value of the magnetic flux through a coil causes an e.m.f. to be induced in that coil.

It should be pointed out that when S (fig. 6.3) is closed, the flux which becomes linked with coil C has also to grow in coil A; consequently, an e.m.f. is induced in A as well as in C. Similarly, when S is opened, the decrease of flux causes an e.m.f. to be induced in both A and C.

The results obtained from the above experiments on electromagnetic induction may now be summarized thus:

(a) When a conductor cuts or is cut by magnetic flux, an e.m.f. is induced in the conductor; or alternatively, when there is a change of magnetic flux passing through a circuit, an e.m.f. is induced in that circuit.
(b) The direction of the induced e.m.f. depends upon the direction of the magnetic flux and upon the direction in which the flux moves relative to the conductor.
(c) The magnitude of the e.m.f. is proportional to the rate at which the conductor cuts or is cut by the magnetic flux; or alternatively, the magnitude of the e.m.f. induced in a circuit is proportional to the rate of change of magnetic flux through the circuit. This last statement is often referred to as *Faraday's Law of Electromagnetic Induction*, although it was not stated in this form by Faraday.

6.3 The transformer

It is only a small step from the apparatus shown in fig. 6.3 to a transformer. The function of the latter is to change the voltage of an alternating-current supply from one value to another. An alternating voltage is maintained across coil A, termed the *primary* winding, and the alternating current through A sets up an alternating flux in the iron core. The variation of this flux causes an alternating e.m.f. to be induced in coil C (termed the *secondary* winding) as well as in coil A; and in the whole of the flux produced by the current in A passes through C, the e.m.f. induced in each turn is the same for the two coils.

Suppose the e.m.f. induced per turn to be, say, 4 V and the number of turns on the primary and secondary windings to be 50 and 500 respectively, then the e.m.f. induced in the primary winding is 200 V and that induced in the secondary winding is 2000 V. The voltage applied to the primary winding is practically equal and opposite to the e.m.f. induced in the primary and is therefore approximately 200 V. Hence such a transformer steps *up* the voltage about ten times.

Had the secondary winding been wound with only five turns, the secondary voltage would have been 20 V and the transformer would therefore step *down* the voltage to roughly a tenth of the voltage applied to the primary winding.

Since the alternating flux induces an e.m.f. in the iron core as well as in the windings, it is necessary to reduce the magnitude of the 'eddy' currents circulating in the core so as to prevent excessive loss of power in the latter. This is done by constructing the core of laminations, about 0.3 to 0.5 mm thick, insulated from one another. The effect of laminating the core is discussed in section 14.2.

6.4   Direction of the induced e.m.f.

The simplest method of determining the direction of the e.m.f. induced or generated in a conductor is to find the direction of the current due to that e.m.f. Thus, in fig. 6.4, AB represents a metal rod with its ends connected

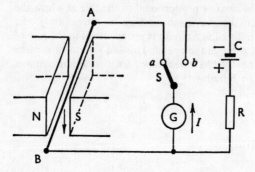

Fig. 6.4   Direction of induced e.m.f.

through a changeover switch S to a moving-coil galvanometer G. With S on side *a*, let us move AB *downwards* between the poles NS of an electromagnet and note the direction of G's deflection. Let us then move S over to *b*, so as to connect G in series with a high resistance R across a cell C. In order that G may again be deflected in the same direction, the polarity of C must be that shown in fig. 6.4; that is, the current through the galvanometer must be in the direction indicated by the arrow alongside G. Hence, the e.m.f. generated in AB must be acting from A towards B when the rod is moved downwards through the magnetic poles NS.

Now arises the problem: how can we remember this relationship in a form that can be easily applied to any other case? Two methods are available for this purpose, namely:

Fleming's* Right-hand Rule *If the first finger of the right hand be pointed in the direction of the magnetic flux, as in fig. 6.5, and if the thumb be pointed in the direction of the motion of the conductor* **relative** *to the magnetic field, then the second finger, held at right angles to both the thumb and the first finger, represents the direction of the e.m.f.* The manipulation

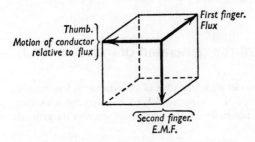

Fig. 6.5 Fleming's Right-hand Rule

of the thumb and fingers and their association with the correct quantity present some difficulty to many students. Each manipulation can only be acquired by experience; and it may be helpful to associate *f*ield or *f*lux with *f*irst finger, *m*otion of the conductor relative to the field with the *m* in thu*m*b and e.m.f. with the *e* in *s*econd finger. If any two of these are correctly applied, the third is correct automatically.

Lenz's Law In 1834, almost immediately after the discovery of induced currents, Heinrich Lenz, a German physicist (1804–65), gave a simple rule, known as Lenz's Law, which can be expressed thus: *The direction of an induced e.m.f. is always such that it tends to set up a current opposing the motion or the change of flux responsible for inducing that e.m.f.*

Let us consider the application of this law to the experiment described in connection with fig. 6.4. The current due to the e.m.f. induced in AB tends to set up a counter-clockwise magnetic flux around the rod, so that the resultant flux in the vicinity of AB is distorted in the opposite direction to that shown in fig. 5.21(c). But such a distorted flux exerts an upward force upon the conductor, trying to oppose its downward movement and therefore trying to prevent that which is responsible for the generation of the e.m.f. Hence, when Lenz's Law is applied to such an example, it is necessary to find the direction of the current which will distort the flux in such a direction as to try and prevent the relative movement of the conductor and the magnetic flux. The direction of such a current is also the direction of the generated e.m.f.

Let us also consider the application of Lenz's Law to the ring shown in fig. 6.3. By applying either the screw or the grip rule given in section 5.6,

---

* John Ambrose Fleming (1894–1945) was Professor of Electrical Engineering at University College, London.

we find that when S is closed and the cell has the polarity shown, the direction of the magnetic flux in the ring is clockwise. Consequently, the current in C must be such as to try to produce a flux in a counterclockwise direction, tending to oppose the growth of the flux due to A, namely the flux which is responsible for the e.m.f. induced in C. But a counter-clockwise flux in the ring would require the current in C to be passing through the coil from X to Y (fig. 6.3). Hence, this must also be the direction of the e.m.f. induced in C.

6.5 Magnitude of the generated or induced e.m.f.

Figure 6.6 represents the elevation and plan of a conductor AA situated in an airgap between poles NS. Suppose AA to be carrying a current, I amperes, in direction shown. By applying either the screw or the grip rule

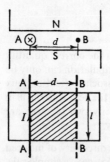

Fig. 6.6 Conductor moved across magnetic field

of section 5.6, it is found that the effect of this current is to strengthen the flux on the right and weaken that on the left of A, so that there is a force of BlI newtons (section 5.11) urging the conductor towards the left, where B is the flux density in teslas and l is the length in metres of the conductor in the magnetic field. Hence, a force of this magnitude has to be applied in the opposite direction to move A towards the right.

The work done in moving conductor AA through a distance d metres to position BB in fig. 6.6 is $(BlI \times d)$ newton metres or joules. If this movement of AA takes place at a uniform velocity in t seconds, the e.m.f. induced in the conductor is constant at, say, E volts. Hence the electrical power generated in AA is IE watts and the electrical energy is IEt watt seconds or joules. Since the mechanical energy expended in moving the conductor horizontally across the gap is all converted into electrical energy, then

$$IEt = BlId$$

$$\therefore \quad E = \frac{Bld}{t} \text{ volts.}$$

But Bld = the total magnetic flux, Φ webers, in the area shown shaded in fig. 6.6, and is therefore the flux cut by the conductor when the latter is moved from AA to BB. Hence

$$E\,[\text{volts}] = \frac{\Phi\,[\text{webers}]}{t\,[\text{seconds}]} \qquad\qquad [6.1]$$

i.e. the e.m.f., in volts, generated in a conductor is equal to the rate (in webers/second) at which the magnetic flux is cutting or being cut by the conductor; and the *weber* may therefore be defined as *that magnetic flux which when cut by a conductor in 1 second generates an e.m.f. of 1 volt.*

Example 6.1 *Calculate the e.m.f. generated in the axle of a car travelling at 90 km/h, assuming the length of the axle to be 1.8 m and the vertical components of the earth's magnetic field to be 40 μT.*

Speed of car = $90 \times 1000/3600 = 25$ m/s.

Vertical component of earth's magnetic field = 40×10^{-6} T,

$\therefore$ rate at which axle cuts magnetic flux

$$= 40 \times 10^{-6}\,[\text{T}] \times 1.8\,[\text{m}] \times 25\,[\text{m/s}]$$

$$= 1800 \times 10^{-6}\,\text{Wb/s}$$

and

e.m.f. generated in axle = 1800×10^{-6} V

$$= 1800\,\mu\text{V}.$$

Example 6.2 *A four-pole generator has a magnetic flux of 12 milliwebers per pole. Calculate the average value of the e.m.f. generated in one of the armature conductors while it is moving through the magnetic flux of one pole, if the armature is driven at 900 rev/min.*

When a conductor moves through the magnetic field of one pole, it cuts a magnetic flux of 12 mWb.

Time taken for a conductor to move through one revolution = $60/900 = (1/15)$ s.

Since the machine has 4 poles, time taken for a conductor to move through the magnetic field of one pole = $(1/4) \times (1/15) = (1/60)$ s,

$\therefore$ average e.m.f. generated in one conductor

$$= (12 \times 10^{-3})\,[\text{Wb}]\ \, 9(1/60)\,[\text{s}] = 0.72\ \text{V}.$$

6.6 Magnitude of e.m.f. induced in a coil

Suppose the magnetic flux through a coil of N turns to be increased by Φ webers in t seconds due to, say, the relative movement of the coil and a magnet (fig. 6.1). Since the magnetic flux cuts each turn, one turn can be regarded as a conductor cut by Φ webers in t seconds; hence, from expression [6.1], the average e.m.f. induced in each turn is Φ/t volts. The current due to this e.m.f., by Lenz's Law, tries to prevent the increase of flux, i.e. tends to set up an opposing flux. Thus, if the magnet NS in fig. 6.1 is moved towards coil C, the flux passing from left to right through the latter is increased. The e.m.f. induced in the coil circulates a current in the direction represented by the dot and cross in fig. 6.7, where—for

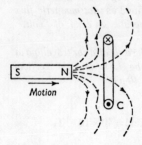

Fig. 6.7 Distortion of magnetic field caused by induced current

simplicity—coil C is represented as one turn. The effect of this current is to distort the magnetic field as shown by the dotted lines, thereby tending to push the coil away from the magnet. By Newton's Third Law of Motion, there must be an equal force tending to oppose the movement of the magnet.

Owing to the fact that the induced e.m.f. circulates a current tending to oppose the increase of flux through the coil, its direction is regarded as negative; hence

average e.m.f. induced in 1 turn $= -\Phi/t$ volts

$\qquad = -$ average rate of change
of flux in webers per
second

and

average e.m.f. induced in coil $= -N\Phi/t$ volts $\qquad$ [6.2]

$\qquad = -$ average rate of change
of flux-linkages per
second.

Next, let us consider the case of the two coils, A and C, shown in fig. 6.3. Suppose that when switch S is closed, the flux in the ring increases by Φ webers in t seconds. Then if coil A has N_1 turns,

average e.m.f. induced in A $= -N_1\Phi/t$ volts.

The minus sign signifies that this e.m.f., in accordance with Lenz's Law, is acting in opposition to the current, trying to prevent its growth.

Hence the induced e.m.f. is acting in opposition to the battery e.m.f., the direction of which is assumed to be positive.

If coil C is wound with N_2 turns and if all the flux produced by coil A passes through C,

average e.m.f. induced in C $= -N_2\Phi/t$ volts.

In this case the minus sign signifies that the e.m.f. circulates a current in such a direction as to tend to set up a flux in opposition to that produced by the current in coil A, thereby delaying the growth of flux in the ring.

Example 6.3 *A magnetic flux of 400 μWb passing through a coil of 1200 turns is reversed in 0·1 s. Calculate the average value of the e.m.f. induced in the coil.*

The magnetic flux has to decrease from $400\,\mu$Wb to zero and then increase to $400\,\mu$Wb in the reverse direction; hence the change of flux in the original direction is $-800\,\mu$Wb.

Substituting in expression (6.2), we have:

$$\text{average e.m.f. induced in coil} = -\frac{1200 \times (-800 \times 10^{-6})}{0.1}$$

$$= 9.6\,\text{V}.$$

This e.m.f. is positive because its direction is the same as the original direction of the current, at first tending to prevent the current decreasing and then tending to prevent it increasing in the reverse direction.

6.7 Summary of important formulae

Average e.m.f. generated in a conductor $= \Phi/t$ volts $\qquad$ [6.1]

where $\Phi = $ flux, in webers, cut in t seconds.

Average e.m.f. induced in a coil of N turns $= -N\Phi/t$ volts $\qquad$ [6.2]

where $\Phi = $ change of flux, in webers, in t seconds.

6.8 Examples

1. Two coils, A and B, are wound on the same iron core. There are 300 turns on A and 2800 turns on B. A current of 4 A through A gives rise to $800\,\mu$Wb in the core. If this current be reversed in 0.02 s, find the average e.m.f.s induced in A and B.
2. A coil of 1500 turns gives rise to a magnetic flux of $2500\,\mu$Wb when

carrying a certain current. If this current be reversed in 0.2 s, find the average e.m.f. induced in the coil.

3. A current of 8 A through a coil of 3000 turns produces a flux of 4 mWb. If this current is reduced to 2 A in 0.1 s calculate the average e.m.f. induced in the coil, assuming the flux to be proportional to the current. What is the direction of the induced e.m.f.?

4. A short coil of 200 turns surrounds the middle of a bar magnet. If the magnet sets up a flux of 80 μWb, find the average e.m.f. induced in the coil when the latter is removed completely from the influence of the magnet in 0.05 s.

5. A coil of 800 turns is wound on an iron core and a certain current produces a flux of 3000 μWb. When the circuit is opened, the residual flux in the iron is 1900 μWb. If this reduction of flux takes place in 0.15 s, calculate the average value of the induced e.m.f.

6. The field winding of a six-pole generator is separately-excited. The six coils are connected in series and each coil is wound with 2500 turns. The flux in each pole is 30 mWb. If the field switch is opened at such a speed that the flux falls to the residual value of 1 mWb in 1/3 s, calculate the average value of the e.m.f. induced in the field winding.

7. An aeroplane having a wing-span of 40 m is flying horizontally at a speed of 600 km/h. Calculate the e.m.f. generated between the tips of the wings, assuming the vertical component of the earth's magnetic field to be 40 μT. Is it possible to measure this e.m.f.?

8. The axle of a certain motor-car is 1.5 metres long. Calculate the e.m.f. generated in it when the car is travelling at 140 km/h. Assume the vertical component of the earth's magnetic field to be 40 μT.

9. A wire, 200 mm long, is moved at a uniform speed of 6 m/s at right angles to its length and to a magnetic field. Calculate the density of the magnetic field if the e.m.f. generated in the wire is 0.4 V.

   If the wire forms part of a closed circuit having a resistance of 0.1 Ω, calculate the force on the wire.

10. Give three practical applications of the mechanical force exerted on a current-carrying conductor in a magnetic field.

    A conductor of active length 0.3 m carries a current of 100 A and lies at right angles to a magnetic field of strength 0.4 T. Calculate the force in newtons exerted on it. If the force causes the conductor to move at a velocity of 10 m/s, calculate (a) the e.m.f. induced in it and (b) the power in watts developed by it.   (E.M.E.U.)

11. Specify the laws of electromagnetic induction and give one application of each law.

    A conductor, 0.1 m in length, moves with uniform velocity of 2 m/s at right angles to itself and to a uniform magnetic field having a flux density of 1 T. Calculate the induced e.m.f. between the ends of the wire.

    Draw a diagram giving a cross-section of the conductor and mark on it the direction of the induced e.m.f., the direction of motion and the direction of the field. Describe the method whereby the direction of the induced e.m.f. is determined.   (N.C.T.E.C.)

12. The flux through a 100-turn coil increases uniformly from zero to 300 μWb in 2 milliseconds. It remains constant at 300 μWb for the third millisecond and then decreases uniformly to zero during the fourth millisecond. Draw to scale graphs representing the variation of the flux and of the e.m.f. induced in the coil.

13. A copper disc, 300 mm in diameter, is rotated at 200 rev/min about a horizontal axis through its centre perpendicular to its plane. If the axis points magnetic north and south, calculate the e.m.f. between the circumference of the disc and the axis. Assume the horizontal component of the flux due to the earth's field to be 18 μT.

14. A six-pole motor has a magnetic flux of 0.08 Wb per pole and the armature is rotating at 700 rev/min. Calculate the average e.m.f. generated per conductor.

15. A four-pole armature is to generate an average e.m.f. of 1.4 V per conductor, the flux pole being 15 mWb. Calculate the speed at which the armature must rotate.

CHAPTER 7

▃▃▃▃▃▃▃▃ Magnetic Circuit

7.1 Introductory

Lines of magnetic flux form closed paths; for instance, in figs. 5.5 and 5.6, the dotted lines represent the flux passing through the iron core and returning through the surrounding air space. Similarly, the dotted lines in fig. 5.17 show that *any one line of flux forms a closed path*, and that in the case of a d.c. machine, part of the flux path is in air and the remainder in iron. The complete closed path followed by a magnetic flux is referred to as a *magnetic circuit*. One of the simplest forms of the magnetic circuit is indicated in fig. 6.3, where the iron ring R provides the path for the magnetic flux.

7.2 Magnetomotive force; magnetic field strength

In section 3.7 it was stated that in an electric circuit an electric current is due to the existence of an electromotive force. By analogy, we may say that in a magnetic circuit the magnetic flux is due to the existence of a *magnetomotive force* (m.m.f.) caused by a current flowing through one or more turns. The value of the m.m.f. is proportional to the current and to the number of turns, and is descriptively expressed in *ampere-turns*; but for the purpose of dimensional analysis, it is expressed in *amperes*, since the number of turns is dimensionless. Hence the unit of magnetomotive force is the *ampere*. If a current of I amperes flows through a coil of N turns, as shown in fig. 7.1, the magnetomotive force is the *total* current linked with the magnetic circuit, namely IN amperes. If the magnetic circuit is homogeneous and of uniform cross-sectional area, the magnetomotive force per metre length of the magnetic circuit is termed the *magnetic field strength* and is represented by the symbol H. Thus, if the mean length of the magnetic circuit of fig. 7.1 is l metres,

$$H = IN/l \text{ amperes/metre} \tag{7.1}$$

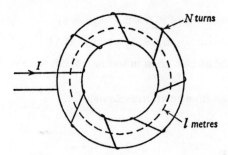

Fig. 7.1 A toroid

7.3 Permeability of free space or magnetic constant

Suppose A in fig. 7.2 to prepresent the cross-section of a long straight

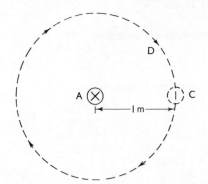

Fig. 7.2 Magnetic field at 1-metre radius due to current in a long straight conductor

conductor, situated in a vacuum and carrying a current of one ampere towards the paper; and suppose the return path of this current to be some considerable distance away from A so that the effect of the return current on the magnetic field in the vicinity of A may be neglected. The lines of magnetic flux surrounding A will therefore be in the form of concentric circles, as already described in section 5.5, and the dotted circle D in fig. 7.2 represents the path of one of these lines of flux at a radius of 1 metre. Since conductor A and its return conductor form 1 turn, the magnetomotive force acting on path D is 1 ampere; and since the length of this line of flux is 2π metres, the magnetic field strength, H, at a radius of 1 m is $1/2\pi$ ampere/ metre.

If B is the flux density in teslas in the region of line D, it follows from expression [5.1] that the force per metre length on a conductor C (parallel to A) carrying 1 ampere at right angles to this flux is given by:

Force, in newtons, per metre length $= B\,[\text{T}] \times 1\,[\text{m}] \times 1\,[\text{A}]$.

But from the definition of the ampere given in section 2.5, this force is 2×10^{-7} newton,

∴ flux density at 1-m radius from conductor carrying 1 A

$$= B = 2 \times 10^{-7}\,\text{T}.$$

Hence,

$$\frac{\text{flux density at C}}{\text{magnetic field strength at C}} = \frac{B}{H} = \frac{2 \times 10^{-7}\,[\text{T}]}{1/2\pi\,[\text{A/m}]}$$

$$= 4\pi \times 10^{-7}\,\text{units.}$$

The ratio B/H for the above condition is termed the *permeability of free space* or *magnetic constant* and is represented by the symbol μ_0. The value of this ratio is almost exactly the same whether the conductor A of fig. 7.2 is assumed to be situated in a vacuum (or free space) or in air or in any other non-magnetic material. Hence,

$$\mu_0 = \frac{B}{H} \text{ for a vacuum and non-magnetic materials}$$

$$= 4\pi \times 10^{-7}\,\text{H/m*} \tag{7.1}$$

and magnetic field strength for non-magnetic materials

$$H = \frac{B}{\mu_0} = \frac{B\,[\text{teslas}]}{4\pi \times 10^{-7}\,[\text{H/m}]} \tag{7.3}$$

Example 7.1 *A coil of 200 turns is wound uniformly over a wooden ring having a mean circumference of 600 mm and a uniform cross-sectional area of 500 mm². If the current through the coil is 4 A, calculate* (a) *the magnetic field strength,* (b) *the flux density and* (c) *the total flux.*

(a) Mean circumference $= 600/1000 = 0.6\,\text{m}$.

∴ $H = IN/l = 4 \times 200/0.6 = 1333\,\text{A/m}$.

(b) From expression [7.2]:

flux density $= \mu_0 H = 4\pi \times 10^{-7} \times 1333$

$$= 0.001\,675\,\text{T}$$

$$= 1675\,\mu\text{T}.$$

---

* It is shown in the footnote on p. 141 that the units of absolute permeability are *henry/metre* (or H/m); e.g. $\mu_0 = 4\pi \times 10^{-7}$ H/m.

(c)   Cross-sectional area $= 500 \times 10^{-6}\,\text{m}^2$

$$\therefore\quad \text{total flux} = 1675\,[\mu\text{T}] \times (500 \times 10^{-6})\,[\text{m}^2]$$

$$= 0.8375\,\mu\text{Wb}.$$

Example 7.2   *Calculate the m.m.f. required to produce a flux of 3 mWb across an airgap 2.5 mm long, having an effective area of 4000 mm².*

Area of airgap $= 4000 \times 10^{-6} = 0.004\,\text{m}^2$,

$$\therefore\quad \text{flux density} = \frac{0.003\,[\text{Wb}]}{0.004\,[\text{m}^2]} = 0.75\,\text{T}.$$

From [7.3],

$$\text{magnetic field strength in tap} = \frac{0.75}{4\pi \times 10^{-7}}$$

$$= 597\,000\,\text{A/m}.$$

Length of gap $= 2.5 \times 10^{-3} = 0.0025\,\text{m}$,

$\therefore$ m.m.f. required to send flux across gap

$$= 597\,000 \times 0.0025 = 1492\,\text{A}.$$

7.4   Relative permeability

In section 5.7 it was shown that the magnetic flux inside a coil is intensified when an iron core is inserted. It follows that if the non-magnetic core of a toroid, such as that shown in fig. 7.1, is replaced by an iron core, the flux produced by a given m.m.f. is greatly increased; and the ratio of the flux density produced in a material to the flux density produced in a vacuum (or in a non-magnetic core) by the same magnetic field strength is termed the *relative permeability* of that material and is denoted by the symbol μ_r. For air, $\mu_r = 1$; but for nickel–iron alloys such as Mumetal (p. 128) it may be as high as 100 000. Graphs representing the value of μ_r are given in figs. 7.6 and 7.7 for various qualities of iron, and it will be seen that for a given quality, the value of μ_r varies over a wide range.

From [7.2], $B = \mu_0 H$ for a non-magnetic material; hence, for a material having a relative permeability μ_r,

$$B = \mu_r \mu_0 H$$

$\therefore$ *absolute permeability* $= B/H = \mu_r \mu_0$.   [7.4]

7.5   Reluctance

Let us consider an iron ring having a cross-sectional area of a square metres and a mean circumference of l metres (fig. 7.1), wound with N turns carrying a current I amperes; then

$$\text{total flux} = \Phi = \text{flux density} \times \text{area} = Ba \qquad [7.5]$$

and

$$\text{m.m.f.} = \text{magnetic field strength} \times \text{length} = Hl \qquad [7.6]$$

Dividing [7.5] by [7.6], we have:

$$\frac{\Phi}{\text{m.m.f.}} = \frac{Ba}{Hl} = \mu_r \mu_0 \times \frac{a}{l}$$

$$\therefore \quad \Phi = \frac{\text{m.m.f.}}{l/\mu_r \mu_0 a} \qquad [7.7]$$

so that

$$\text{m.m.f.}/\Phi = l/\mu_r \mu_0 a$$

$$= \textit{reluctance} \text{ of magnetic circuit.}$$

This expression is similar in form to:

$$\text{e.m.f.}/I = \rho l/a$$

for the electric circuit. The denominator, $l/\mu_r \mu_0 a$, in expression [7.7] is similar in form to $\rho l/a$ for the resistance of a conductor except that the absolute permeability, $\mu_r \mu_0$, for the magnetic material corresponds to the reciprocal of the resistivity, namely the conductivity of the conducting material.

Since the m.m.f. is equal to the number of amperes acting on the magnetic circuit,

$$\therefore \quad \text{magnetic flux} = \frac{\text{m.m.f.}}{\text{reluctance}} = \frac{\text{no. of amperes}}{\text{reluctance}} \qquad [7.8]$$

where

$$\text{reluctance} = l/\mu_r \mu_0 a \qquad [7.9]$$

$$= l/\mu_0 a \text{ for non-magnetic materials.}$$

The symbol for reluctance is S and from expression [7.8] it is seen that reluctance can be expressed in amperes/weber.

7.6   Comparison of the electric and magnetic circuits

It is helpful to tabulate side by side the various electric and magnetic quantities and their relationships. These are given in Table 7.1.

Table 7.1

Electric circuit		Magnetic circuit	
Quantity	Unit	Quantity	Unit
E.M.F.	volt	M.M.F.	ampere
—	—	Magnetic field strength	ampere/metre
Current	ampere	Magnetic flux	weber
Current density	ampere/m^2	Magnetic flux density	tesla
Resistance	ohm	Reluctance	ampere/weber
$\left[= \rho \cdot \dfrac{l}{a} \right]$		$\left[= \dfrac{1}{\mu_r \mu_0} \cdot \dfrac{l}{a} \right]$	
Current = e.m.f./resistance		Flux = m.m.f./reluctance	

One important difference between the electric and magnetic circuits is the fact that energy must be supplied to *maintain* the flow of electricity in a circuit, whereas the magnetic flux, once it is set up, does not require any further supply of energy. For instance, once the flux produced by a current in a solenoid has attained its steady value, the energy subsequently absorbed by that solenoid is all dissipated as heat due to the resistance of the winding.

Example 7.3   *A mild-steel ring having a cross-sectional area of 500 mm^2 and a mean circumference of 400 mm has a coil of 200 turns wound uniformly around it. Calculate (a) the reluctance of the ring and (b) the current required to produce a flux of 800 μWb in the ring.*

(a)   Flux density in ring $= \dfrac{800 \times 10^{-6}\,[\text{Wb}]}{500 \times 10^{-6}\,[\text{m}^2]} = 1.6\,\text{T}.$

From fig. 7.7, the relative permeability of mild steel for a flux density of 1.6 T is about 370; hence, from expression [7.9]:

reluctance of ring $= \dfrac{0.4}{370 \times 4\pi \times 10^{-7} \times 500 \times 10^{-6}}$

(b)   From expression [7.7],

$$800 \times 10^{-6} = \frac{\text{m.m.f.}}{1.725 \times 10^6}$$

$$\therefore \quad \text{m.m.f.} = 1380\,\text{A}$$

and

magnetizing current $= 1380/200 = 6.9\,\text{A}.$

Alternatively, from expression [7.4],

$$\text{magnetic field strength} = \frac{B}{\mu_r \mu_0} = \frac{1.6}{370 \times 4\pi \times 10^{-7}}$$

$$= 3450 \,\text{A/m},$$

$$\therefore \quad \text{m.m.f.} = 3450 \,[\text{A/m}] \times 0.4 \,[\text{m}] = 1380 \,\text{A}$$

and

$$\text{magnetizing current} = 1380/200 = 6.9 \,\text{A}.$$

7.7 Composite magnetic circuit

Suppose a magnetic circuit to consist of two specimens of iron, A and B, arranged as in fig. 7.3. If l_1 and l_2 are the mean lengths in metres of the

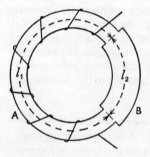

Fig. 7.3 Composite magnetic circuit

magnetic circuits of A and B respectively, a_1 and a_2 their cross-sectional areas in square metres and μ_1 and μ_2 their relative permeabilities, then

$$\text{reluctance of A} = \frac{l_1}{\mu_1 \mu_0 a_1}$$

and

$$\text{reluctance of B} = \frac{l_2}{\mu_2 \mu_0 a_2}$$

Since A and B are magnetically in series, i.e. since the same flux passes through the two specimens, assuming the magnetic leakage (section 7.9) to be negligible,

$$\text{total reluctance of magnetic circuit} = \frac{l_1}{\mu_1 \mu_0 a_1} + \frac{l_2}{\mu_2 \mu_0 a_2}$$

and

$$\text{total flux} = \Phi = \frac{\text{m.m.f. of coil}}{\text{total reluctance}}$$

$$= \frac{IN}{\dfrac{l_1}{\mu_1\mu_0 a_1} + \dfrac{l_2}{\mu_2\mu_0 a_2}}. \qquad [7.10]$$

7.8 Determination of the magnetization curve for an iron ring

Figure 7.4 shows an iron ring C of uniform cross-section, uniformly

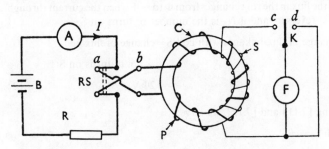

Fig. 7.4 Determination of the magnetization curve for an iron ring

wound with a coil P, thereby eliminating magnetic leakage. Coil P is connected to a battery B through a reversing switch RS, an ammeter A and a variable resistor R. Another coil S, which need not be distributed around the ring, is connected through a two-way switch K to fluxmeter F which is a special type of permanent-magnet moving-coil instrument. Current is led into and out of the moving coil of F by fine wires or ligaments so arranged as to exert negligible control over the position of the moving coil. When the flux in the ring is varied, the e.m.f. induced in S sends a current through the fluxmeter and produces a deflection that is proportional to the change of flux-linkages in coil S.

The current through coil P is adjusted to a desired value by means of R and switch RS is then reversed several times to bring the iron into a *cyclic* condition, i.e. into a condition such that the flux in the ring reverses from a certain value in one direction to the same value in the reverse direction. During this operation, switch K should be on *d*, thereby short-circuiting the fluxmeter. With switch RS on, say, *a*, switch K is moved over to *c*, the current through P is reversed by moving RS quickly over to *b* and the fluxmeter deflection is noted.

If

N_P = number of turns on coil P,

l = mean circumference of the ring, in metres

and

I = current through P, im amperes,

magnetic field strength = $H = IN_P/l$ amperes/metre.

If

θ = fluxmeter deflection when current through P is reversed

and

c = fluxmeter constant

= no. of weber-turns per unit of scale deflection,

change of flux-linkages with coils S = $c\theta$.   [7.11]

If the flux in the ring changes from Φ to $-\Phi$ when the current through coil P is reversed, and if N_S is the number of turns on S,

change of flux-linkages with coil S = change of flux

× no. of turns on S

= $2\Phi N_S$.   [7.12]

Equating [7.11] and [7.12], we have:

$2\Phi N_S = c\theta$

so that

$$\Phi = \frac{c\theta}{2N_S} \text{ webers.}$$

If a = cross-sectional area of ring in square metres.

Flux density in ring = $B = \Phi/a$

$$= \frac{c\theta}{2aN_S} \text{ teslas.} \qquad [7.13]$$

The test is performed with different values of the current; and from the data, a graph representing the variation of flux density with magnetic field strength can be plotted. The graphs in fig. 7.5 show the relationship between the flux density and the magnetic field strength obtained for different qualities of iron; and the graphs in figs. 7.6 and 7.7 represent the corresponding values of the relative permeability plotted against the magnetic field strength and the flux density respectively. In actual practice it is found that the magnetic property of different specimens of iron of the same chemical composition may vary considerably, being very dependent upon heat and mechanical treatments. The data for mild steel, wrought iron and sheet steel are so similar that they can be represented by a common graph. 'Stalloy' is an alloy of iron and silicon commonly used in the construction of transformers and a.c. machines.

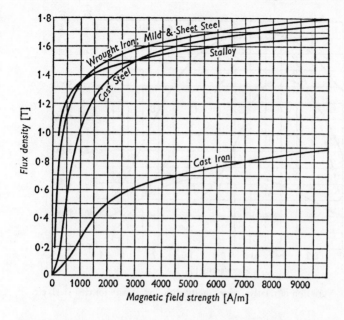

Fig. 7.5 Variation of flux density with magnetic field strength for cast steel, etc.

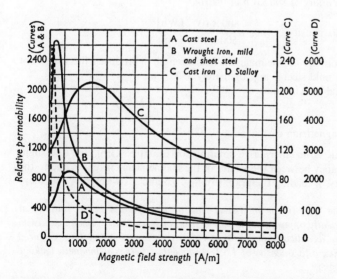

Fig. 7.6 Variation of relative permeability with magnetic field strength for cast steel, etc.

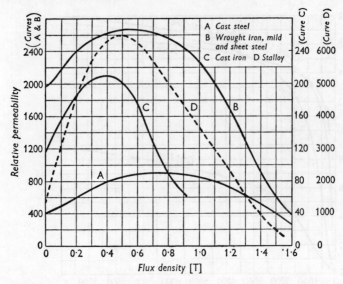

Fig. 7.7 Variation of relative permeability with flux density for cast steel, etc.

Example 7.4 *A mild-steel ring having a cross-sectional area of 500 mm²
and a mean circumference of 400 mm has a coil of 200 turns wound uniformly
around it. Using the data given in fig. 7.5, calculate the current required to
produce a flux of 800 µWb in the ring.*

$$\text{Flux density in ring} = \frac{800 \times 10^{-6}\,[\text{Wb}]}{500 \times 10^{-6}\,[\text{m}^2]} = 1.6\,\text{T}.$$

From fig. 7.5, the magnetic field strength to produce a flux density of
1.6 T in mild steel is approximately 3500 A/m.

∴ total m.m.f. required = 1400/200 = 7 A.

and

magnetizing current = 1400/200 = 7 A.

This is an alternative solution for example 7.3(b), and the slight
discrepancy between the values of the magnetizing current is due to the
difficulty of reading the graphs of figs. 7.5 and 7.7 with great accuracy.

Example 7.5 *If the mild-steel ring of example 7.4 is cut radially at
diametrically opposite points and a sheet of brass 0.5 mm thick inserted in
each gap, as indicated in fig. 7.8, calculate the magnetomotive force required
to produce a flux density of 1.6 T. Assume that there is no magnetic leakage
or fringing.*

From example 7.4, the magnetomotive force required for the steel is
1400 A.

Magnetic Circuit

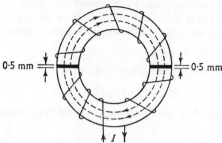

Fig. 7.8 Magnetic circuit for example 7.5

From expression [7.3], the magnetic field strength for the non-magnetic gaps $= 1.6/(4\pi \times 10^{-7}) = 1.274 \times 10^6$ A/m.

$\therefore$ total m.m.f. for the two gaps $= 1.274 \times 10^6$ [A/m] $\times 0.001$ [m]

$$= 1274 \text{ A}.$$

Hence   total m.m.f. $= 1400 + 1274 = 2674$ A.

Example 7.6   *A core is built of steel laminations stamped out as shown in fig. 7.9, in which the lower figure represents a sectional plan at XX. The width, w, of each side limb is half that of the centre limb. The iron in the latter has a cross-sectional area of 900 mm² and is wound with 150 turns of insulated wire. If the mean length of the flux path is 350 mm, find the current required to set up a flux of 1.2 mWb in the centre limb.*

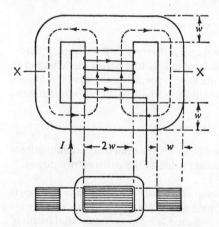

Fig. 7.9 Magnetic circuit for example 7.6

A current in the direction indicated by the arrowheads on the conductors in fig. 7.9 produces a magnetic flux which is upwards in the centre limb. This flux divides, one half returning via one outer limb and the other half via the other limb, as represented by the dotted lines. Since the area of each outer limb is half that of the centre limb, we may assume the flux density to be the same in every part of the magnetic circuit. Hence,

$$\text{flux density} = \frac{\text{total flux in centre limb}}{\text{cross-sectional area of centre limb}}$$

$$= 0.0012\,[\text{Wb}]/0.0009\,[\text{m}^2] = 1.333\,\text{T}.$$

From fig. 7.5 the magnetic field strength to produce this flux density in sheet steel $\simeq 900\,\text{A/m}$,

$$\therefore \quad \text{total m.m.f. required} = 900\,[\text{A/m}] \times 0.35\,[\text{m}] = 315\,\text{A}$$

and

$$\text{magnetizing current} = 315/150 = 2.1\,\text{A}.$$

It will be noticed that the flux paths in the side limbs are in parallel and that the m.m.f. which sends the flux through one side limb must be responsible for sending flux through the other side limb. This is a case where it may be helpful to compare the magnetic circuit of fig. 7.9 with the equivalent electric circuit shown in fig. 7.10. The current I from battery B

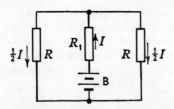

Fig. 7.10   Electric circuit equivalent to fig. 7.9

divided equally through the two equal resistances R, R and the p.d. across the right-hand R is exactly the same p.d. as that across the left-hand R.

If R_1 be the resistance of the middle portion of the circuit, it follows from Kirchhoff's Second Law that for both the right- and the left-hand loops,

$$\text{total e.m.f. of battery} = I \times R_1 + \tfrac{1}{2}I \times R.$$

Similarly, in the magnetic circuit of fig. 7.9, the total magnetomotive force is that required to send flux round either of the paths shown dotted.

Example 7.7   *The magnetic circuit of a two-pole d.c. generator is shown in fig. 7.11. The armature A is of the smooth-core type, i.e. it has no slots, the armature winding being on the surface. Each gap G has a radial length of 4 mm and a cross-sectional area of 6000 mm². Each pole core P has a radial length of 70 mm and a cross-sectional area of 5000 mm² and is made of wrought iron. The yoke Y is of cast steel, having a cross-sectional area of 1500 mm². The estimated mean length of the flux paths in the yoke is 420 mm*

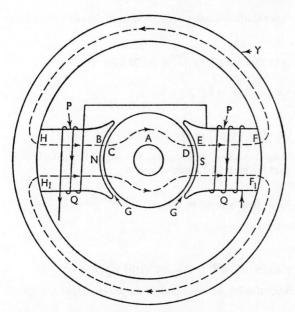

Fig. 7.11 Magnetic circuit for example 7.7

*and that in the armature is 100 mm. The cross-sectional area of each side of
the armature core is 1200 mm². Find the number of turns to be wound on each
pole in order that 3 A may set up a flux of 3 mWb per pole. Assume the flux in
the pole cores and yoke to be the same as that in the armature.*

The simplest procedure is to consider the path taken by any one line
of flux, follow that line round the whole of its magnetic circuit and
calculate the magnetomotive force required for each component of that
circuit. For instance, let us start at point B in fig. 7.11 and follow line
BCDEFHB, then:

total m.m.f. = sum of the m.m.f.s for
BC, CD, DE, EF, FH and HB.

It is evident from fig. 7.11 that the same m.m.f. will also set up flux round
loop H_1F_1.
Path BC.

Flux density in gap = 0.003 [Wb]/0.006 [m²] = 0.5 T.

From expression [7.3],

$$\text{value of } H \text{ for gap} = \frac{0.5 \times 10^7}{4\pi} = 0.398 \times 10^6 \,\text{A/m}$$

∴ m.m.f. to send flux from **B** to **C**

$$-(0.398 \times 10^{-6})\,[\text{A/m}] \times (4 \times 10^{-3})\,[\text{m}] = 1592\,\text{A}.$$

Path CD. Since the flux divides equally between the two sides of the core,

$$\text{flux density in armature} = \frac{0.003}{0.0012 \times 2} = 1.25\,\text{T}.$$

From graph for sheet steel in fig. 7.5,

$$H \simeq 700\,\text{A/m}$$

∴   m.m.f. to send flux from C to $D = 700 \times 0.1 = 70\,\text{A}.$

Path DE. Magnetomotive force required to send the flux from D to E is the same as that for path BC, namely, 1592 A.

Path EF.

Flux density in pole core $= 0.003/0.005 = 0.6\,\text{T}.$

From graph for wrought iron in fig. 7.5,

$$H \simeq 200\,\text{A/m}$$

∴   m.m.f. to send flux from E to $F = 200 \times 0.07 = 14\,\text{A}.$

Path FH. Since the flux divides equally between the upper and lower yokes,

$$\text{flux density in yoke} = \frac{0.003}{0.0015 \times 2} = 1.0\,\text{T}.$$

From graph for cost steel in fig. 7.5,

$$H \simeq 900\,\text{A/m}$$

∴   m.m.f. to send flux from F to $H = 900 \times 0.42 = 378\,\text{A}.$

Path HB. Magnetomotive force required to send the flux from H to B is the same as that for path EF, namely 14 A. Hence,

total m.m.f. for the whole magnetic circuit

$$= 1592 + 70 + 1592 + 14 + 378 + 14 = 3660\,\text{A}.$$

Table 7.2 shows the above calculations conveniently tabulated. Since the current is to be 3 A,

total no. of turns on coils QQ $= 3660/3 = 1220.$

These can be arranged as 610 turns on each pole core.

7.9 Magnetic leakage and fringing

Suppose *dd* in fig. 7.12 to represent a metal ring symmetrically situated relative to the airgap in the iron ring, and suppose the magnetizing winding to be concentrated over a short length of the core. As far as ring *dd*

Table 7.2

Components of magnetic circuit	Total area (m²)	Total length (m)	Flux density (T)	Amperes per metre (A/m)	Total m.m.f. (A)
Airgaps	0.006	0.008	0.5	398 000	3184
Armature	0.0024	0.1	1.25	700	70
Pole cores	0.005	0.14	0.6	200	28
Yoke	0.003	0.42	1.0	900	378

Total m.m.f. for whole magnetic circuit = 3660 A

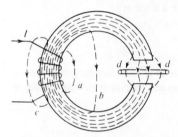

Fig. 7.12 Magnetic leakage and fringing

is concerned, the flux passing through it can be regarded as the *useful* flux and that which returns by such paths as *a*, *b* and *c* is *leakage* flux.

The useful flux passing across the gap tends to bulge outwards as shown roughly in fig. 7.12, thereby increasing the effective area of the gap and reducing the flux density in the gap. This effect is referred to as *fringing*; and the longer the airgap, the greater is the fringing.

The distinction between useful and leakage fluxes may be more obvious if we consider an electrical machine. For instance, fig. 7.13 shows two poles of a six-pole d.c. machine. The armature slots have, for simplicity, been omitted. Some of the dotted lines do not enter the armature core and thus do not assist in generating an e.m.f. in the armature winding; consequently, they represent leakage flux. On the other hand, some of the flux passes between the pole tips and the armature core, as shown in fig. 7.13, and is referred to as fringing flux. Since this fringing flux is cut by the armature conductors, it forms part of the useful flux.

From figs. 7.12 and 7.13 it is seen that the effect of leakage flux is to increase the total flux through the exciting winding, and

$$\text{leakage factor} = \frac{\text{total flux through exciting winding}}{\text{useful flux}} \qquad [7.14]$$

The value of the leakage factor for electrical machines is usually about 1.15 to 1.25.

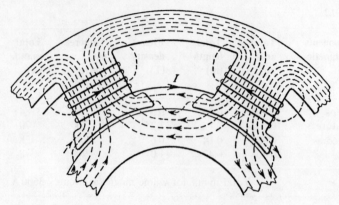

Fig. 7.13   Magnetic leakage and fringing in a machine

Example 7.8   *A magnetic circuit is made up of steel laminations shaped as in fig. 7.14. The width of the iron is 40 mm and the core is built up to a depth*

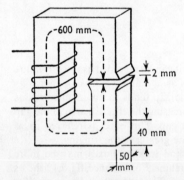

Fig. 7.14   Magnetic circuit for example 7.8

of 50 mm, of which 8 per cent is taken up by insulation between the laminations. The gap is 2 mm long and the effective area of the gap is 2500 mm². The coil is wound with 800 turns. If the leakage factor is 1.2, calculate the magnetizing current required to produce a flux of 2.5 mWb across the airgap.

Flux density in airgap = 0.0025 [Wb]/0.0025 [m²] = 1 T,

$$\therefore \quad \text{value of } H \text{ for gap} = \frac{1}{4\pi \times 10^{-7}} = 796\,000 \text{ A/m}$$

and   m.m.f. for gap = 796 000 [A/m] × 0.002 [m]

$$= 1592 \text{ A}.$$

Total flux through coil = flux in gap × leakage factor

$$= 0.0025 \times 1.2 = 0.003 \, \text{Wb}.$$

Since only 92 per cent of the cross-section of the core consists of iron,

∴ area of iron in core = 40 × 50 × 0.92

$$= 1840 \, \text{mm}^2 = 0.001 \, 84 \, \text{m}^2$$

and

flux density in core = 0.003 [Wb]/0.001 84 [m²] = 1.63 T.

From fig. 7.5, corresponding magnetic field strength for steel laminations ≃ 4000 A/m.

It will be evident from fig. 7.12 that when there is magnetic leakage, the flux density is not uniform over the whole length of the iron core. It is impossible, however, to allow for the variation of magnetic field strength due to this variation of flux density; and the usual practice is to assume that the magnetic field strength estimated for the region of maximum density applies to the whole of the iron core, thereby erring on the safe side. Hence

m.m.f. for iron core = 4000 [A/m] × 0.6 [m] = 2400 A

and

total m.m.f. = 1592 + 2400 = 3992 A

∴ magnetizing current = 3992/800 ≃ 5 A.

7.10 Hysteresis

If we take a closed iron ring which has been completely demagnetized* and measure the flux density with increasing values of the magnetic field strength, the relationship between the two quantities is represented by curve OAC in fig. 7.15. If the value of *H* is then reduced, the flux density follows curve CD; and when the value of *H* has been reduced to zero, the flux density remaining in the iron is OD and is referred to as the *remanent flux density*.

If the magnetic field strength is then increased in the reverse direction, the flux density decreases, until at some value OE, the flux has been reduced to zero. The magnetic field strength OE required to wipe out the residual magnetism is termed the *coercive force*. Further increase of *H* causes the flux density to grow in the reverse direction as represented by curve EF. If the reversed magnetic field strength OL is adjusted to the

---

* The simplest method of demagnetizing is to reverse the magnetic current a large number of times, the maximum value of the current at each reversal being reduced until it is ultimately zero.

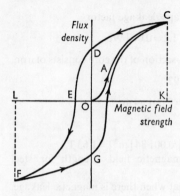

Fig. 7.15 Hysteresis loop

same value as the maximum field strength OK in the initial direction the final flux density LF is the same as KC.

If the magnetic field strength is varied backwards from OL to OK, the flux density follows a curve FGC similar to curve CDEF, and the closed figure CDEFGC is termed the *hysteresis** loop.

It is a remarkable fact that for most magnetic materials, ranging from the hardest of steels, such as those used in the construction of permanent magnets, to the most easily magnetizable materials, such as nickel-iron alloys, the ratio of the residual flux density OD to the maximum flux density KC usually lies between about 0.6 and 0.75. On the other hand, the value of the *coercive force* varies enormously for different materials. Thus, for 'Alcomax' (an alloy of iron, aluminium, nickel, cobalt and copper) the coercive force may be as high as 80 000 amperes/metre; whereas for 'Mumetal' (nickel, iron, copper and molybdenum), the value is about 3 amperes/metre.

7.11 Current-ring theory of magnetism

It may be relevant at this point to consider why the presence of iron in a current-carrying coil increases the value of the magnetic flux and why magnetic hysteresis occurs in iron. As long ago as 1823, André-Marie Ampère—after whom the unit of current was named—suggested that the increase in the magnetic flux might be due to electric currents circulating within the molecules of the iron. Subsequent discoveries have confirmed this suggestion and the following brief explanation may assist in giving some idea of the current-ring theory of magnetism.

---

* 'Hysteresis' is a Greek word meaning 'lagging behind'.

An atom consists of a nucleus of positive electricity surrounded, at distances large compared with their diameter, by electrons, which are charges or negative electricity (section 9.2). The electrons revolve in orbits around the nucleus, and each electron also spins around its own axis—somewhat like a gyroscope—and the magnetic characteristics of iron appear to be due mainly to this electron spin. The movement of an electron around a circular path is equivalent to a minute current flowing in a circular ring. In an iron atom, four more electrons spin round in one direction than in the reverse direction, and the axes of spin of these electrons are parallel with one another; consequently, the effect is equivalent to four current rings producing magnetic flux in a certain direction.

The iron atoms are grouped together in *domains*, each about 0.1 mm in width; and in any one domain the magnetic axes of all the atoms are parallel with one another. In an unmagnetized bar of iron, the magnetic axes of different domains are in various directions so that their magnetizing effects cancel one another.

When an unmagnetized bar of iron is moved into a current-carrying solenoid, there are sudden tiny increments of the magnetic flux as the magnetic axes of the various domains are orientated so that they coincide with the direction of the m.m.f. due to the solenoid, thereby increasing the magnitude of the flux. This phenomenon is known as the *Barkhausen effect* and can be demonstrated by winding a search coil on the iron bar and connecting it through an amplifier to a loudspeaker. The sudden increments of flux due to successive orientation of the various domains, while the iron bar is being moved into the solenoid, induce e.m.f. impulses in the search coil and the effect can be heard as a rustling noise.

It follows that when a current-carrying solenoid has an iron core, the magnetic flux can be regarded as consisting of two components: (a) the flux produced by the solenoid without an iron core, (b) the flux due to ampere-turns equivalent to the current rings formed by the spinning electrons in the orientated domains. This component reaches its maximum value when all the domains have been orientated so that their magnetic axes are in the direction of the magnetic flux. The iron is then said to be saturated.

This alignment of the domains has a certain amount of stability, depending upon the quality of the iron. Consequently the iron may retain much of its magnetism after the external magnetomotive force has been removed, as already discussed in section 7.10, the remanent flux being maintained by the m.m.f. due to the electronic current-rings in the iron. A permanent magnet can therefore be regarded as an electromagnet, the relatively large flux being due to the high value of the *inherent m.m.f.* (i.e. coercive force × length of magnet) retained by the steel after it has been magnetized.

A disturbance in the alignment of the domains necessitates the expenditure of energy in taking the specimen of iron through a cycle of magnetization. This energy appears as heat in the specimen and is referred to as *hysteresis loss*.

7.12 Summary of important formulae

Magnetomotive force $= F = IN$ amperes

Magnetic field strength $= H = IN/l$ amperes/metre $\qquad$ [7.1]

Permeability of free space or magnetic constant

$\quad = \mu_0 = B/H$ for vacuum or non-magnetic materials

$\qquad = 4\pi \times 10^{-7}$ henrys/metre $\qquad$ [7.2]

Magnetic field strength for non-magnetic materials

$$= \frac{B\,[\text{teslas}]}{(4\pi \times 10^{-7})\,[\text{H/m}]} \qquad [7.3]$$

Absolute permeability $= B/H = \mu_r\mu_0$ $\qquad$ [7.4]

$$\text{Magnetic flux} = \Phi = \frac{IN}{\text{reluctance}} \qquad [7.8]$$

$$\text{Reluctance} = S = l/\mu_r\mu_0 a. \qquad [7.9]$$

For a composite magnetic circuit,

$$\Phi = \frac{IN}{\dfrac{l_1}{\mu_1\mu_0 a_1} + \dfrac{l_2}{\mu_2\mu_0 a_2} + \cdots} \qquad [7.10]$$

$$\text{Leakage factor} = \frac{\text{total flux}}{\text{useful flux}}. \qquad [7.14]$$

7.13 Examples

Data of B/H, when not given in question, should be taken from fig. 7.5.

1. A long straight conductor, situated in air, is carrying a current of 200 A, the return conductor being some distance away. Calculate (a) the magnetic field strength and (b) the flux density at a radius of 100 mm.
2. The flux density at a distance of 60 mm from the centre of a long straight conductor C is 0.02 T. Assuming the return conductor to be some distance away, calculate the current in conductor C.
3. Two straight parallel conductors, situated in air and spaced 100 mm between centres, carry current of 1000 A in opposite directions. Calculate (a) the magnetic field strength and (b) the flux density at points on a straight line joining the two conductors (i) midway between the conductors and (ii) 30 mm from one conductor.

Also, calculate the magnetic field strength and the flux density at the same two points if the conductors are each carrying 1000 A in the same direction.

4. The airgap in a certain magnetic circuit is 1.2 mm long and 2500 mm² in cross-section. Calculate (a) the reluctance of the gap, (b) the m.m.f. required to send a flux of 800 μWb across the gap and (c) the magnetic field strength in the gap.

5. A cast-iron ring having a cross-sectional area of 500 mm² and a mean diameter of 150 mm is wound with 600 turns. Calculate the current required to set up a flux of 350 μWb.

6. An iron magnetic circuit has a uniform cross-sectional area of 400 mm² and a length of 700 mm. For a flux density of 1.5 T, calculate (a) the reluctance of the circuit and (b) the total m.m.f. required. Assume the relative permeability for this flux density to be 2500.

7. An iron magnetic circuit has a uniform cross-sectional area of 750 mm² and a length of 200 mm. A coil of 100 turns is wound uniformly over the circuit. With a current of 1 A, the total flux is 450 μWb; and with 3 A, the flux is 900 μWb. Calculate the values of (a) the magnetic field strength and (b) the relative permeability of the iron for each value of flux.

8. A coil is wound uniformly with 300 turns over an iron ring having a mean circumference of 400 mm and a cross-section of 500 mm². If the coil has a resistance of 8 Ω and is connected across a 20-V d.c. supply, calculate (a) the m.m.f. of the coil, (b) the magnetic field strength, (c) the total flux and (d) the reluctance of the ring. Assume the value of the relative permeability to be 900.

9. A coil P of 200 turns is wound uniformly on an iron ring having a cross-sectional area of 400 mm² and a mean diameter of 300 mm. Another coil Q of 20 turns, wound on the ring, is connected to a fluxmeter having a constant of 150 microweber-turns per division. When a current of 1.5 A through P is reversed, the fluxmeter deflection is 78 divisions. Calculate (a) the flux density in the ring and (b) the corresponding value of the relative permeability.

10. Write a brief description of the fluxmeter method as employed to determine the magnetization curve of a sample of iron in the form of a ring.

    A coil of 250 turns is wound uniformly on an iron ring whose cross-sectional area is 500 mm² and whose mean diameter is 300 mm. A second coil of 20 turns wound on the ring is connected to a fluxmeter whose constant is 180 microweber-turns per division. When a current of 2 A is reversed in the 250-turn coil, the fluxmeter deflection is 90 divisions. Calculate the flux density in the ring for a current of 2 A and the corresponding value of the relative permeability.

11. A wooden ring with a mean circumference of 800 mm and a cross-sectional area of 600 mm² is uniformly wound with a coil P of 400 turns. Another coil Q of 800 turns, wound on the ring, is connected to a fluxmeter having a constant of 300 microweber-turns per division. Calculate the fluxmeter deflection when a current of 5 A through P is reversed.

12. A magnetic circuit has a uniform cross-sectional area of 500 mm² and a mean length of 800 mm. The B/H curve for the material is given below:

H (A/m)	100	200	300	400	500	600
B (T)	0.76	1.1	1.23	1.3	1.34	1.27

    Calculate the m.m.f. required to produce a flux density of 1.2 T and also the relative permeability of the material at this density.

13. (a) The flux density in air at a point 40 mm from the centre of a long straight conductor A is 0.03 T. Assuming that the return conductor is a considerable distance away, calculate the current in A.

    (b) In a certain magnetic circuit, having a length of 500 mm and a cross-sectional area of 300 mm², a magnetomotive force of 200 A produces a flux of 400 μWb. Calculate (i) the reluctance of the magnetic circuit and (ii) the relative permeability of the core.

(E.M.E.U.)

14. An iron ring having a mean circumference of 300 mm and a cross-sectional area of 400 mm² is wound with a coil of 80 turns. From the following data calculate the current required to set up a flux of 520 μWb:

Flux density, teslas	1.0	1.2	1.4
Amperes/metre	350	600	1250

15. If a radial saw-cut, 0.7 mm wide, is made in the ring of Q. 14, calculate the current required to set up the same flux of 520 μWb. Neglect any magnetic leaking and fringing.

16. An iron ring has internal and external diameters of 100 mm and 120 mm respectively and the axial thickness of the ring is 15 mm. The ring is uniformly wound with 300 turns. Calculate the current required to set up a flux of 200 μWb, given the following data:

Flux density (T)	1.2	1.3	1.4	1.5
Relative permeability	2000	1560	1150	750

    If a radial gap, 0.3 mm wide, is made in the ring, calculate (a) the current required to maintain the same flux of 200 μWb and (b) the corresponding reluctance of the magnetic circuit.

17. A U-shaped wrought-iron electromagnet has an armature of the same material across the ends, leaving an airgap of 0.15 mm between each end and the armature. The cross-sectional area of the core and armature is 600 mm² and the total length of iron path is 0.5 m. The electromagnet is wound with 300 turns. Calculate the current required to set up a flux of 750 μWb.

18. One portion of a wrought-iron ring has a cross-sectional area of 400 mm² and an effective length of 200 mm, the corresponding values for the remainder of the ring being 300 mm² and 120 mm respectively. The ring is wound with 150 turns. Calculate the current required to set up a flux of 480 μWb.

19. The two halves of a ring of uniform cross-section consist of cast iron and cast steel respectively. The cross-sectional area is 350 mm² and the mean circumference of the ring is 0.6 m. The butt joints between the two halves are equivalent to a single radial airgap 0.2 mm wide. Calculate the m.m.f. necessary to set up a flux of 200 μWb.

20. A mild-steel core has the dimensions shown in fig. 7.16. The cross-

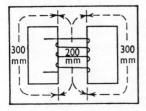

Fig. 7.16

sectional area of the centre limb is 1500 mm² and that of each side limb is 900 mm². A coil of 200 turns is wound on the centre limb. Calculate the current required to set up a magnetic flux on 2.4 mWb in the centre limb. Neglect any magnetic leakage.

21. A magnetic system is made of wrought iron arranged as in fig. 7.17. The centre limb has a cross-sectional area of 600 mm² and each of the outer limbs has a cross-sectional area of 400 mm². Each airgap has an area of 400 mm² and a length of 2 mm. If the coil is wound with 500 turns, calculate the current required to set up a flux of 900 μWb in the centre limb, assuming no magnetic leakage. The mean lengths of the various magnetic paths are indicated on the diagram.

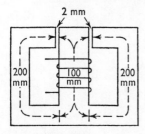

Fig. 7.17

22. An electromagnet with its armature has an iron length of 400 mm and a cross-sectional area of 500 mm². There is a total airgap of 1.8 mm. Assuming a leakage factor of 1.2, calculate the magnetomotive force required to produce a flux of 400 μWb in the armature. Points on the *B/H* curve are as follows:

Flux density (T)	0.8	1.0	1.2
Magnetic field strength (A/m)	800	1000	1600

23. A circular cast-steel lifting magnet has a cross-sectional area of 8000 mm² and a mean magnetic path length of 600 mm. When it is used to lift certain steel plates, the reluctance of which is negligible, there is an effective gap of 0.8 mm at each pole. Calculate the m.m.f. required to establish a flux of 10 mWb in the airgaps. Assume a leakage factor of 1.15.

CHAPTER 8
▰ Inductance in a d.c. Circuit

8.1 Inductive and non-inductive circuits

Let us consider what happens when a coil C (fig. 8.1) and a resistor R, connected in parallel, are switched across a battery B. C consists of a large number of turns wound on an iron core D (or it may be the field winding of a generator or motor), and R may be an element of an electric heater connected in series with a *centre-zero* ammeter A_2.

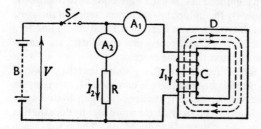

Fig. 8.1   Inductive and non-inductive circuits

When switch S is closed it is found that the current I_2 through R increases instantly to its final value, whereas the current i_1* through C takes an appreciable time to grow—as indicated in fig. 8.2. The final value, I_1, is equal to battery voltage V/resistance of coil C. In fig. 8.2, I_2 has been shown a little larger than I_1, but this is of no importance.

When S is opened, current through C decreases comparatively slowly, but the current through R instantly reverses its direction and becomes the same current as i_1; in other words, the current of C is circulating round R.

Let us now consider the reason for the difference in the behaviour of the currents in C and R.

The growth of current in C is accompanied by an increase of flux— shown dotted—in the iron core D. But it has been pointed out in section 6.4 that any change in the flux linked with a coil is accompanied by an

---

* A small letter is used to represent the instantaneous value of a varying quantity.

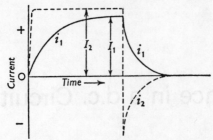

Fig. 8.2 Variation of switch-on and switch-off currents

e.m.f. induced in that coil, the direction of which—according to Lenz's Law—is always such as to oppose the change responsible for inducing the e.m.f., namely the growth of current in C. In other words, the induced e.m.f. is acting in opposition to the current and therefore to the applied voltage.

In circuit R, the flux is so small that the induced e.m.f. is negligible.

When switch S is opened, the current in both C and R tend to decrease; but any decrease of i_1 is accompanied by a decrease of flux in D and therefore by an e.m.f. induced in C in such a direction as to oppose the decrease of i_1. Consequently the induced e.m.f. is now acting in the same direction as the current. But it is evident from fig. 8.1 that after S has been opened, the only return path for C's current is that via R; hence the reason why i_1 and i_2 are now one and the same current.

If the experiment is repeated without R, it is found that the growth of i_1 is unaffected; but when S is opened there is considerable arcing at the switch due to the maintenance of current across the gap by the e.m.f. induced in C. The more quickly S is opened, the more rapidly does the flux in D collapse and the greater is the e.m.f. induced in C. This is the reason why it is dangerous to break quickly the full excitation of an electromagnet such as the field winding of a d.c. machine. One method of avoiding this risk is to connect a *discharge resistor* R permanently in parallel with the winding, as in fig. 8.1; but this involves a waste of energy in R while the winding is in circuit. This waste of energy can be avoided by the use of a special switch such as that shown in fig. 8.3, where the blade B, pivoted at

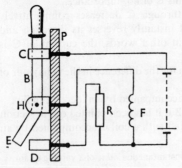

Fig. 8.3 Field-discharge switch

H, carries an extension E. Hinge H and jaws C and D are fixed to a board P of insulating material. When the switch is being opened, E makes contact with D before B breaks contact with C, thereby connecting R and F momentarily in parallel across the source. After B has broken contact with C, resistor R and winding F form a closed circuit and the current decreases relatively slowly so that there is no risk of a dangerously high e.m.f. being induced in F. The resistance of the discharge resistor is usually about the same as that of the field winding.

Any circuit in which a change of current is accompanied by a change of flux, and therefore by an induced e.m.f., is said to be *inductive* or to possess *self inductance* or merely *inductance*. It is impossible to have a perfectly *non-inductive* circuit, i.e. a circuit in which no flux is set up by a current; but for most purposes a circuit which is not in the form of a coil may be regarded as being practically non-inductive—even the open helix of an electric fire is almost non-inductive. In cases where the inductance has to be reduced to the smallest possible value—for instance, in resistance boxes—the wire is bent back on itself, as shown in fig. 8.4, so that the

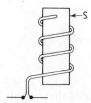

Fig. 8.4 Non-inductive coil

magnetizing effect of one conductor is neutralized by that of the adjacent conductor. The wire can then be coiled round an insulator S without increasing the inductance.

8.2 Unit of inductance

The unit of inductance is termed the *henry*, in commemoration of a famous American physicist, Joseph Henry (1797–1878), who, quite independently, discovered electromagnetic induction within a year after it had been discovered in this country by Michael Faraday in 1831. *A circuit has an inductance of 1 henry (or 1 H) if an e.m.f. of 1 volt is induced in the circuit when the current varies uniformly at the rate of 1 ampere per second.* If either the inductance or the rate of change of current be doubled, the induced e.m.f. is doubled. Hence we may generalize by saying that if a circuit has an inductance of L henrys and if the current *increase* from i_1 to i_2 amperes in t seconds:

$$\text{average rate of change of current} = \frac{i_2 - i_1}{t} \text{ amperes/second,}$$

and

average induced e.m.f. $= -L \times$ rate of change of current

$$= -L \times \frac{i_2 - i_1}{t} \text{ volts.} \quad [8.1]$$

The minus sign indicates that the direction of the induced e.m.f. is opposite to that of the current increase. It also indicates that energy is being *absorbed* from the electric circuit and stored as magnetic energy in the coil.

When the current is decreasing, the induced e.m.f. tends to prevent the decrease of the current and its direction is therefore the same as that of the current. Energy is then being *supplied* from the magnetic circuit to the electric circuit, so that the coil is acting as a generator* of electrical energy.

Example 8.1   *If the current through a coil having an inductance of 0.5 H is reduced from 5 A to 2 A in 0.05 s, calculate the average value of the e.m.f. induced in the coil.*

Average rate of change of current

$$= \frac{2 - 5}{0.05} = -60 \text{ amperes/second.}$$

From [8.1]

average e.m.f. induced in coil $= -0.5 \times (-60) = 30 \text{ V.}$

The direction of the induced e.m.f. is the same as that of the current opposing its decrease.

8.3 Inductance in terms of flux-linkages per ampere

Suppose a current of I amperes through a coil of N turns to produce a flux of Φ webers, and suppose the reluctance of the magnetic circuit to remain constant so that the flux is proportional to the current. Also, suppose the inductance of the coil to be L henrys.

If the current is increased from zero to I amperes in t seconds,

average rate of change of current $= I/t$ amperes/second,

$\therefore$ average e.m.f. induced in coil $= -LI/t$ volts   [8.2]

In section 6.6, it was explained that the value of the e.m.f., in volts,

---

* In section 16.1 it is shown that in a generator the current and the generated e.m.f. are in the same direction, so that electric power is being *supplied* by the generator; whereas in an electric motor, the direction of the generated e.m.f. is opposite to that of the current, so that electric power is *absorbed* by the motor.

induced in a coil is equal to the rate of change of flux-linkages per second. Hence, when the flux increases from zero to Φ webers in t seconds,

average rate of change of flux $= \Phi/t$ webers/second

and

average e.m.f. induced in coil $= -N\Phi/t$ volts   [8.3]

Equating expressions [8.2] and [8.3], we have:

$-LI/t = -N\Phi/t$

$\therefore \quad L = N\Phi/I$ henrys   [8.4]

$=$ flux-linkages/ampere.

The term 'flux-linkages' merely means the product of the flux in webers and the number of turns through which that flux passes or with which the flux is linked. Thus if a coil of 20 turns has a flux of 0.1 weber through it, the flux-linkages $= 0.1 \times 20$ weber-turns.

Expression [8.4] gives us an alternative method of defining the unit of inductance, namely: *A coil possesses an inductance of 1 henry if a current of 1 ampere through the coil produces a flux-linkage of 1 weber-turn.*

Example 8.2 *A coil of 800 turns is wound on a wooden former and a current of 5 A through the coil produces a magnetic flux of 200 μWb. Calculate (a) the inductance of the coil and (b) the average value of the e.m.f. induced in the coil when the current is reversed in 0.2 s.*

(a) Since the relative permeability of wood is unity, the flux is proportional to the current. Hence, from [8.4], we have:

$$\text{inductance} = \frac{200 \times 10^{-6} \times 800}{5} = 0.032 \, \text{H}.$$

(b) The current changes from 5 A to -5 A in 0.2 s,

$\therefore$ average rate of change of current $= -(5 \times 2)/0.2$

$= -50 \, \text{A/s}.$

Hence, from expression [8.2],

average e.m.f. induced in coil $= -0.032 \times (-50)$

$= 1.6 \, \text{V}.$

Alternatively, since the flux changes from $200 \, \mu$Wb to $-200 \, \mu$Wb in 0.2 s,

average rate of change of flux $= -(200 \times 10^{-6} \times 2)/0.2$

$= -0.002 \, \text{Wb/s}.$

Hence, from expression [8.3],

average e.m.f. induced in coil $= -800 \times (-0.002)$

$= 1.6 \, \text{V}.$

The sign is positive because the e.m.f. is acting in the direction of the original current, at first trying to prevent the current decreasing to zero and then opposing its growth in the reverse direction.

8.4 Factors determining the inductance of a coil

Let us first consider a coil uniformly wound on a *non-magnetic* ring of uniform section—similar to that of fig. 7.1. From expression [7.2] it follows that the flux density, in teslas, in such a ring is $4\pi \times 10^{-7} \times$ amperes/metre. Consequently, if l be the length of the magnetic circuit, in metres, and a its cross-sectional area, in square metres, then for a coil of N turns with a current I amperes:

$$H = IN/l \text{ amperes/metre}$$

and

$$\text{total flux} = \Phi = Ba = \mu_0 Ha$$
$$= 4\pi \times 10^{-7} \times (IN/l)a \text{ webers.}$$

Substituting for Φ in expression [8.4] we have:

$$\text{inductance} = L = 4\pi \times 10^{-7} \times aN^2/l \text{ henrys.} \qquad [8.5]$$

Hence the inductance is proportional to the square of the number of turns and to the cross-sectional area, and is inversely proportional to the length of the magnetic circuit.

If the coil is wound on a closed iron circuit, such as a ring, the problem of defining the inductance of such a coil becomes involved due to the fact that the variation of flux is no longer proportional to the variation of current. Suppose the relationship between the magnetic flux and the magnetizing current to be as shown in fig. 8.5; then if the iron has initially no residual magnetism, an increase of current from zero to OA causes the flux to increase from zero to AC, but when the current is subsequently reduced to zero, the decrease of flux is only DE. If the current is then increased to OG in the reverse direction, the change of flux is EJ. Consequently, we can have an infinite number of inductance values, depending upon the particular variation of current that we happen to consider.

Since we are usually concerned with the effect of inductance in an a.c. circuit, where the current varies from a maximum in one direction to the same maximum in the reverse direction, it is convenient to consider the value of the inductance as being the ratio of the change of flux-linkages to the change of current when the latter is reversed. Thus, for the case shown in fig. 8.5:

$$\text{inductance of coil} = \frac{DJ}{AG} \times \text{number of turns.}$$

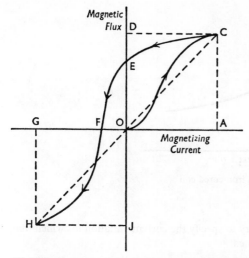

Fig. 8.5 Variation of magnetic flux with magnetizing current for a closed iron circuit

This value of inductance is the same as if the flux varied linearly along the dotted line COH in fig. 8.5.

If μ_r represents the value of the relative permeability corresponding to the maximum value AC of the flux, then the inductance of the iron-cored coil, as defined above, is μ_r times that of the same coil with a non-magnetic core. Hence, from expression [8.5], we have:

inductance of iron-cored coil (for reversal of flux)

$$= 4\pi \times 10^{-7} \times \frac{aN^2}{l} \times \mu_r \text{ henrys.*} \qquad [8.6]$$

The variations of relative permeability with magnetic field strength for various qualities of iron are shown in fig. 7.6; hence it follows from expression [8.6] that as the value of an alternating current through a coil having a closed iron circuit is increased, the value of the inductance increases to a maximum and then decreases, as shown in fig. 8.6. It will now be evident that when the value of the inductance of such a coil is

---

* From expression [8.6], $L\,[\text{henrys}] = \mu_0\mu_r \times a\,[\text{metres}^2] \times N^2/l\,[\text{metres}]$

$$\therefore \quad \text{absolute permeability} = \mu_0\mu_r = \frac{L\,[\text{henrys}] \times l\,[\text{metres}]}{N^2 \times a\,[\text{metres}^2]}$$

$$= \frac{Ll}{N^2a} \text{ henrys/metre,}$$

i.e., the units of absolute permeability are *henrys/metre* (or H/m); e.g. $\mu_0 = 4\pi \times 10^{-7}$ H/m.

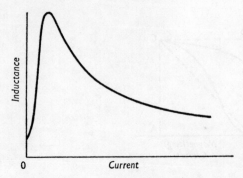

Fig. 8.6   Inductance of an iron-cored coil

stated, it is also necessary to specify the current variation for which that value has been determined.

Example 8.3   *A ring of 'Stalloy' stampings having a mean circumference of 400 mm and a cross-sectional area of 500 mm² is wound with 200 turns. Calculate the inductance of the coil corresponding to a reversal of a magnetizing current of (a) 1 A, (b) 10 A.*

(a) $H = \dfrac{1\,[\text{A}] \times 200\,[\text{turns}]}{0.4\,[\text{m}]} = 500\,\text{A/m}.$

From fig. 7.5:

   corresponding flux density $= 1.22\,\text{T}$

∴   total flux $= 1.22\,[\text{T}] \times 0.0005\,[\text{m}^2] = 0.000\,61\,\text{Wb}.$

From expression [8.4]

   $\text{inductance} = \dfrac{0.000\,61\,[\text{Wb}] \times 200\,[\text{turns}]}{1.0\,[\text{A}]} = 0.122\,\text{H}.$

(b) $H = \dfrac{10\,[\text{A}] \times 200\,[\text{turns}]}{0.4\,[\text{m}]} = 5000\,\text{A/m}.$

From fig. 7.5:

   corresponding flux density $= 1.58\,\text{T}$

∴   total flux $= 1.58 \times 0.0005 = 0.000\,79\,\text{Wb}$

and

   $\text{inductance} = (0.000\,79 \times 200)/10 = 0.0158\,\text{H}.$

Example 8.4 *If a coil of 200 turns be wound on a non-magnetic core having the same dimensions as the 'Stalloy' ring of example 8.3, calculate its inductance.*

From expression [8.6] we have:

$$\text{inductance} = \frac{4\pi \times 10^{-7}\,[\text{H/m}] \times 0.0005\,[\text{m}^2] \times (200)^2\,[\text{turns}^2]}{0.4\,[\text{m}]}$$

$$= 0.000\,062\,8\ \text{H}$$

$$= 0.0628\ \text{mH}$$

$$= 62.8\ \mu\text{H}.$$

A comparison of this inductance with the values obtained in example 8.3 for a coil of the same dimensions shows why an iron core is used when a large inductance is required.

8.5 Iron-cored inductor in a d.c. circuit

An inductor (i.e. a piece of apparatus used primarily because it possesses inductance) is frequently used in the output circuit of a rectifier to smooth out any variation (or ripple) in the direct current (section 17.7). If the inductor were made with a closed iron circuit, the relationship between rhe flux and the magnetizing current would be represented by curve OBD in fig. 8.7. It will be seen that if the current increases from OA to OC, the flux

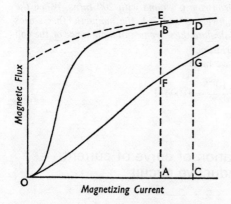

Fig. 8.7 Effect of inserting an airgap in an iron core

increases from AB to CD. If this increase takes place in t seconds, then:

average induced e.m.f.

$$= -\text{number of turns} \times \text{rate of change of flux}$$

$$= -N \times (CD - AB)/t \text{ volts.} \qquad [8.7]$$

Let L_1 be the *incremental* inductance of the coil over this range of flux variation, i.e. the effective value of the inductance when the flux is not proportional to the magnetizing current and varies over a relatively small range, then

$$\text{average induced e.m.f.} = -L_1 \times (\text{OC} - \text{OA})/t \text{ volts.} \qquad [8.8]$$

Equating [8.7] and [8.8], we have:

$$L_1 \times (\text{OC} - \text{OA})/t = N \times (\text{CD} - \text{AB})/t$$

$$\therefore \quad L_1 = N \times \frac{\text{CD} - \text{AB}}{\text{OC} - \text{OA}} \qquad [8.9]$$

$$= N \times \text{average slope of curve BD.}$$

From fig. 8.7 it is evident that the slope is very small when the iron is saturated. This effect is accentuated by hysteresis; thus if the current is reduced from OC to OA, the flux decreases from CD only to AE, so that the effective inductance is still further reduced.

If a short radial airgap were made in the iron ring, the flux produced by current OA would be reduced to some value AF. For the reduced flux density in the iron, the total m.m.f. required for the iron and the gap is approximately proportional to the flux; and for the same increase of current, AC, the increase of flux = CG − AF. As (CG − AF) may be much greater than (CD − AB), we have the curious result that the effective inductance of an iron-cored coil in a d.c. circuit may be increased by the introduction of an airgap.

Example 8.5   *A laminated iron ring is wound with 200 turns. When the magnetizing current varies between 5 and 7 A, the magnetic flux varies between 760 and 800 μWb. Calculate the incremental inductance of the coil over this range of current variation.*

From expression [8.8] we have:

$$L_1 = 200 \times \frac{(800 - 760) \times 10^{-6}}{(7 - 5)} = 0.004 \text{ H.}$$

8.6 Graphical derivation of curve of current growth in an inductive circuit

In section 8.1 the growth of current in an inductive circuit was discussed qualitatively; we shall now consider how to derive the curve showing the growth of current in a circuit of known resistance and inductance (assumed constant).

When dealing with an inductive circuit it is convenient to separate the effects of inductance and resistance by representing the inductance L as an inductor or coil having no resistance and the resistance R as a resistor

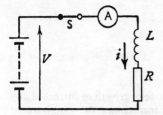

Fig. 8.8 Inductive circuit

having no inductance, as shown in fig. 8.8. It is evident from the latter that the current ultimately reaches a steady value I (fig. 8.9), where $I = V/R$.

Let us consider any instant A during the growth of the current. Suppose the current at that instant to be i amperes, represented by AB in fig. 8.9. The corresponding p.d. across R is Ri volts. Also at that instant the

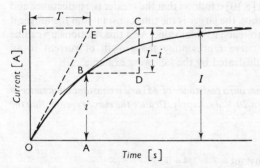

Fig. 8.9 Growth of current in an inductive circuit

rate of change of the current is given by the slope of the curve at B, namely the slope of the tangent to the curve at B.

But the slope of $\text{BC} = \text{CD}/\text{BD} = (I-i)/\text{BD}$ amperes/second. Hence

e.m.f. induced in L at instant $A = -L \times$ rate of change of current

$$= -L \times (I-i)/\text{BD volts}.$$

The total applied voltage V is absorbed partly in providing the voltage drop across R^* and partly in neutralizing the e.m.f. induced in L: i.e.

$$V = Ri + L \times (I-i)/\text{BD}.$$

---

* If R were negligibly small, then

$$V = L \times \text{rate of change of current}$$

$\therefore$ rate of change of current $= V/L$ amperes/second.

Substituting RI for V, we have:

$$RI = Ri + L \times (I-i)/\mathrm{BD}$$

$$\therefore \quad R(I-i) = L \times (I-i)/\mathrm{BD}$$

hence $\mathrm{BD} = L/R$.

In words, this expression means that the rate of growth of current at any instant is such that if the current continued increasing at that rate, it would reach its maximum value of I amperes in L/R seconds. Hence this period is termed the *time constant* of the circuit and is usually represented by the symbol T: i.e.

time constant $= T = L/R$ seconds.   [8.10]

Immediately after switch S is closed, the rate of growth of the current is given by the slope of the tangent OE drawn to the curve at the origin; and if the current continued growing at this rate, it would attain its final value in time FE $= T$ seconds.

From expression [8.10] it follows that the greater the inductance and the smaller the resistance, the larger is the time constant and the longer it takes for the current to reach its final value. Also this relationship can be used to deduce the curve representing the growth of current in an inductive circuit, as illustrated by the following example.

Example 8.6   *A coil having a resistance of $4\,\Omega$ and a constant inductance of 2 H is switched across a 20-V d.c. supply. Deduce the curve representing the growth of the current.*

From [8.10],

time constant $= T = 2/4 = 0.5$ s.

Final value of current $= I = 20/4 = 5$ A.

With the horizontal and vertical axes suitably scaled, as in fig. 8.10, draw a horizontal dotted line at the level of 5 A. Along this line mark off a period MN $= T = 0.5$ s, and join ON.

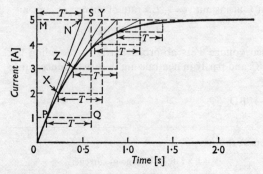

Fig. 8.10   Graph for example 8.6

Take any point P relatively near the origin and draw a horizontal dotted line $PQ = T = 0.5$ s, and at Q draw a vertical dotted line QS. Join PS.

Repeat the operation from a point X on PS, Z on XY, etc.

A curve touching OP, PX, XZ, etc., represents the growth of the current. The greater the number of points used in the construction, the more accurate is the curve.

8.7 Energy stored in a magnetic field

Suppose the current in a coil having a *constant* inductance L henrys to grow at a uniform rate from zero to I amperes in t seconds, as represented by straight line OA in fig. 8.11(a), and then to remain constant at I amperes. During the t seconds, the rate of increase of current is I/t

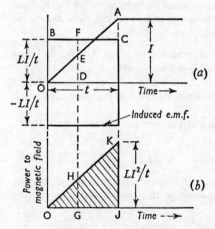

Fig. 8.11 Energy absorbed by a magnetic field

amperes/second and the e.m.f. induced in the coil therefore remains constant at $-LI/t$ volts. Hence the component of the applied voltage to neutralize this induced e.m.f. also remains constant at LI/t volts and is represented by the horizontal line BC in fig. 8.11(a). Immediately the current reaches I amperes, its rate of change becomes zero and the induced e.m.f. instantly falls to zero.

At instant D, the current is DE and the applied voltage to neutralize the induced e.m.f. is DF. Consequently the power supplied to the magnetic field at that instant is (DE × DF) watts and is represented by GH in fig. 8.11(b). It is therefore obvious that the power absorbed by the magnetic field increases uniformly from zero to LI^2/t watts as represented by the straight line OK.

Hence, average power absorbed by magnetic field $=\frac{1}{2}LI^2/t$ watts and total energy absorbed by magnetic field

$$= \text{average power} \times \text{time}$$

$$=(\tfrac{1}{2}LI^2/t) \times t = \tfrac{1}{2}LI^2 \text{ joules} \qquad\qquad [8.11]$$

and is represented by the area of the shaded triangle OJK in fig. 8.11(b).

When an inductive circuit is opened, the current has to die away and the magnetic energy has to be dissipated. If there is no resistor in parallel with the circuit, the energy is largely dissipated as heat in the arc at the switch. With a parallel resistor as described in section 8.1, the energy is dissipated as heat generated by the decreasing current in the total resistance of the circuit in which that current is flowing.

Example 8.7   *A coil has a resistance of 5 Ω and an inductance of 1.2 H. The current through the coil is increased uniformly from zero to 10 A in 0.2 s, maintained constant for 0.1 s and then reduced uniformly to zero in 0.3 s. Plot graphs representing the variation with time of* (a) *the current,* (b) *the induced e.m.f.,* (c) *the p.d.s across the resistance and the inductance,* (d) *the resultant applied voltage and* (e) *the power to and from the magnetic field. Assume the coil to be wound on a non-metallic core.*

The variation of current is represented by graph A in fig. 8.12; and since the p.d. across the resistance is proportional to the current, this p.d. increases from zero to (10 A × 5 Ω), namely 50 V, in 0.2 s, remains constant at 50 V for 0.1 s and then decreases to zero in 0.3 s, as represented by graph B.

During the first 0.2 s, the current is increasing at the rate of 10/0.2, namely 50 A/s,

∴   corresponding induced e.m.f. $= -50 \times 1.2 = -60$ V.

During the following 0.1 s, the induced e.m.f. is zero; and during the last 0.3 s, the current is decreasing at the rate of $-10/0.3$, namely -33.3 A/s,

∴   corresponding induced e.m.f. $= -(-33.3 \times 1.2) = 40$ V.

The variation of the induced e.m.f. is represented by the uniformly dotted graph C in fig. 8.12. Since the p.d. applied across the inductance has to neutralize the induced e.m.f., its variation is represented by the chain-dotted graph D which is exactly similar to graph C except that the signs are reversed.

The resultant voltage applied to the coil is obtained by adding graphs B and D; thus the resultant voltage increases uniformly from 60 V to 110 V during the first 0.2 s, remains constant at 50 V for the next 0.1 s and then changes uniformly from 10 V to -40 V during the last 0.3 s, as shown by graph E in fig. 8.12.

The power supplied to the magnetic field increases uniformly from zero to (10 A × 60 V), namely 600 W, during the first 0.2 s. It is zero during the next 0.1 s. Immediately the current begins to decrease, energy is being

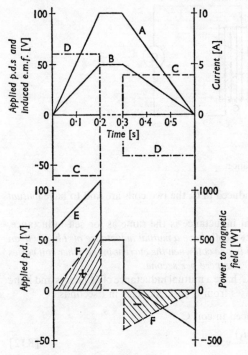

Fig. 8.12 Graphs for example 8.7

returned from the magnetic field to the electric circuit, and the power decreases uniformly from $(-40\,V \times 10\,A)$, namely $-400\,W$, to zero as represented by graph F.

The positive shaded area enclosed by graph F represents the energy $(= \frac{1}{2} \times 600 \times 0.2 = 60\,J)$ absorbed by the magnetic field during the first 0.2 s; and the negative shaded area represents the energy $(= \frac{1}{2} \times 400 \times 0.3 = 60\,J)$ returned from the magnetic field to the electric circuit during the last 0.3 s. The two areas are obviously equal in magnitude, i.e. all the energy supplied to the magnetic field is returned to the electric circuit.

8.8 Mutual inductance

If two coils A and C are placed relatively to each other as in fig. 8.13 then when S is closed, some of the flux produced by the current in A becomes linked with C and the e.m.f. induced in C circulates a momentary current through galvanometer G. Similarly, when S is opened, the collapse of the flux induces an e.m.f. in the reverse direction in C. Since a change of current in coil A is accompanied by a change of flux linked with coil C and

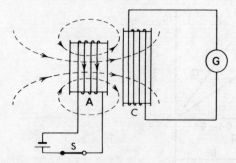

Fig. 8.13   Mutual inductance

therefore by an e.m.f. induced in C, the two coils are said to have *mutual inductance.*

The unit of mutual inductance is the same as for self inductance, namely the *henry;* and *two coils have a mutual inductance of 1 henry if an e.m.f. of 1 volt is induced in one coil when the current in the other coil varies uniformly at the rate of 1 ampere per second.*

If two coils, A and C, have a mutual inductance of M henrys and if the current in coil A increases from i_1 to i_2 amperes in t seconds:

average e.m.f. induced in coil C

$$= -M \times \frac{(i_2 - i_1)}{t} \text{ volts} \qquad [8.12]$$

$$= -M \times \text{average rate of change of current in coil A.}$$

The minus sign indicates that the e.m.f. induced in coil C tends to circulate a current in such a direction as to oppose the increase of flux due to the growth of current in coil A (see section 6.6).

If Φ_1 and Φ_2 represent the flux in webers linked with coil C due to currents i_1 and i_2 amperes respectively in coil A (termed the *primary*), and if N_2 represents the number of turns on coil C (termed the *secondary*),

$$\text{average e.m.f. induced in C} = -\frac{\Phi_2 - \Phi_1}{t} \cdot N_2 \text{ volts.} \qquad [8.13]$$

Equating expressions [8.12] and [8.13], we have:

$$M\left(\frac{i_2 - i_1}{t}\right) = \frac{\Phi_2 - \Phi_1}{t} \cdot N_2$$

$$\therefore \quad M = \frac{\Phi_2 - \Phi_1}{i_2 - i_1} \cdot N_2 \text{ henrys}$$

$$= \frac{\text{change of flux-linkages with secondary}}{\text{change of current in primary}}. \qquad [8.14]$$

If the relative permeability of the magnetic circuit remains constant, the flux linked with coil C is directly proportional to the current in coil A, so that

$$M = \frac{\text{flux-linkages with secondary}}{\text{current in primary}} = \Phi_2 N_2 / I_1. \quad [8.15]$$

If two coils have self-inductances L_1 and L_2 respectively and if the mutual inductance between the coils is M, then

$$\frac{M}{\sqrt{(L_1 L_2)}} = coupling\ coefficient \quad [8.16]$$

This term is much used in radio work to denote the degree of coupling between two coils; thus, if the coils are close together, most of the flux produced by current in one coil passes through the other and the coils are said to be *tightly* coupled. If the coils are well apart, only a small fraction of the flux is linked with the secondary, and the coils are said to be *loosely* coupled.

Example 8.8 *Two coils have a mutual inductance of 0.3 H. If the current in one coil is varied from 5 to 2 A in 0.4 s, calculate: (a) the average e.m.f. induced in the other coil; (b) the change of flux linked with the latter, assuming that it is wound with 200 turns.*

(a) Average rate of change of current in one coil

$$= \frac{(2-5)\,[\text{A}]}{0.4\,[\text{s}]} = -7.5\,\text{A/s}$$

$\therefore$ average e.m.f. induced in other coil

$$= -0.3\,[\text{H}] \times (-7.5)\,[\text{A/s}] = 2.25\,\text{V}.$$

(b) If Φ be the change of flux, in webers, linked with the second coil:

$$\text{average e.m.f. induced in that coil} = \frac{\Phi \times \text{number of turns}}{\text{time in seconds}}$$

i.e. $2.25 = (\Phi \times 200)/0.4$

$$\therefore \quad \Phi = 0.0045\,\text{Wb}.$$

8.9 Summary of important formulae

Induced e.m.f. $= -L \times$ rate of change of current.   [8.1]

Self inductance (in henrys) $= L =$ flux-linkages/ampere   [8.4]

Inductance of coil (in henrys) $= 4\pi \times 10^{-7} \times \dfrac{aN^2}{l} \times \mu_r$ [8.6]

Time constant of inductive circuit

$$= T = \frac{L\,[\text{henrys}]}{R\,[\text{ohms}]}\ \text{seconds} \quad [8.10]$$

Magnetic energy stored in an inductor

having constant inductance $= \frac{1}{2}LI^2$ joules   [8.11]

Mutual inductance (in henrys)

$$= \frac{\text{flux-linkages with secondary}}{\text{current in primary}}$$ [8.15]

Coupling coefficient $= \dfrac{M}{\sqrt{(L_1 L_2)}}$.   [8.16]

8.10 Examples

1. A 1500-turn coil surrounds a magnetic circuit which has a reluctance of 6×10^6 A/Wb. What is the inductance of the coil?
2. Calculate the inductance of a circuit in which 30 V are induced when the current varies at the rate of 200 A/s.
3. At what rate is the current varying in a circuit having an inductance of 50 mH when the induced e.m.f. is 8 V?
4. What is the value of the e.m.f. induced in a circuit having an inductance of 700 μH when the current varies at a rate of 5000 A/s?
5. A certain coil is wound with 50 turns and a current of 8 A produces a flux of 200 μWb. Calculate (a) the inductance of the coil and (b) the average e.m.f. induced when the current is reversed in 0.2 s.
6. A toroid is wound with 300 turns on a plastic ring having a cross-sectional area of 400 mm² and a mean circumference of 350 mm. Calculate (a) the inductance of coil and (b) the induced e.m.f. when the current is reduced at the rate of 200 A/s.
7. A coil wound with 500 turns has a resistance of 2 Ω. It is found that a current of 3 A produces a flux of 500 μWb. Calculate (a) the inductance and the time constant of the coil and (b) the average e.m.f. induced in the coil when the flux is reversed in 0.3 s.
8. If the coil referred to in Question 7 is switched across a 10-V d.c. supply, derive a curve showing the growth of the current, assuming the inductance to remain constant.
9. What is meant by (a) self inductance and (b) mutual inductance? Define the unit in which each is measured.

    If a coil of 150 turns is linked with a flux of 0.01 Wb when carrying a current of 10 A, calculate the inductance of the coil. If this current is uniformly reversed in 0.1 s, calculate the induced e.m.f.

    (N.C.T.E.C.)
10. Explain the term *inductance*, name the unit used and state how it is defined.

    A coil has an inductance of 5 H and its resistance is negligible. When it is connected to a 15-V d.c. supply, determine the average rate at which the current increases and hence find the current 4 s after connection to the supply. If this current is interrupted in 0.5 s, what average e.m.f. will be induced in the coil? (U.E.I.)

11. A coil having an inductance of 2 H and a resistance of 5 Ω is switched across a 20-V d.c. supply. Derive the curve of current growth in the coil and estimate the value of the current after 0.5 s.

12. The field winding of a d.c. machine has a resistance of 80 Ω and an inductance of 90 H. It has a resistor of 70 Ω connected permanently in parallel. With a current of 6 A in the field winding, the combined circuits are disconnected from the supply. Derive a curve showing the variation of the field current with time.

13. A coil having a resistance of 20 Ω and an inductance of 4 H is connected across a 50-V d.c. supply. Determine (a) the initial rate of growth of the current, (b) the value of the current after 0.1 2, (c) the time required for the current to grow to 1.5 A, and (d) the energy stored in the magnetic field when the current has attained its steady value.

14. The current through a certain coil is increased uniformly from 0 to 10 A in 0.1 s. It is then increased uniformly from 10 to 20 A in 0.3 s and subsequently decreased uniformly from 20 A to zero in 0.4 s. If the inductance of the coil remains constant at 2 H and if the resistance is negligible, plot graphs to scale representing the variation of (a) the current, (b) the e.m.f. induced in the coil and (c) the voltage applied to the coil.

15. A certain circuit has a resistance of 10 Ω and a constant inductance of 3 H. The current through this circuit is increased uniformly from 0 to 5 A in 0.6 s, maintained constant at 5 A for 0.1 s and then reduced uniformly to zero in 0.3 s. Draw to scale graphs representing the variation of (a) the current, (b) the induced e.m.f. and (c) the resultant applied voltage.

16. The current in a circuit consisting of an inductor of 0.1 H which has a resistance of 50 Ω is increasing uniformly from zero at a rate of 5 A/s. Draw to scale graphs of this current and of the voltage across the circuit for the first 1/100 of a second.

    Caclculate the energy stored in the inductor when the current through the inductor is 0.1 A. What is the value of the voltage across the circuit at that instant?

17. A coil consists of two similar sections wound on a common core. Each section has an inductance of 0.1 H. Calculate the inductance of the coil when the sections are connected (a) in series, (b) in parallel.

18. A coil, having a resistance of 8 Ω and an inductance of 3 H, is connected across a 20-V d.c. supply. Calculate the energy stored in the magnetic field when the current has attained its final value.

19. If two coils have a mutual inductance of 400 μH, calculate the e.m.f. induced in one coil when the current in the other coil varies at a rate of 30 000 A/s.

20. If an e.m.f. of 5 V is induced in a coil when the current in an adjacent coil varies at a rate of 80 A/s, what is the value of the mutual inductance of the two coils?

21. If the mutual inductance between two coils is 0.2 H, calculate the average e.m.f. induced in one coil when the current in the other coil is increased from 0.5 to 3 A in 0.05 s.

22. If the toroid of Question 6 has a second winding of 80 turns wound over the plastic ring and inside the first winding of 300 turns, calculate the mutual inductance.

23. When a current of 2 A through a coil P is reversed, a deflection of 36 divisions is obtained on a fluxmeter connected to a coil Q. If the fluxmeter constant is 150 microweber-turns/division, what is the value of the mutual inductance of coils P and O?

24. Two coils, A and B, of 600 and 100 turns respectively, are wound uniformly around a wooden ring having a mean circumference of 300 mm. The cross-sectional area of the ring is 400 mm^2. Calculate (a) the mutual inductance of the coils and (b) the average e.m.f. induced in coil B when a current of 2 A in coil A is reversed in 0.01 s.

25. Two coils have self inductances of 10 mH and 20 mH respectively. If the mutual inductance is 12 mH, what is the value of the coupling coefficient?

26. Two similar coils, each having a self inductance of 400 μH, are so spaced that the coupling coefficient is 0.2. Calculate the value of the mutual inductance.

CHAPTER 9

Electrostatics

9.1 Electrification by friction

The fact that amber, when rubbed, acquires the property of attracting light bodies was referred to over 2500 years ago by a Greek philosopher, named Thales, as a phenomenon that was quite familiar even at that time; but it was not until about A.D. 1600 that Dr Gilbert of Colchester discovered that other bodies, such as glass, could also be electrified by rubbing.

If a glass rod be rubbed* with silk and placed in a stirrup hung from a wooden stand (fig. 9.1), and if the end of another similarly rubbed glass rod

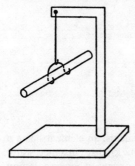

Fig. 9.1 Attraction and repulsion between electric charges

be brought near one end of the suspended rod, they are found to repel each other. But if an ebonite rod rubbed with fur is brought near the suspended glass rod, attraction takes place. Similarly, if an excited ebonite rod be supported on the stirrup, the approach of a similar ebonite rod causes repulsion, whereas an excited glass rod causes attraction.

The glass and the ebonite appear to be charged with different kinds of electricity; and the above experiments show that bodies charged with the

---

* These electrostatic experiments are difficult to perform satisfactorily if the atmosphere is humid, but they do form a very simple and useful introduction to the idea of positive and negative charges of electricity and to the existence of forces of repulsion between like charges and of attraction between unlike charges.

same kind of electricity repel, while bodies charged with opposite kinds of electricity attract one another.

About 1750 Benjamin Franklin, in America, suggested that electricity was some form of fluid which passed from one body to the other when they were rubbed together, and that in the case of glass rubbed with silk, the electric fluid passed from the silk into the glass so that the glass contained a 'plus' or 'positive' amount of electricity. On the other hand, when ebonite was rubbed with fur, the electric fluid passed from the ebonite to the fur, leaving the ebonite with a 'minus' or 'negative' amount of electricity. Franklin's one-fluid theory has long since been discarded, but his convention still remains; thus, the glass is said to be positively charged and the ebonite negatively charged, and it is this convention which governs the signs used for the terminals of batteries and direct-current generators.

We may now summarize the results of the above experiments by saying that like charges repel and unlike charges attract each other, as shown in fig. 9.2, where the arrows represent the directions of the forces on the charges.

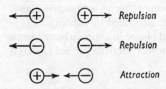

Fig. 9.2   Repulsion and attraction between electric charges

9.2   Structure of the atom

Every material is made up of one or more elements, an element being a substance composed entirely of atoms of the same kind; for instance, water is a combination of the elements hydrogen and oxygen, whereas common salt is a combination of the elements sodium and chlorine. The atoms of different elements differ in their structure, and this accounts for different elements possessing different characteristics.

Every atom consists of a relatively massive core or nucleus carrying a positive charge, around which *electrons* move in orbits at distances that are great compared with the size of the nucleus. Each *electron* has a mass of 9.11×10^{-31} kg and a *negative* charge, $-e$, equal to 1.602×10^{-19} C. The nucleus of every atom except that of hydrogen consists of *protons* and *neutrons*. Each *proton* carries a *positive* charge, e, equal in magnitude to that of an electron and its mass is 1.673×10^{-27} kg, namely 1836 times that of an electron. A *neutron*, on the other hand, carried *no* resultant charge and its mass is approximately the same as that of a proton. Under normal conditions, an atom is neutral, i.e. the total negative charge on its electrons is equal to the total positive charge on the protons.

The atom possessing the simplest structure is that of hydrogen—it consists merely of a nucleus of one proton together with a single electron which may be thought of as revolving in an orbit, of about 10^{-10} m diameter, around the proton, as in fig. 9.3(a).

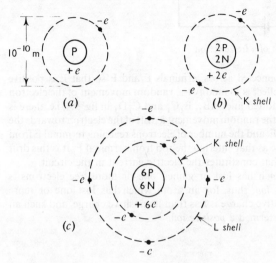

Fig. 9.3 Hydrogen, helium and carbon atoms

Figure 9.3(b) shows the arrangement of a helium atom. In this case, the nucleus consists of two protons and two neutrons, with two electrons orbiting in what is termed the K *shell*. The nucleus of a carbon atom has six protons and six neutrons and therefore carries a positive charge **6e**. This nucleus is surrounded by six orbital electrons, each carrying a negative charge of $-\mathbf{e}$, two electrons being in the K shell and four in the L shell, and their relative positions may be imagined to be as shown in fig. 9.3(c).

The further away an electron is from the nucleus, the smaller is the force of attraction between that electron and the positive charge on the nucleus; consequently the easier it is to detach such an electron from the atom. When atoms are packed tightly together, as in a metal, each outer electron experiences a small force of attraction towards neighbouring nuclei, with the result that such an electron is no longer bound to any individual atom, but can move at random within the metal. These electrons are termed *free* or *conduction* electrons and only a slight external influence is required to cause them to drift in a desired direction.

The full lines AB, BC, CD, etc., in fig. 9.4 represent paths of the random movement of one of these free electrons in a metal rod when there is *no* p.d. across terminals EF; i.e. the electron is accelerated in direction AB until it collides with an atom with the result that it may rebound in direction BC, etc. Different free electrons move in different directions so as to maintain the electron density constant throughout the metal; in other words, there is no resultant drift of electrons towards either E or F.

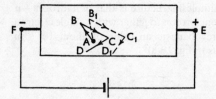

Fig. 9.4   Movement of a free electron

If a cell is connected across terminals E and F so that E is positive relative to F, the effect is to modify the random movement of the electron as shown by the dotted lines AB_1, B_1C_1 and C_1D_1 in fig. 9.4, i.e. there is superimposed on the random movement a drift of the electron towards the positive terminal E; and the number of electrons reaching terminal E from the rod is the same as that entering the rod from terminal F. It is this drift of the electrons that constitutes the electric current in the circuit.

An atom which has lost or gained one or more free electrons is referred to as an *ion*; thus, for an atom which has lost one or more electrons, its negative charge is less than its positive charge, and such an atom is therefore termed a *positive ion*.

9.3   Movement of electrons in a conductor

In section 9.2 it was explained how the drift of electrons in a desired direction can be produced by introducing a source of electromotive force into the circuit. In fig. 9.5, DE represents an enlarged view of a metal rod

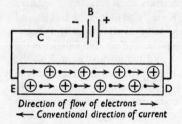

Direction of flow of electrons $\longrightarrow$
$\longleftarrow$ Conventional direction of current

Fig. 9.5   Movement of electrons in a conductor

forming part of a closed circuit which includes a battery B. The circles with crosses represent positive ions, namely atoms which have lost one or more of their outermost electrons. These ions are locked in the structure of the metal and are therefore unable to move. The small black circles represent the free electrons moving from left to right. This procession or drift of the electrons takes place round the whole circuit, including battery B i.e. the number of electrons emerging per second from the negative terminal of B is exactly the same as that entering the positive terminal per second. It follows that an electric current in a metal conductor consists of a movement of electrons from a point at the *lower* potential to a point at the *higher* potential, namely in the opposite direction to that taken as the conventional direction of the current. The latter was universally adopted long before the discovery of the electron and so we continue to say that an electric current flows from a point at the *higher* potential to that at the *lower* potential.

Since each electron carries a negative charge of 1.602×10^{-19} coulomb, it follows that when the current in a circuit is 1 ampere (or 1 coulomb/second), the number of electrons passing any given point must be such that:

$$1.602 \times 10^{-19} \times \text{no. of electrons/second} = 1 \text{ coulomb/second}$$

$$\therefore \quad \text{no. of electrons/second} = 6.24 \times 10^{18}$$

i.e. when the current in a circuit is 1 ampere, electrons are passing any given point of the circuit at the rate of 6.24×10^{18} per second.

9.4 Capacitor

Two metal plates, separated by an insulator, constitute a *capacitor**, namely an arrangement which has the capacity of storing electricity as an excess of electrons on one plate and a deficiency on the other.

The most common type of capacitor used in practice consists of two strips of metal foil, represented by full lines in fig. 9.6, separated by strips of waxed paper, shown dotted, these strips being wound spirally, forming—in effect—two very large metal surfaces near to each other. The whole

Fig. 9.6   Paper-insulated capacitor

assembly is thoroughly soaked in hot paraffin wax. In radio receivers, some of the capacitors consist of two sets of metal vanes, one of which is fixed and the other set is so arranged that the vanes can be moved into and out of the space between the fixed vanes without touching the latter.

A charged capacitor may be regarded as a reservoir of electricity and its action can be demonstrated by connecting a capacitor of, say, 20 microfards in series with a resistor R, a centre-zero microammeter A

---

* Recommended by the British Standards Institution, but the term 'condenser' is still widely used.

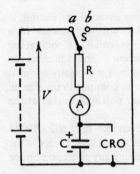

Fig. 9.7 Capacitor charged and discharged through a resistor

and a two-way switch S, as in fig. 9.7. A cathode ray oscilloscope (section 18.23) is connected across C. If R has a resistance of, say, 1 megohm, it is found that when S is closed on *a*, the deflection on A rises immediately to its maximum value and then falls off to zero, as indicated by curve A in fig. 9.8. At the same time, the p.d. across C grows in the manner shown by curve M. When S is moved over to *b*, the current again rises immediately to the same maximum but in the reverse direction, and then falls off as shown by curve B. Curve N shows the corresponding variation of p.d. across C.

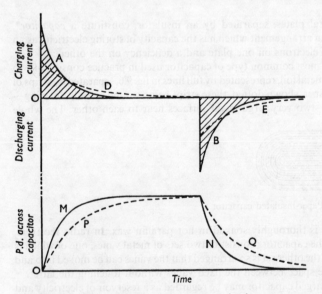

Fig. 9.8 Charging and discharging currents and p.d.s

If the experiment is repeated with a resistance of, say, 2 megohms it is found that the initial current, both on charging and on discharging, is halved, but it takes about twice as long to fall off, as shown by the dotted curves D and E. Curves P and Q represent the corresponding variation of the p.d. across C during charge and discharge respectively.

The shaded area between curve A and the horizontal axis in fig. 9.8 represents the product of the average charging current (in amperes) and the time (in seconds), namely the quantity of electricity (in coulombs) required to charge the capacitor to a p.d. of V volts. Similarly the shaded area enclosed by curve B represents the same quantity of electricity obtainable from the capacitor during discharge.

9.5 Hydraulic analogy of a capacitor

The operation of charging and discharging a capacitor may be more easily understood if we consider the hydraulic analogy given in fig. 9.9, where P represents a piston operated by a rod R, and D is a rubber diaphragm stretched across a cylindrical chamber C. The cylinders are connected by pipes E and are filled with water.

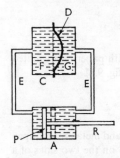

Fig. 9.9 Hydraulic analogy of a capacitor

When no force is being exerted on P, the diaphragm is flat, as shown dotted, and the piston is in position A. If P is pushed towards the left, water is withdrawn from G and forced into F, and the diaphragm is in consequence distended, as shown by the full line. The greater the force applied to P, the greater is the amount of water displaced. But the rate at which this displacement takes place depends upon the resistance offered by pipes E; thus the smaller the cross-sectional area of the pipes, the longer is the time required for the steady state to be reached. The force applied to P is analogous to the e.m.f. of the battery, the quantity of water displaced corresponds to the charge, the rate at which the water passes any point in the pipes corresponds to the current, and the cylinder G with its elastic diaphragm is the analogue of the capacitor.

When the force exerted on P is removed, the distended diaphragm forces water out of F back into G and the piston is pushed back to its original position A. The strain energy stored in the diaphragm due to its distension is converted into heat by the frictional resistance. The effect is similar to the discharge of the capacitor through a resistor.

No water can pass from F to G through the diaphragm so long as the latter remains intact but if the diaphragm is strained excessively it bursts, just as the insulation in a capacitor is punctured when the terminal p.d. becomes excessive.

9.6 Types of capacitors

Capacitors may be divided into the following five main groups according to the nature of the dielectric:

(a) Air capacitors

This type usually consists of one set of fixed plates and another set of movable plates, and is mainly used for radio work where it is required to vary the capacitance.

(b) Paper capacitors

The electrodes consist of metal foils interleaved with paper impregnated with wax or oil and rolled into a compact form (fig. 9.6).

(c) Mica capacitors

This type consists either of alternate layers of mica and metal foil clamped tightly together or of thin films of silver sputtered on the two sides of a mica sheet. Owing to its relatively high cost, this type is mainly used in high-frequency circuits when it is necessary to reduce to a minimum the loss in the dielectric.

(d) Ceramic capacitors

The electrodes consist of metallic coatings (usually silver) on the opposite faces of a thin disc or plate of ceramic material such as the hydrous silicate of magnesia or talc—somewhat similar to that used by tailors for marking cloth. This type of capacitor is mainly used in high-frequency circuits subject to wide variation of temperature.

(e) Electrolytic capacitors

The type most commonly used consists of two aluminium foils, one with an oxide film and one without, the foils being interleaved with a material such as paper saturated with a suitable electrolyte, for example, ammonium borate. The aluminium-oxide film is formed on the one foil by passing it through an electrolytic bath of which the foil forms the positive electrode. The finished unit is assembled in a container—usually of aluminium—and hermetically sealed. The oxide film acts as the dielectric; and as its thickness in a capacitor suitable for a working voltage of 100 V is only about 0.15 micrometre, a very large capacitance is obtainable in a relatively small volume.

The main disadvantages of the electrolytic capacitor are: (a) the insulation resistance is comparatively low and (b) it is only suitable for circuits where the voltage applied to the capacitor never reverses its direction. Electrolytic capacitors are mainly used where very large capacitances are required, e.g. for reducing the ripple in the voltage wave obtained from a rectifier, section 23.4.

9.7 Relationship between the charge and the applied voltage

The method described in section 9.4 for determining the charge on a capacitor is instructive, but is unsuitable for accurate measurement. A much better method is to discharge the capacitor through a ballistic galvanometer,* since the deflection of the latter is proportional to the charge.

Let us charge a capacitor C (fig. 9.10) to various voltages by means of a slider on a resistor R connected across a battery B, S being on a; and for each voltage, note the deflection of G when C is discharged through it by

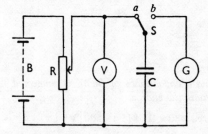

Fig. 9.10 Measurement of charge by ballistic galvanometer

---

* A ballistic galvanometer has a moving coil suspended between the poles of a permanent magnet, but the coil is wound on a non-metallic former so that there is little damping when the coil is connected in series with a high resistance. The first deflection or 'throw' is proportional to the number of coulombs discharged through the galvanometer if the duration of the discharge is short compared with the time of one oscillation of the coil.

moving S over to *b*. Thus, if θ is the first deflection or 'throw' observed when the capacitor, charged to a p.d. of *V* volts, is discharged through G, and if *k* is the ballistic constant of G in coulombs per unit of first deflection, then:

charge on $C = Q = k\theta$ coulombs.

It is found that for a given capacitor,

$$\frac{\text{charge on C [coulombs]}}{\text{p.d. across C [volts]}} = \text{a constant.} \qquad [9.1]$$

9.8 Capacitance

The property of a capacitor to store an electric charge when its plates are at different potentials is referred to as its *capacitance*.

The unit of capacitance is termed the *farad* (symbol F)—a curtailment of 'Faraday'—and may be defined as *the capacitance of a capacitor which requires a p.d. of 1 volt to maintain a charge of 1 coulomb on that capacitor*.

It follows from expression [9.1] and from the definition of the farad that:

$$\frac{\text{charge [coulombs]}}{\text{applied p.dp [volts]}} = \text{capacitance [farads]}$$

or, in symbols $Q/V = C$

$$\therefore \quad Q = CV. \qquad [9.2]$$

In practice, the farad is found to be inconveniently large and the capacitance is usually expressed in *microfarads* (μF) or in *picofarads* (pF), where

1 microfarad $= 10^{-6}$ farad

and

1 picofarad $= 10^{-12}$ farad.

Example 9.1 *A capacitor having a capacitance of 80 μF is connected across a 500-V d.c. supply. Calculate the charge.*

From expression [9.2], charge $= 80 \times 10^{-6}$ [F] $\times 500$ [V]

$$= 0.04 \, \text{C}.$$

9.9 Capacitors in parallel

Suppose two capacitors, having capacitance C_1 and C_2 farads respectively, to be connected in parallel (fig. 9.11) across a p.d. of *V* volts. The

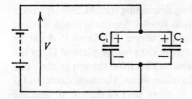

Fig. 9.11 Capacitors in parallel

charge on C_1 is Q_1 coulombs and that on C_2 is Q_2 coulombs, where:

$$Q_1 = C_1 V \qquad \text{and} \qquad Q_2 = C_2 V.$$

If we were to replace C_1 and C_2 by a single capacitor of such capacitance C farads that the same total charge of $(Q_1 + Q_2)$ coulombs would be produced by the same p.d., then $Q_1 + Q_2 = CV$.

Substituting for Q_1 and Q_2, we have:

$$C_1 V + C_2 V = CV$$

$$\therefore \quad C = C_1 + C_2. \tag{9.3}$$

Hence *the resultant capacitance of capacitors in parallel is the arithmetic sum of their respective capacitances.*

9.10 Capacitors in series

Suppose C_1 and C_2 in fig. 9.12 to be two capacitors connected in series with suitable centre-zero ammeters A_1 and A_2, a resistor R and a two-way switch S. When S is put over to *a*, A_1 and A_2 are found to indicate exactly

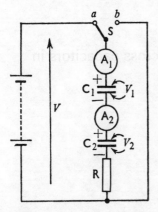

Fig. 9.12 Capacitors in series

the same charging current, each reading decreasing simultaneously from a maximum to zero, as already shown in fig. 9.8. Similarly, when S is put over to b, A_1 and A_2 indicate similar discharges. It follows that during charge the displacement of electrons from the positive plate of C_1 to the negative plate of C_2 is exactly the same as that from the upper plate (fig. 9.12) of C_2 to the lower plate of C_1. In other words, the displacement of Q coulombs of electricity is the same in every part of the circuit and the charge on each capacitor is therefore Q coulombs.

If V_1 and V_2 be the corresponding p.d.s across C_1 and C_2 respectively, then from expression [9.2],

$$Q = C_1 V_1 = C_2 V_2 \qquad [9.4]$$

so that

$$V_1 = Q/C_1, \qquad \text{and} \qquad V_2 = Q/C_2.$$

If we were to replace C_1 and C_2 by a single capacitor of capacitance C farads such that it would have the same charge Q coulombs with the same p.d. of V volts, then:

$$Q = CV, \qquad \text{or} \qquad V = Q/C.$$

But it is evident from fig. 9.12 that $V = V_1 + V_2$. Substituting for V, V_1 and V_2, we have:

$$\frac{Q}{C} = \frac{Q}{C_1} + \frac{Q}{C_2}$$

$$\therefore \quad \frac{1}{C} = \frac{1}{C_1} + \frac{1}{C_2}. \qquad [9.5]$$

Hence *the reciprocal of the resultant capacitance of capacitors connected in series is the sum of the reciprocals of their respective capacitances.*

9.11 Distribution of voltage across capacitors in series

From expression [9.4],

$$\frac{V_2}{V_1} = \frac{C_1}{C_2}. \qquad [9.6]$$

But

$$V_1 + V_2 = V$$

$$\therefore \quad V_2 = V - V_1.$$

Substituting for V_2 in [9.6], we have:

$$\frac{V - V_1}{V_1} = \frac{C_1}{C_2}$$

$$\therefore \quad \frac{V}{V_1} = \frac{C_1}{C_2} + 1 = \frac{C_1 + C_2}{C_2}$$

Hence

$$V_1 = V \times \frac{C_2}{C_1 + C_2} \qquad [9.7]$$

and

$$V_2 = V \times \frac{C_1}{C_1 + C_2}. \qquad [9.8]$$

Example 9.2 *Three capacitors have capacitances of 2 μF, 4 μF and 8 μF respectively. Find the total capacitance when they are connected* (a) *in parallel,* (b) *in series.*

(a) From expression [9.3]

total capacitance $= 2 + 4 + 8 = 14\ \mu F$.

(b) If C be the resultant capacitance in microfarads when the capacitors are in series, then from (9.5):

$$\frac{1}{C} = \frac{1}{2} + \frac{1}{4} + \frac{1}{8} = 0.5 + 0.25 + 0.125 = 0.875$$

$$\therefore \quad C = 1.143\ \mu F.$$

Example 9.3 *If two capacitors having capacitances of 6 μF and 10 μF respectively are connected in series across a 200-V supply, find* (a) *the p.d. across each capacitor,* (b) *the charge on each capacitor.*

(a) Let V_1 and V_2 be the p.d.s across the 6-μF and 10-μF capacitors respectively; then, from expression [9.7],

$$V_1 = 200 \times \frac{10}{6 + 10} = 125\ \text{volts}$$

and

$$V_2 = 200 - 125 = 75\ \text{volts}.$$

(b) Charge on each capacitor = charge on C_1

$$6 \times 10^{-6}\ [\text{F}] \times 125\ [\text{V}]$$

$$= 0.000\,75\ \text{coulomb}.$$

9.12 Relationship between the capacitance and the dimensions of a capacitor

It follows from expression [9.3] that if two similar capacitors are connected in parallel, the capacitance is double that of one capacitor. But the effect of connecting two similar capacitors in parallel is merely to double the area of each plate. In the same way, the effect of connecting, say, five similar capacitors in parallel is to increase the area of the plates five times and to increase the capacitance fivefold. In general, we may therefore say that the capacitance of a capacitor is proportional to the area of the plates.

On the other hand, if two similar capacitors are connected in series, it follows from expression [9.5] that the capacitance is halved. We have, however, doubled the thickness of the insulation between the plates that are connected to the supply. If, say, five similar capacitors are connected in series, the thickness of the insulation between the outermost plates is increased fivefold, but the capacitance is reduced to a fifth of that of one capacitor. Hence we may say in general that the capacitance of a capacitor is inversely proportional to the distance between the plates; and the above relationships may be summarized thus:

$$\text{capacitance} \propto \frac{\text{area of plates}}{\text{distance between plates}}.$$

9.13 Electric field strength and electric flux density

Let us consider a capacitor consisting of metal plates, M and N, in a glass enclosure G, shown chain-dotted in fig. 9.13, from which all the air has been removed. Let a be the area in square metres of one side of each plate and let d be the distance in metres between the plates. Let Q be the charge in coulombs due to a p.d. of V volts between the plates.

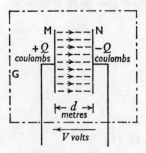

Fig. 9.13 A parallel-plate capacitor

Suppose the dimensions of the plates to be so large compared with the distance betwen them that we may assume negligible fringing of the electric flux, i.e. all the lines of electric flux may be assumed to pass straight across from M to N, as shown by the dotted lines in fig. 9.13.

The electric field strength (the term *electric force* is now obsolete) in the region between the two plates M and N is the *potential drop per unit length or potential gradient,** namely V/d volts/metre; and the *direction* of the electric field strength at any point is the direction of the mechanical force on a positive charge situated at that point, namely from the positively-charged plate M towards the negatively-charged plate N in fig. 9.13. The symbol for electric field strength is E†, i.e.

$$\mathbf{E} = V/d \text{ volts/metre.} \tag{9.9}$$

In SI, *one* unit of electric flux is assumed to emanate from a positive charge of 1 coulomb and to enter a negative charge of 1 coulomb. Hence, if the charge on plates M and N is Q coulombs, electric flux between M and $N = \Psi = Q$ coulombs and

$$\text{electric flux density} = \mathbf{D} = Q/a \text{ coulombs/metre}^2 \tag{9.10}$$

where a = area or dielectric, in square metres, at right angles to direction of electric flux.

From expressions [9.9] and [9.10],

$$\frac{\text{electric flux density}}{\text{electric field strength}} = \frac{\mathbf{D}}{\mathbf{E}} = \frac{Q}{a} \div \frac{V}{d} = \frac{Q}{V} \times \frac{d}{a} = \frac{Cd}{a}.$$

In electromagnetism, the ratio of the magnetic flux density in a vacuum to the magnetic field strength is termed the 'permeability of free space' or 'magnetic constant' and is represented by μ_0. Similarly, in electrostatics, the ratio of the electric flux density in a vacuum to the electric field strength is termed the *permittivity of free space* or *electric constant* and is represented by ε_0. Hence,

$$\varepsilon_0 = Cd/a$$

or $\quad C = \varepsilon_0 a/d.$ [9.11]

The effect of filling the space between M and N (fig. 9.13) with air at atmospheric pressure is to increase the capacitance by 0.06 per cent compared with the value when the space is completely evacuated; hence for all practical purposes, expression [9.11] can be applied to capacitors having air dielectric.

The value of ε_0 can be determined experimentally by charging a capacitor, of known dimensions and with air dielectric, to a p.d. of V volts

---

* The potential gradient necessary to break down a dielectric—usually by producing a tiny hole through it—is termed the *dielectric strength* of that material and is generally expressed in kilovolts per millimetre of thickness of the specimen.
† Bold type is used to enable the symbols **E** and **D** for electric field strength and electric flux density respectively to be easily distinguished from the letters E and D used to represent other quantities.

and then discharging it through a ballistic galvanometer having a ballistic constant k coulombs/unit deflection. If the deflection is θ divisions,

$$Q = CV = k\theta$$

$$\therefore \quad \varepsilon_0 = C \cdot \frac{d}{a} = \frac{k\theta}{V} \cdot \frac{d}{a}.$$

From carefully conducted tests it has been found that the value of ε_0 is 8.85×10^{-12} F/m (see footnote on p. 171).

Hence the capacitance of a parallel-plate capacitor with vacuum or air dielectric is given by:

$$C = \frac{8.85 \times 10^{-12} \, [\text{F/m}] \times a \, [\text{metres}^2]}{d \, [\text{metres}]} \text{ farads.} \qquad [9.12]$$

It may be mentioned at this point that there is a definite relationship between μ_0, ε_0 and the velocity of light and other electromagnetic waves; thus,

$$\frac{1}{\mu_0 \varepsilon_0} = \frac{1}{4\pi \times 10^{-7} \times 8.85 \times 10^{-12}} = 8.99 \times 10^{16} = (2.998 \times 10^8)^2$$

But the velocity of light $= 2.998 \times 10^8$ metres/second

$$\therefore \quad \text{velocity of light in metres/second} = \frac{1}{\sqrt{(\mu_0 \varepsilon_0)}} \qquad [9.13]$$

This relationship was discovered by Prof. Clerk Maxwell in 1865 and enabled him to predict the existence of radio waves about twenty years before their effect was demonstrated experimentally by Prof. H. Hertz (see page 190).

9.14 Relative permittivity

If the experiment described in section 9.13 is performed with a sheet of glass filling the space between plates M and N, it is found that the value of the capacitance is greatly increased; and the ratio of the capacitance of a capacitor having a certain material as dielectric to the capacitance of that capacitor with vacuum (or air) dielectric is termed the *relative permittivity* of that material and is represented by the symbol ε_r. (The term 'dielectric constant' is now obsolete.) Values of the relative permittivity of some of the most important insulating materials are given in Table 9.1.

From expression [9.11], it follows that if the space between the metal plates of the capacitor in fig. 9.13 is filled with a dielectric having a relative permittivity ε_r,

$$\text{capacitance} = C = \frac{\varepsilon_0 \varepsilon_r a}{d} \text{ farads} \qquad [9.14]$$

$$= \frac{8.85 \times 10^{-12} \varepsilon_r a}{d} \text{ farads}$$

Table 9.1

Material	Relative permittivity
Air	1.0006
Paper (dry)	2–2.5
Bakelite	4.5–5.5
Glass	5–10
Rubber	2–3.5
Mica	3–7
Porcelain	6–7

and charge due to a p.d. of V volts $= Q = CV$

$$= \frac{\varepsilon_0 \varepsilon_r aV}{d} \text{ coulombs,}$$

$\therefore \quad \dfrac{\text{electric flux density}}{\text{electric field strength}} = \dfrac{\mathbf{D}}{\mathbf{E}} = \dfrac{Q}{a} \div \dfrac{V}{d} = \dfrac{Qd}{Va} = \varepsilon_0 \varepsilon_r$

$$= \varepsilon = \text{absolute permittivity.*} \qquad [9.15]$$

This expression is similar in form to expression [7.4] deduced for the magnetic circuit, namely:

$$\frac{\text{magnetic flux density}}{\text{magnetic field strength}} = \frac{B}{H} = \mu_0 \mu_r.$$

9.15 Capacitance of a multi-plate capacitor

Suppose a capacitor to be made up of n parallel plates, alternate plates being connected together as in fig. 9.14.

Fig. 9.14 Multi-plate capacitor

* From expression [9.14]

$$\text{absolute permittivity} = \varepsilon_0 \varepsilon_r = \frac{C\,[\text{farads}] \times d\,[\text{metres}]}{a\,[\text{metres}^2]}$$

$$= Cd/a \text{ farads/metre}$$

i.e., the units of absolute permittivity are *farads/metre* or (F/m); e.g. $\varepsilon_0 = 8.85 \times 10^{-12}$ F/m.

Let

a = area of *one* side of each plate in square metres

d = thickness of dielectric in metres

and

ε_r = relative permittivity of the dielectric.

Figure 9.14 shows a capacitor with seven plates, four being connected to A and three to B. It will be seen that each side of the three plates connected to B is in contact with the dielectric, whereas only one side of each of the outer plates is in contact with it. Consequently, the useful surface area of each set of plates is $6a$ square metres. For n plates, the useful area of each set is $(n-1)a$ square metres;

$$\therefore \quad \text{capacitance} = \frac{\varepsilon_0 \varepsilon_r (n-1)a}{d} \text{ farads}$$

$$= \frac{8.85 \times 10^{-12} \varepsilon_r (n-1)a}{d} \text{ farads.} \qquad [9.16]$$

Example 9.4   *A capacitor is made with seven metal plates connected as in fig. 9.14 and separated by sheets of mica having a thickness of 0.3 mm and a relative permittivity of 6. The area of one side of each plate is 50 000 mm². Calculate the capacitance in microfarads.*

Using expression [9.16], we have $n=7$, $a=0.05\,\text{m}^2$, $d=0.003$ m and $\varepsilon_r=6$.

$$\therefore \quad C = \frac{(8.85 \times 10^{-12})\,[\text{F/m}] \times 6 \times (6 \times 0.05)\,[\text{m}^2]}{0.0003\,[\text{m}]}$$

$$= 0.0531 \times 10^{-6}\,\text{F} = 0.0531\,\mu\text{F}.$$

Example 9.5   *A p.d. of 400 V is maintained across the terminals of the capacitor of example 9.4. Calculate (a) the charge, (b) the electric field strength or potential gradient and (c) the electric flux density in the dielectric.*

(a)   Charge $= Q = CV = 0.0531\,[\mu\text{F}] \times 400\,[\text{V}]$

$$= 21.24 \text{ microcoulombs.}$$

(b)   Electric field strength or potential gradient

$$= V/d = 400\,[\text{V}]/0.0003\,[\text{m}] = 1\,333\,000\,\text{V/m}$$

$$= 1333\,\text{kV/m}.$$

(c)   Electric flux density $= Q/a = 21.24\,[\mu\text{C}]/(0.05 \times 6)\,[\text{m}^2]$

$$= 70.8 \text{ microcoulombs/metre}^2.$$

9.16 Comparison of electrostatic and electromagnetic terms

Table 9.2 compares the terms and symbols used in electrostatics with the corresponding terms and symbols used in electromagnetism.

Table 9.2

Electrostatics term	Symbol	Electromagnetism term	Symbol
Electric flux	Ψ	Magnetic flux	Φ
Electric flux density	$\mathbf{D}$	Magnetic flux density	B
Electric field strength	$\mathbf{E}$	Magnetic field strength	H
Electromotive force	E	Magnetomotive force	F
Electric potential difference	V	Magnetic potential difference	—
Permittivity of free space	ε_0	Permeability of free space	μ_0
Relative permittivity	ε_r	Relative permeability	μ_r
Absolute permittivity $= \dfrac{\text{electric flux density}}{\text{electric field strength}}$		Absolute permeability $= \dfrac{\text{magnetic flux density}}{\text{magnetic field strength}}$	
i.e. $\varepsilon_0\varepsilon_r = \varepsilon = \mathbf{D}/\mathbf{E}$		i.e. $\mu_0\mu_r = \mu = B/H$	

9.17 Charging and discharging currents of a capacitor

Consider a capacitor being charged through a resistance R, as in the circuit described in fig. 9.7. Initially the current is high and then decreases with time to become eventually zero. When it is zero the capacitor is fully charged.

Initially there is no charge on the capacitor and thus, since $V = Q/C$ (equation [9.2]), there is no potential difference across the capacitor. Therefore the source potential difference must be entirely across the resistance. Thus the initial current is V/R, where V is the applied potential difference.

When the capacitor begins to acquire charge then the potential difference across it increases. This results in a decrease in the potential difference across the resistor and a decrease in current. When the capacitor is fully charged there is no current in the circuit and thus no potential difference across the resistor (potential difference $= IR$). Hence the potential difference across the capacitor must be the same as the applied potential difference V; all the potential difference is across the capacitor.

Figure 9.15 shows how the current and potential differences change with time during the charging of the capacitor. At any instant of time if you added the p.d. across the resistor to the p.d. across the capacitor the answer obtained would be V, the applied potential difference.

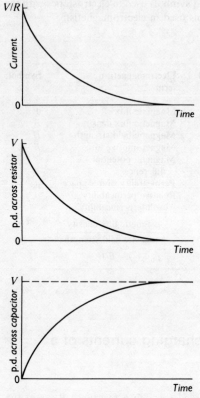

Fig. 9.15 Charging a capacitor

If we now consider the capacitor being discharged through a resistor R, then the current is initially high and decreases with time to eventually become zero. The initially fully charged capacitor has a potential difference V across it. This is applied across the resistor and so causes a current V/R to flow initially. The current flows in the opposite direction to that which occurred when the capacitor was being charged. As the capacitor discharges so the potential difference across it decreases, the p.d. being proportional to the charge. Hence the p.d. across the resistance decreases and so does the current. Figure 9.16 shows how the current and the potential differences change with time during the discharging of the capacitor. If at any instant of time we added the p.d. across the resistor to the p.d. across the capacitor the answer obtained would be zero, there being no source of e.m.f. in the circuit.

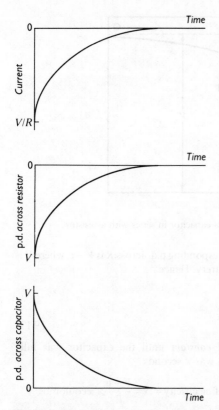

Fig. 9.16 Discharging a capacitor

9.18 Graphical derivation of curve of voltage across a capacitor connected in series with a resistor across a d.c. supply

In section 9.4 we derived the curves of the voltage across a capacitor during charging and discharging from the readings on a cathode ray connected across the capacitor. We will now consider how these curves can be derived graphically from the values of the capacitance, the resistance and the applied voltage. At the instant when S is closed on *a* (fig. 9.17), there is no p.d. across C; consequently the whole of the voltage is applied across R and the initial value of the charging current $= I = V/R$.

The growth of the p.d. across C is represented by the curve in fig. 9.17. Suppose *v* to be the p.d. across C and *i* to be the charging current *t* seconds

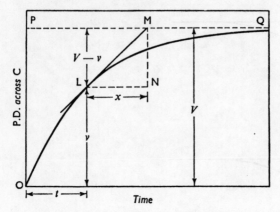

Fig. 9.17   Growth of p.d. across a capacitor in series with a resistor

after S is put over to *a*. The corresponding p.d. across R is $V - v$, where V is the terminal voltage of the battery. Hence

$$iR = V - v$$

and

$$i = (V - v)/R. \qquad\qquad [9.17]$$

If this current remained *constant* until the capacitor was fully charged, and if the time taken was x seconds:

$$\text{corresponding quantity of electricity} = ix = \frac{V - v}{R} \times x \text{ coulombs.}$$

With a constant charging current, the p.d. across C would have increased uniformly up to V volts, as represented by the tangent LM drawn to the curve at L.

But the charge added to the capacitor also

$$= \text{increase of p.d.} \times C = (V - v) \times C.$$

Hence

$$\frac{V - v}{R} \times x = C(V - v)$$

and

$$x = CR = \text{the time constant, } T, \text{ of the circuit.} \qquad [9.18]$$

The construction of the curve representing the growth of the p.d. across a capacitor is therefore similar to that described in section 8.6 for the growth of current in an inductive circuit. Thus, OA in fig. 9.18 represents the battery voltage V, and AB the time constant T. Join OB, and from a point D fairly near the origin draw $DE = T$ seconds and draw EF perpendicularly. Join DF, etc. Draw a curve such that OB, DF, etc., are tangents to it.

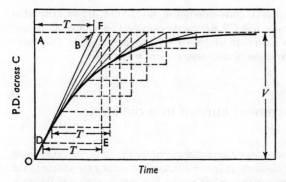

Fig. 9.18   Growth of p.d. across a capacitor in series with a resistor

From expression (9.17) it is evident that the instantaneous value of the charging current is proportional to $(V-v)$, namely the vertical distance between the curve and the horizontal line PQ in fig. 9.17. Hence the shape of the curve representing the charging current is the inverse of that of the p.d. across the capacitor and is the same for both charging and discharging currents (assuming the resistance to be the same), and its construction is illustrated by the following example.

Example 9.6   *A 20-μF capacitor is charged to a p.d. of 400 V and then discharged through a circuit having a resistance of 100 000 Ω. Derive a curve representing the discharge current.*

From expression [9.18],

$$\text{time constant} = 100\,000\,[\Omega] \times \frac{20}{1\,000\,000}\,[\text{F}] = 2\,\text{s}.$$

Initial value of discharge current

$$= \frac{V}{R} = \frac{400\,[\text{V}]}{100\,000\,[\Omega]} = 0.004\,\text{A} = 4\,\text{mA}.$$

Hence draw OA in fig. 9.19 to represent 4 mA and OB to represent 2 s.

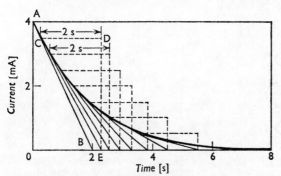

Fig. 9.19   Discharge current, example 9.16

Join AB. From a point C corresponding to, say, 3.5 mA, draw CD equal to 2 s and DE vertically. Join CE. Repeat the construction at intervals of, say, 0.5 mA and draw a curve to which AB, CE, etc., are tangents. This curve represents the variation of discharge current with time.

9.19 Displacement current in a dielectric

In the preceding sections we have considered the charging current as being the movement of electrons in the conductors connecting the source to the plates of the capacitor; e.g. when switch S is put over to *a* in fig. 9.12, electrons flow from the positive plate of C_1 via the battery to the negative plate of C_2, and the same number of electrons flow from the upper plate of C_2 to the lower plate of C_1. This current is referred to as *conduction* current.

Let us consider the capacitor of fig. 9.13, in which the metal plates M and N are in an evacuated glass enclosure G. There are no electrons in the space between the plates and therefore there cannot be any movement of electrons in this space when the capacitor is being charged. We know, however, that an electric field is being set up and that energy is being stored in the space between the plates; in other words, the space between the plates of a charged capacitor is in a state of electrostatic strain.

We do not know the exact nature of this strain (any more than we know the nature of the strain in a magnetic field), but Prof. Clerk Maxwell, in 1865, introduced the concept that any *change* in the electric flux in any region is equivalent to an electric current in that region, and he called this electric current a *displacement* current, to distinguish it from the *conduction* current referred to above.

This displacement current produces a magnetic field exactly as if it had been a conduction current. For instance, when a capacitor having circular parallel metal plates M and N (fig. 9.20) is being charged by a current *i* flowing in the direction shown, a magnetic field is created in the space between the plates, as indicated by the concentric dotted lines. The

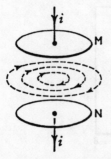

Fig. 9.20   Magnetic field due to displacement current

plane of these concentric circles is parallel to the plates.

This magnetic field disappears as soon as the displacement current ceases, i.e. as soon as the charge on the capacitor ceases to increase. When the capacitor is discharged, the magnetic field re-appears in the reverse direction and again disappears when the discharge ceases. In other words, *the magnetic field is set* up only when the electric field is undergoing a change of intensity*. Hence, when a capacitor is being charged or discharged, we can say that the current is *continuous* around the *whole* circuit, being in the form of *conduction current* in the wires and *displacement current* in the dielectric of the capacitor. This means that we can apply Kirchhoff's First Law to plate M of the capacitor of fig. 9.20 by saying that the conduction current entering the plate is equal to the displacement current leaving that plate.

9.20 Energy involved in moving an electric charge between two points differing in potential

If the p.d. between two points, A and B, of a closed circuit is V volts, A being at a higher potential than B, and if the current flowing from A towards B is I amperes, then:

power absorbed by length AB of circuit $= IV$ watts.

If the current remains constant for t seconds,

quantity of electricity $= Q = It$ coulombs

and electrical energy *absorbed* by length AB of circuit

$= IVt$ joules

$= QV$ joules.

In the case of a source of electricity, such as a cell or a generator, the conventional direction of current through the source is from the negative terminal to the positive terminal. Hence, for a quantity of electricity, Q coulombs, generated at an e.m.f. E volts, we have

electrical energy *generated* $= QE$ joules.

In the cell, this energy represents an expenditure of chemical energy, whereas in the generator, it involves an expenditure of mechanical energy.

An electron carries a negative charge of 1.6×10^{-19} coulomb, consequently the work done in moving an electron from a point A to a point B, when the potential of A is 1 volt *above* that of B, is

---

* Conversely, an electric field appears whenever a magnetic field changes in intensity, and it is this reciprocity between electric and magnetic fields that forms the basis of electromagnetic theory dealing with the radiation of electromagnetic waves.

1.6×10^{-19} joule. This amount of energy is referred to as an *electron-volt*, a term that has acquired considerable importance in the field of electronics, i.e.

$$1 \text{ electron-volt} = 1.6 \times 10^{-19} \text{ joule.} \qquad [9.19]$$

If the potential of the anode of a vacuum diode (section 23.4) is 100 V above that of the cathode, the electrical energy *absorbed* when an electron travels from the cathode to the anode is 100 electron-volts or 1.6×10^{-17} joule. Hence, for an anode current of 5 mA and an anode–cathode p.d. of 100 V,

number of electrons passing per second from cathode to anode

$$= 6.24* \times 10^{18} \times 5 \times 10^{-3}$$

$$= 31.2 \times 10^{15}$$

$\therefore$ energy absorbed per second $= 31.2 \times 10^{15} \times 100$ electron-volts

$$= 3.12 \times 10^{18} \text{ electron-volts}$$

$$= 3.12 \times 10^{18} \times 1.6 \times 10^{-19} \text{ joule}$$

$$= 0.5 \text{ joule.}$$

Alternatively,

power absorbed $= 0.005 \times 100 = 0.5$ watt or joule/second

$$= 0.5/(1.6 \times 10^{-19}) \text{ electron-volts/second}$$

$$= 3.12 \times 10^{18} \text{ electron-volts/second.}$$

The electrical energy thus absorbed is converted into kinetic energy possessed by the accelerated electrons when they reach the anode. The bombardment of the anode by the electrons then converts this kinetic energy into heat.

9.21 Energy stored in a capacitor

If a capacitor having capacitance C farads is charged at a constant rate of I amperes for t seconds, as in fig. 9.21, the charge is It coulombs. If the final p.d. across the capacitor is V volts, the charge is also CV coulombs.

$\therefore \quad CV = It, \quad$ and $\quad V = It/C.$

The p.d. across C is therefore proportional to the duration of the charge and is represented by the thick line in fig. 9.21. It follows that:

average p.d. across C during charging $= \frac{1}{2}V$ volts

---

* It was explained in section 9.3 that 1 ampere corresponds to 6.24×10^{18} electrons/second.

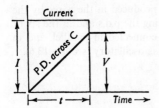

Fig. 9.21 Charging of a capacitor

and

average power to C during charging $= I \times \frac{1}{2}V$ watts

∴ energy supplied to C during charging

= average power × time

$= \frac{1}{2}IVt$ joules

$= \frac{1}{2}V \times CV$ joules

i.e.

electrostatic energy stored in $C = \frac{1}{2}CV^2$ joules.   [9.20]

Example 9.7 *A 50-μF capacitor is charged from a 200-V supply. After being disconnected it is immediately connected in parallel with a 30-μF capacitor. Find: (a) the p.d. across the combination; (b) the electrostatic energies before and after the capacitors are connected in parallel.*

(a) From [9.2], charge $= 50 \times 10^{-6}$[F] $\times 200$[V]

$= 0.01$ C.

When the capacitors are connected in parallel, the total capacitance is 80 μF, and the charge of 0.01 coulomb is divided between the two capacitors: hence

0.01 [C] $= 80 \times 10^{-6}$ [F] × p.d.

∴ p.d. across capacitors $= 125$ V.

(b) From [9.20] it follows that when the 50-μF capacitor is charged to a p.d. of 200 V:

electrostatic energy $= \frac{1}{2} \times (50 \times 10^{-6})$ [F] $\times (200)^2$ [V²]

$= 1$ joule.

With the capacitors in parallel:

total electrostatic energy $= \frac{1}{2} \times (80 \times 10^{-6})$ [F] $\times (125)^2$ [V²]

$= 0.625$ joule.

It is of interest to note that there is a reduction in the energy stored in the capacitors. This loss appears as heat produced in the wires by the circulating current responsible for equalizing the p.d.s and in the spark that may occur when the capacitors are connected in parallel, and in electromagnetic radiation if the discharge is oscillatory (section 13.6).

9.22 Summary of important formulae

Q [coulombs] $= C$ [farads] $\times V$ [volts]

1 microfarad $= 10^{-6}$ farad and 1 picofarad $= 10^{-12}$ farad.

For capacitors in parallel,

$$C = C_1 + C_2 + \qquad\qquad\qquad [9.3]$$

For capacitors in series,

$$\frac{1}{C} = \frac{1}{C_1} + \frac{1}{C_2} + \qquad\qquad\qquad [9.5]$$

For C_1 and C_2 in series,

$$V_1 = V \cdot \frac{C_2}{C_1 + C_2} \qquad\qquad\qquad [9.7]$$

and

$$V_2 = V \cdot \frac{C_1}{C_1 + C_2} \qquad\qquad\qquad [9.8]$$

Electric field strength $= \mathbf{E} = V/d$ volts/metre $\qquad [9.9]$

Electric flux density $= \mathbf{D}$

$$= Q/a \text{ coulombs/square metre.} \qquad [9.10]$$

$$\frac{\text{Electric flux density}}{\text{Electric field strength}} = \frac{\mathbf{D}}{\mathbf{E}} = \varepsilon_0 \varepsilon_r \qquad\qquad [9.15]$$

$$= \text{absolute permittivity}$$

where

$\varepsilon_0 =$ permittivity of free space or electric constant

$$= 8.85 \times 10^{-12} \text{ F/m}$$

$$\frac{1}{\sqrt{(\mu_0 \varepsilon_0)}} = 2.998 \times 10^8 \text{ m/s}$$

$$= \text{velocity of electromagnetic waves.} \qquad [9.13]$$

Relative permittivity of a material

$$= \frac{\text{capacitance of capacitor with that material as dielectric}}{\text{capacitance of same capacitor with vacuum dielectric}}$$

Capacitance of parallel-plate capacitor with n plates

$$= \frac{\varepsilon_0 \varepsilon_r (n-1)a}{d} \text{ farads.} \qquad [9.16]$$

Charging current of capacitor in amperes

$= $ rate of change of charge in coulombs/second

$= C \text{ [farads]} \times$ rate of change of p.d. in volts/second.

For R and C in series,

time constant in seconds $= T = R \text{ [ohms]} \times C \text{ [farads]}$ $\qquad [9.18]$

1 electron-volt $= 1.6 \times 10^{-19}$ joule $\qquad [9.19]$

Energy stored in capacitor $= \frac{1}{2}CV^2$ joules. $\qquad [9.20]$

9.23 Examples

1. A capacitor is formed of two metallized paper sheets, each side of which has an area of 0.2 m^2. The two sheets are separated by paper 0.05 mm thick having a relative permittivity of 2.5. Calculate the capacitance.

2. A 1-microfarad capacitor is constructed of two strips of metal foil separated by paper dielectric 0.02 mm thick, wound spirally. The paper has a relative permittivity of 3.0. The width of each metal strip is 80 mm. Find the length, in metres, of metal foil required.

3. A capacitor consists of two metal plates, each having an area of 0.06 m^2, spaced 2 mm apart. The whole of the space between the plates is occupied by a dielectric having a relative permittivity of 5. A p.d. of 400 V is maintained between the two plates. Calculate (a) the capacitance in picofarads, (b) the charge in microcoulombs and (c) the electric flux density and the electric field strength in the dielectric.

4. A capacitor consists of two parallel metal plates, each $200 \text{ mm} \times 150 \text{ mm}$, separated by air dielectric. What is the spacing between the plates if the capacitance is 200 pF?

5. A capacitor consists of two metal plates, each 100 mm square, placed parallel and 3 mm apart. The space between the plates is occupied by a plate of insulating material 3 mm thick. The capacitor is charged to 300 V.

    (a) If the metal plates were isolated from the 300-V supply and the insulating plate is removed, what would be the voltage between the metal plates?

(b) If the spacing between the metal plates were then increased to 6 mm, what would be the voltage between them?

Assume throughout that the insulation is perfect.

6. A slab of insulating material, 4 mm thick, is inserted between the plates of a parallel-plate capacitor. To restore the capacitance to its original value, it is necessary to increase the spacing between the plates by 2 mm. Calculate the relative permittivity of the slab.

7. A capacitor is charged to a p.d. of 20 V and then discharged through a ballistic galvanometer having a ballistic constant of 0.1 μC per scale division. If the first deflection or 'throw' of the galvanometer is 83 divisions, what is the value of the capacitance in microfarads?

8. A capacitor consists of two metal plates, each 200 mm × 200 mm, spaced 1 mm apart, the dielectric being air. The capacitor is charged to a p.d. of 100 V and then discharged through a ballistic galvanometer having a ballistic constant of 0.0011 μC per scale division. The amplitude of the first deflection is 32 divisions. Calculate the value of ε_0.

Also calculate the electric field strength and the electric flux density in the air dielectric when the terminal p.d. is 100 V.

9. When the capacitor of Question 8 is immersed in oil, charged to a p.d. of 30 V and then discharged throough the same galvanometer, the first deflection is 27 divisions. Calculate the relative permittivity of the oil.

Also calculate the electric field strength and the electric flux density in the oil when the terminal p.d. is 30 V. What is the value of the energy stored in the capacitor?

10. A capacitor consists of two flat plates, each 0.2 m × 0.2 m, separated by an air space 2 mm wide. The capacitor is charged off a 400-V d.c. supply, and a sheet of glass, 0.3 m × 0.3 m and 2 mm thick, is placed between the plates immediately they are disconnected from the supply. Determine (a) the capacitance with air dielectric, (b) the capacitance with glass dielectric, assuming the relative permittivity of glass to be 6, (c) the p.d. across the capacitor after the glass is inserted, (d) the charge on the capacitor and (e) the energy stored in the capacitor before and after the glass plate is inserted.

11. A parallel-plate capacitor has a capacitance of 300 pF. It has nine plates, each 40 mm × 30 mm, separated by mica having a relative permittivity of 5. Find the thickness of the mica.

12. Calculate the capacitance, in picofarads, of a capacitor having seven parallel plates separated by a dielectric, 0.15 mm thick. The area of one side of each plate is 2000 mm^2 and the relative permittivity of the dielectric is 3.

13. What are the different values of capacitance which can be obtained with two capacitors of 0.1 μF and 0.2 μF capacitance?

14. Two capacitors, having capacitances of 10 μF respectively, are connected in series across a 200-V d.c. supply. Calculate (a) the charge on each capacitor, (b) the p.d. across each capacitor and (c) the capacitance of a single capacitor that would be equivalent to these

two capacitors in series.

15. Two capacitors, having capacitances of 10 μF and 15 μF respectively are connected in parallel and a capcitor of 8 μF is connected in series. Find the capacitance of a single capacitor that is equivalent to the above combination.

16. Three capacitors of 2, 3 and 6 μF are connected in series across a 500-V d.c. supply. Calculate (a) the charge on each capacitor, (b) the p.d. across each capacitor and (c) the energy stored in the 6-μF capacitor.

17. Three capacitors of identical dimensions have dielectrics of relative permittivity 1, 3 and 5 respectively. They are connected in series across a 400-V d.c. supply. Calculate the p.d. across each capacitor.

18. The energy stored in a certain capacitor when connected to a 400-V d.c. supply is 0.3 joule. Calculate (a) the capacitance and (b) the charge on the capacitor.

19. A variable capacitor having a capacitance of 800 pF is charge to a p.d. of 100 V. The plates of the capacitor are then separated until the capacitance is reduced to 200 pF. What is the change of p.d. across the capacitor? Also, what is the energy stored in the capacitor when its capacitance is (a) 800 pF and (b) 200 pF?

20. A 20-μF capacitor is charged off a 60-V d.c. supply. After being disconnected from the supply, it is immediately connected across an uncharged 5-μF capacitor. Calculate (a) the p.d. across the parallel capacitors, (b) the charge on each capacitor and (c) the total electrostatic energies before and after the capacitors are connected in parallel.

21. A 10-μF capacitor charged to 500 V is connected across an uncharged 4-μF capacitor. Calculate (a) the voltage across the two capacitors and (b) the stored energy before and after they are connected together.

22. The voltage applied across a 10-μF capacitor is varied thus: the p.d. is increased uniformly from 0 to 600 V in 2 s. It is then maintained constant at 600 V for 1 s and subsequently decreased uniformly to zero in 5 s. Plot a graph showing the variation of current during these 8 s and calculate (a) the charge and (b) the energy stored in the capacitor when the terminal voltage is 600 V.

23. A 10-μF capacitor is connected in series with a 50-kΩ resistor across a 50-V d.c. supply. Derive curves showing how the charging current and the p.d. across the capacitor vary with time.

24. A 20-μF capacitor has an insulation resistance of 40 MΩ. The capacitor is charged to a p.d. of 500 V and then disconnected from the supply. Plot graphs showing the variation of current and of p.d. with time after disconnection. From these graphs determine the values of the current and the p.d. after an interval of 10 minutes.

25. A 0.5-μF capacitor, charged to a p.d. of 100 V, is discharged through a 2-MΩ resistor. Calculate (a) the initial value of the discharge current; (b) the initial rate of decay of the voltage across the capacitor and (c) the energy dissipated in the resistor.

26. A 0.1-μF capacitor is charged to a p.d. of 5 V. Calculate (a) the charge in micro-coulombs and (b) the number of electrons displaced.

27. When the current in a wire is 600 A, what is the number of electrons per second passing a given point of the wire?

28. If the number of electrons passing per second through an ammeter is 7×10^{16}, what is the ammeter reading?

29. If there are 3×10^{15} electrons passing per second between two metal surfaces and if the p.d. between the surfaces is 200 V, calculate the energy absorbed in 20 min (a) in joules and (b) in electron-volts.

30. In a thermionic valve, the current between the filament and the anode is 3 mA and the p.d. is 120 V. Calculate (a) the number of electrons per second passing from the filament to the anode and (b) the energy absorbed in 5 min in (i) electron-volts and (ii) joules.

CHAPTER 10

Alternating Voltage and Current

10.1 A sine wave

Since alternating voltages and currents are often represented by sine waves, it is very desirable that students should plot sine waves to scale so as to become familiar with their exact shape. This may be done either by reference to the sine values or graphically, as in fig. 10.1. Thus,

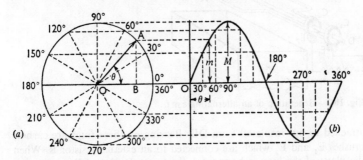

Fig. 10.1 Construction of a sine wave

with any convenient radius OA, draw a circle and insert radii every, say, 30°. On the right, mark off a horizontal scale in degrees and draw the dotted vertical lines to meet the horizontal projections from the corresponding points on the circle. A curve drawn through the various intersections is a sine wave; thus for any angle θ in fig. 10.1(a):

$$\sin \theta = \frac{\text{perpendicular}}{\text{hypotenuse}} = \frac{AB}{OA}$$

$$\therefore \quad AB = OA \times \sin \theta.$$

---

* A good sine wave can be drawn with the aid of ordinates spaced 30° apart on the horizontal axis; thus, $\sin 0° = 0$, $\sin 30° = 0.5$, $\sin 60° = 0.866$, $\sin 90° = 1.0$, $\sin 120° = 0.866$, $\sin 150° = 0.5$, etc.

In fig. 10.1(b), M is the maximum value of the alternating quantity and is also referred to as the *peak* or the *crest* value. If m is the instantaneous value of the alternating quantity, $m = M \sin \theta$.

10.2 Generation of an alternating e.m.f.

Figure 10.2 shows a loop AB carried by a spindle DD rotated at a constant speed in an anticlockwise direction in a uniform magnetic field due to poles NS. The ends of the loop are brought out to two slip-rings C_1 and C_2,

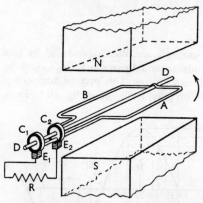

Fig. 10.2   Generation of an alternating e.m.f.

attached to but insulated from DD. Bearing on these rings are carbon brushes E_1 and E_2 which are connected to an external resistor R. When the plane of the loop is horizontal, the two sides A and B are moving parallel to the direction of the magnetic flux; therefore no flux is being cut and no e.m.f. is being generated in the loop.

In fig. 10.3(a), the vertical dotted lines represent lines of magnetic flux and loop AB is shown after it has rotated through an angle θ from the horizontal position, namely the position of zero e.m.f. Suppose the peripheral speed of each side of the loop to be v metres per second; then at the instant shown in fig. 10.3, this peripheral speed can be represented by the length of a line AL drawn at right angles to the plane of the loop. We can resolve AL into two components AM and AN, perpendicular and parallel respectively to the direction of the magnetic flux, as shown in fig. 10.3(b). Since

$$\angle \text{MLA} = 90° - \angle \text{MAL} = \angle \text{MAO} = \theta,$$

$$\therefore \qquad \text{AM} = \text{AL} \sin \theta = v \sin \theta.$$

The e.m.f. generated in A is due entirely to the component of the speed

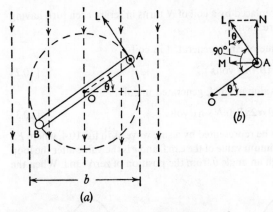

Fig. 10.3 Instantaneous value of generated e.m.f.

perpendicular to the magnetic field. Hence, if B is the flux density in webers per square metre and if l is the length in metres of each of the parallel sides A and B of the loop, it follows from expression [6.1] that:

e.m.f. generated in one side of loop $= Blv \sin \theta$ volts

and

total e.m.f. generated in loop $= 2Blv \sin \theta$ volts   [10.1]

i.e. the generated e.m.f. is proportional to $\sin \theta$. When $\theta = 90°$, the plane of the loop is vertical and both sides of the loop are cutting the magnetic flux at the maximum rate, so that the generated e.m.f. is then at its maximum value E_m. From expression [10.1], it follows that when $\theta = 90°$, $E_m = 2Blv$ volts.

If

$b =$ breadth of the loop in metres

and

$n =$ speed of rotation in revolutions/second,

then

$v = \pi bn$ metres/second

and

$E_m = 2Bl \times \pi bn$ volts

$\quad\ = 2\pi BAn$ volts

where

$A = lb =$ area of loop in square metres.

If the loop is replaced by a coil of N turns in series, each turn having an area of A square metres,

maximum value of e.m.f. generated in coil

$$= E_m = 2\pi B A n N \text{ volts} \qquad [10.2]$$

an instantaneous value of e.m.f. generated in coil

$$= e^* = E_m \sin \theta = 2\pi B A n N \sin \theta \text{ volts} \qquad [10.3]$$

This e.m.f. can be represented by a sine wave as in fig. 10.4, where E_m represents the maximum value of the e.m.f. and e is the value after the loop has rotated through an angle θ from the position of zero e.m.f. When the

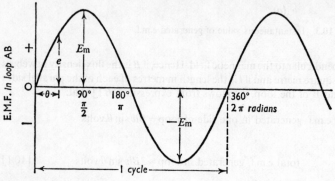

Fig. 10.4 Sine wave of e.m.f.

loop has rotated through 180° or π radius, the e.m.f. is again zero. When θ is varying between 180° and 360° (π and 2π radians), side A of the loop is moving towards the right in fig. 10.2 and is therefore cutting the magnetic flux in the opposite direction to that during the first half-revolution. Hence, if we regard the e.m.f. as positive while θ is varying between 0 and 180°, it is negative while θ is varying between 180° and 360°; i.e. when θ varies between 180° and 270°, the value of the e.m.f. increases from zero to $-E_m$ and then decreases to zero as θ varies between 270° and 360°. Subsequent revolutions of the loop merely produce a repetition of the e.m.f. wave.

Each repetition of a variable quantity, recurring at equal intervals, is termed a *cycle*, and the duration of one cycle is termed its *period* (or *periodic time*). The number of such cycles that occur in one second is termed the *frequency* of that quantity. The unit of frequency is the *hertz* (Hz), in memory of Heinrich Rudolf Hertz, who in 1888, was the first to demonstrate experimentally the existence and properties of electro-magnetic radiation predicted by Maxwell in 1865.

---

* Small letters are used to represent instantaneous values and capital letters represent definite values such as maximum, average or r.m.s. values. Capital I and V without any subscript represent r.m.s. values.

Example 10.1 *A coil of 100 turns is rotated at 1500 rev/min in a magnetic field having a uniform density of 0.05 T, the axis of rotation being at right angles to the direction of the flux. The area per turn is 4000 mm². Calculate (a) the frequency, (b) the period, (c) the maximum value of the generated e.m.f. and (d) the value of the generated e.m.f. when the coil has rotated through 30° from the position of zero e.m.f.*

(a) Since the e.m.f. generated in the coil undergoes one cycle of variation when the coil rotates through one revolution,

$\therefore$ frequency = no. of cycles/second

$\qquad\qquad$ = no. of revolutions/second

$\qquad\qquad$ = 1500/60 = 25 Hz.

(b) $\qquad$ Period = time of 1 cycle

$\qquad\qquad\qquad$ = 1/25 = 0.04 s.

(c) From expression [10.2],

$\qquad E_m = 2\pi \times 0.05\,[\text{T}] \times 0.004\,[\text{m}^2] \times 25\,[\text{rev/s}] \times 100\,[\text{turns}]$

$\qquad\quad = 3.14\,\text{V}.$

(d) For $\theta = 30°$, $\sin 30° = 0.5$,

$\therefore \quad e = 3.14 \times 0.5 = 1.57\,\text{V}.$

In practice, the e.m.f. generated in a conductor of an alternating-current generator (or alternator) is seldom sinusoidal. For instance, let us consider a four-pole machine, a portion of which is shown in fig. 10.5, and suppose AB to represent a full-pitch coil, i.e. a coil of such width that when side A is opposite the centre of a N pole, the other side B is opposite the centre of an adjacent S pole. The gaps under the poles are assumed to be of

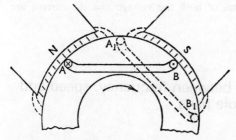

Fig. 10.5 Generation of an alternating e.m.f.

uniform length, so that the flux is uniformly distributed over the pole faces—as indicated by the short dotted lines. Beyond the pole-tips, however, the density of the fringing flux decreases rapidly owing to the greater length of the paths of the magnetic flux between the poles and the armature.

Since the value of the e.m.f. generated in AB is proportional to the rate at which these conductors cut the magnetic flux, it follows that while A and B are moving under the pole-faces, the rate of cutting the magnetic flux remains constant and the e.m.f. is represented by the straight line CD in fig. 10.6. After the coil has passed beyond the pole-faces, the e.m.f. decreases

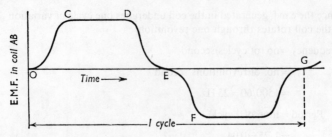

Fig. 10.6   Waveform of e.m.f. generated in fig. 10.5

rapidly, as indicated by DE, until when the conductors are midway between the poles—as shown by A_1B_1 in fig. 10.5—the e.m.f. is zero. As the conductors move into the fields of the succeeding poles, the e.m.f. generated in the coil grows in the reverse direction, as indicated by EF in fig. 10.6. Hence the e.m.f. generated in the coil varies in the manner shown in that figure; and the variation which takes place during interval OG is repeated indefinitely, so long as the speed is kept constant.

The sharpness of the corners at C and D in fig. 10.6 can be reduced by rounding off the pole-tips in fig. 10.5, thereby giving a more gradual variation of the flux density in those regions. Also, in an actual machine, the conductors are distributed in slots around the periphery of the armature, and it is found that the effect of this distribution is to make the waveform of the resultant e.m.f. nearly sinusoidal. In most a.c. calculations, the waveforms of both the voltage and the current are assumed to be sinusoidal.

10.3 Relationship between frequency, speed and number of pole pairs

Let us consider the four-pole machine of fig. 10.5. When conductor A is moving across the flux entering the armature from pole N, the direction of the e.m.f. is, by the Right-Hand Rule (section 6.4), towards the paper. When A is moving across the flux entering pole S from the armature, the direction of the e.m.f. is outwards from the paper; i.e. when the conductor passes a pair of poles, the e.m.f. varies through one cycle. Hence for a four-pole machine, the number of cycles of e.m.f. generated during one revolution is 2, namely the number of pairs of poles. It follows that if a machine has *p* pairs of poles and if the speed of rotation is *n* revolutions/second,

$$\text{frequency} = f = \text{no of cycles/second}$$

$$= \text{no. of cycles/rev} \times \text{no. of revs/second}$$

$$= pn \text{ hertz} \qquad\qquad [10.4]$$

Thus, if a two-pole machine has to generate an e.m.f. having a frequency of 50 Hz, then from expression [10.4],

$$50 = 1 \times n$$

$$\therefore \quad \text{speed} = 50 \text{ revolutions/second}$$

$$= 3000 \text{ rev/min.}$$

Since it is not possible to have fewer than two poles, the highest speed at which a 50-Hz alternator can be operated is 3000 rev/min.

10.4 Average or mean value of an alternating current or voltage

Let us first consider the case of a current having a waveform which is not sinusoidal. For instance, fig. 10.7(*a*) is typical of the waveform of the

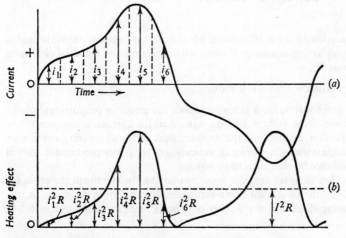

Fig. 10.7   Average and r.m.s. values

current taken by a transformer on no load. The base line for the first half-cycle is divided into, say, 6 equal parts and the mid-ordinates i_1, i_2, etc., are drawn and measured. Then:

average value of current over half a cycle

$$= I_{av} = \frac{i_1 + i_2 + i_3 + i_4 + i_5 + i_6}{6}$$

In general, if n equally-spaced mid-ordinates $i_1, i_2, ..., i_n$ are taken over either the positive or the negative half-cycle:

average value of current over half a cycle

$$= I_{av} = \frac{i_1 + i_2 + \cdots + i_n}{n} \qquad [10.5]$$

The larger the number of ordinates used, the more accurate is the result. With symmetrical waves it would be useless to add together the mid-ordinates over a whole cycle because the result would be zero.

If the above calculation is repeated with a sine wave, it is found that:

average value $= 0.637^* \times$ maximum value $\qquad [10.6]$

It is sometimes necessary to determine the average value of an alternating current or voltage; but in general the root-mean-square value, referred to in the next section, is far more important.

10.5 Root-mean-square value of an alternating current or voltage

In section 3.6 it was shown that when a current I amperes flows through a circuit having resistance R ohms with a difference of potential of V volts across it:

power $= V^2/R = I^2 R$ watts.

In other words, for a given resistance the power is proportional to the square of the voltage and the square of the current; consequently the important quantity in the measurement of an alternating current or voltage is that which gives an indication of the power produced in a given resistance by that current or voltage.

Let us again consider a current having the waveform shown in fig. 10.7(a). If this current flows through a circuit having resistance R, the heating effect of i_1 is $i_1^2 R$, that of i_2 is $i_2^2 R$, etc., as shown in fig. 10.7(b).

---

* By the aid of integral calculus, it can be shown that this value is $2/\pi = 0.637$.

Furthermore, the heating effect is positive during both half-cycles of the current wave; hence:

average heating effect over half a cycle

$$= \frac{i_1^2 R + i_2^2 R + \cdots + i_6^2 R}{6}.$$

In general, if there are n equally-spaced mid-ordinates in half a cycle, then:

average heating effect over half a cycle

$$= \frac{i_1^2 R + i_2^2 R + \cdots + i_n^2 R}{n}.$$

Suppose I amperes to be the value of the *direct current* through the *same* resistance R to produce a steady heating effect equal to the average heating effect of the alternating current, and thus to produce the same quantity of heat in half a cycle: then

$$I^2 R = \frac{i_1^2 R + i_2^2 R + \cdots + i_n^2 R}{n}.$$

$$\therefore \quad I = \sqrt{\left(\frac{i_1^2 + i_2^2 + \cdots + i_n^2}{n} \right)} \qquad [10.7]$$

= square *root* of the *mean* of the *squares* of the current

= root-mean-square (or r.m.s.) value of the current.

It will therefore be noted that the *root-mean-square* (or *r.m.s.*) *value of an alternating current is measured in terms of the direct current that produces the same heating effect in the same resistance*.

If the construction described above be applied to a sine wave of current, it is found that the variation of the heating effect follows a curve which is symmetrical* about the dotted horizontal line in fig. 10.8. In virtue of this symmetry the shaded areas above the dotted line are equal to those below the dotted line. Hence, if I_m be the maximum or peak value of the alternating current, the average heating effect over a cycle (or half a cycle) is half the maximum heating effect, namely $\frac{1}{2} I_m^2 R$ watts. If I amperes be the value of the *direct* current to give the same heating effect in the same resistance, then:

$$I^2 R = \frac{1}{2} I_m^2 R$$

$$\therefore \quad I = 0.707 I_m. \qquad [10.8]$$

---

* The symmetry can also be proved by trigonometry: thus, $\sin^2 \theta = \frac{1}{2} - \frac{1}{2} \cos 2\theta$. In words, this means that the square of a sine wave may be regarded as being made up of two components: (a) a constant quantity equal to half the maximum value of the $\sin^2 \theta$ curve; and (b) a cosine curve having twice the frequency of the $\sin \theta$ curve. From fig. 10.8 it is seen that the curve of the heating effect undergoes two cycles of change during one cycle of current.

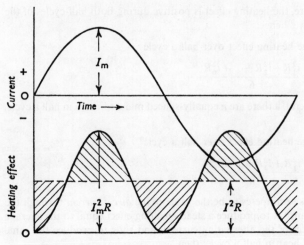

Fig. 10.8   R.m.s. value of a sinusoidal current

The following simple experiment may be found useful in illustrating the significance of the r.m.s. value of an alternating current. A metal-filament lamp L (fig. 10.9) is connected to an a.c. supply by closing switch S on contact a and the brightness of the filament is noted. The arm of S is then moved to b and the slider on resistor R is adjusted to give the same brightness.* The reading on a moving-coil ammeter A then gives the value

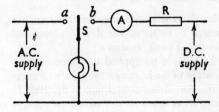

Fig. 10.9   An experiment to demonstrate the r.m.s. value of an alternating current

of the direct current that produces the same heating effect as that produced by the alternating current. If the reading on ammeter A is, say, 0.3 ampere when equality of brightness has been attained, the r.m.s. value of the alternating current is 0.3 ampere.

This experiment can be usefully extended† by connecting the Y-plates of a cathode-ray oscilloscope directly across lamp L (fig. 10.9) and

---

* For more precise adjustment, an illuminance meter can be placed at a convenient distance from the lamp and R adjusted to give the same reading when S is moved over from a to b.

† This extension was suggested by W. A. Turner and P. Godfrey in *Bulletin of Electrical Engineering Education*, No. 3, 1949.

measuring the deflections obtained with the direct and alternating voltages giving equal lamp brightness. The value (peak deflection with alternating voltage)/(deflection with direct voltage) gives the ratio (peak value)/(r.m.s. value) of the alternating voltage. If the supply voltage is sinusoidal, the value obtained experimentally should be compared with the theoretical figure of 1.414.

10.6 Form factor and peak or crest factor of a wave

The *form factor* is the ratio of the r.m.s. value to the average value of the wave; thus for a sine wave:

$$\text{form factor} = \frac{0.707 \times \text{maximum value}}{0.637 \times \text{maximum value}}$$

$$= 1.11. \tag{10.9}$$

The *peak* or *crest factor* is the ratio of the peak (or maximum) value to the r.m.s. value of the wave; thus for a sine wave:

$$\text{peak or crest factor} \frac{\text{maximum value}}{0.707 \times \text{maximum value}}$$

$$= 1.414. \tag{10.10}$$

10.7 Measurement of the r.m.s. values of alternating currents and voltages

If an ammeter or a voltmeter is to read the r.m.s. value, its action must depend upon some effect which is proportional to the square of the quantity to be measured; and the instruments principally used for this purpose can be grouped thus:

(a) moving-iron ammeters and voltmeters (section 18.6),
(b) thermal ammeters and voltmeters (section 18.7),
(c) electrodynamic (or dynamometer) ammeters and voltmeters (section 18.8), and
(d) electronic voltmeters (section 18.9).

10.8 Summary of important formulae

Instantaneous value of e.m.f. generated in a coil rotating in a uniform magnetic field

$$e = E_m \sin \theta$$

$$= 2\pi BAnN \sin \theta \text{ volts} \qquad [10.3]$$

$$f = np \text{ hertz} \qquad [10.4]$$

For n equidistant mid-ordinates over half a cycle,

$$\text{average value} = \frac{i_1 + i_2 + \cdots + i_n}{n} \qquad [10.5]$$

and

$$\text{r.m.s. value} = \sqrt{\left(\frac{i_1^2 + i_2^2 + \cdots + i_n^2}{n}\right)}. \qquad [10.7]$$

For sinusoidal waves,

$$\text{average value} = 0.637 \times \text{maximum value} \qquad [10.6]$$

and

$$\text{r.m.s. value} = 0.707 \times \text{maximum value} \qquad [10.8]$$

$$\text{Form factor} = \frac{\text{r.m.s. value}}{\text{average value}} \qquad [10.9]$$

$$= 1.11 \text{ for a sine wave.}$$

$$\text{Peak or crest factor} = \frac{\text{peak or maximum value}}{\text{r.m.s. value}} \qquad [10.10]$$

$$= 1.414 \text{ for a sine wave.}$$

10.9 Examples

1. Calculate the speed at which an eight-pole alternator must be driven in order that it may generate an e.m.f. having a frequency of 60 Hz.
2. An alternator driven by an internal-combustion engine at 375 rev/min is required to generate an alternating e.m.f. having a frequency of 50 Hz. Calculate the number of poles for which the alternator must be wound.
3. If a four-pole alternator is driven at a speed of 1800 rev/min, calculate the frequency of the generated e.m.f.
4. A coil of 200 turns is rotated at 600 rev/min in a magnetic field having a uniform density of 5000 μT, the axis of rotation being at right angles

to the direction of the field. The mean area per turn is $3000\,\text{mm}^2$. Calculate (a) the frequency and the period, (b) the r.m.s. and the maximum values of the generated e.m.f.

5. A circular coil of 60 turns, carried by a spindle placed at right angles to a magnetic field of uniform density, is rotated at a uniform speed. The mean diameter of the coil is 300 mm. Calculate the speed in order that the frequency of the generated e.m.f. may be 40 Hz and find the density of the magnetic field if the r.m.s. value of the generated e.m.f. is 36 V. What are the average and the maximum values of the e.m.f.?

   What should be the direction of the spindle relative to that of the magnetic field if no e.m.f. is to be generated when the coil is rotated?

6. What is meant by *electromagnetic induction*?

   A length of stiff copper wire is made into a rectangle measuring $240\,\text{mm} \times 100\,\text{mm}$. It is rotated at a uniform speed of 600 rev/min about one of its longer sides as an axis, this side lying in and at right angles to a uniform magnetic field of 0.5 T. Calculate the average e.m.f. induced in the coil.   (U.L.C.I.)

7. Sketch the waveform of the e.m.f. generated by rotating a coil in a uniform magnetic field. Show how to obtain the r.m.s. and average values and explain the meaning of the term *form factor*.

   Calculate the maximum value of the e.m.f. generated in a coil which is rotating at 50 rev/s in a uniform magnetic field of 0.8 T. The coil is wound on a square former having sides 50 mm in length, and is wound with 300 turns.   (U.E.I.)

8. Draw to an adequate scale the sine wave representing a current having an amplitude of 20 A. By means of the mid-ordinate method, find (a) the average value and (b) the r.m.s. value of the current.

9. Draw sine waves to represent a voltage having a peak value of 200 V and a current having a peak value of 50 A, the current being assumed to lag the voltage by 30°.

10. A sinusoidal current has an r.m.s. value of 10 A at a frequency of 50 Hz. Find the duration of 1 cycle in milliseconds. Plot the current wave over one cycle, using a vertical scale of 1 cm to represent 2 A and a horizontal scale of 1 cm to represent 2 ms. From this graph, find (a) the value of the current 3 ms after it has passed through its zero value and (b) the time taken for the current to grow from zero to 6 A. Check these values mathematically from the expression: $i = I_m \sin \theta$.

11. What is meant by the r.m.s. value of an alternating current and why is this value used?

    A sinusoidal alternating current has a maximum value of 2 A. Draw to scale the curve of current over half a cycle and obtain graphically the r.m.s. value of the current.   (N.C.T.E.C.)

12. The time between the positive peak of an alternating voltage wave and the first succeeding negative peak is 0.02 s. What is the frequency? Plot this voltage wave and obtain the r.m.s. value from your graph, assuming the wave to be sinusoidal and 200-V peak value. If this voltage is applied to a 57.6-Ω resistor, calculate the power in watts.

                                                                        (U.E.I.)

13. What is understood by the r.m.s. value of an alternating current? A sinusoidal current has a maximum value of 5 A. Plot the current to scale and determine graphically its r.m.s. value. If the current flows for 15 minutes through a 20-Ω resistor, what energy, in joules, is dissipated in the resistor?   (N.C.T.E.C.)

14. An alternating current had the following values for half-cycle:

Angle in radians	0	$\pi/9$	$\pi/6$	$\pi/3$	$\pi/2$	$2\pi/3$	$5\pi/6$	$8\pi/9$	π
Current in amperes	0	5	20	35	40	35	20	5	0

These current values are joined by straight lines. Obtain the r.m.s. value of this current from your graph. If the time of one complete cycle is 0.015 s, what is the frequency?   (U.E.I.)

15. An alternating current has the following values (in amperes) at equal intervals of time: 0, 2, 3, 3.5, 4, 4.4, 4.5, 3, 0, -2, -3, etc. Draw to scale the waveform over one cycle and determine the average and the r.m.s. values of the current.

16. What is meant by the *root mean square* value of an alternating current? Explain why this value is used. An alternating current has the following values over one-half of a cycle:

Time in milliseconds	0	1	2	3	4	5	6	7	8	9	10
Current in amperes	0	2.5	3.4	3.1	3.3	3.6	3.3	3.1	3.4	2.5	0

Determine the r.m.s. value and the frequency of the current wave.

                                                                 (N.C.T.E.C.)

17. An alternating current varies in the following manner during one-half-cycle: During the first quarter-cycle, it increases uniformly from 0 to 10 A; during the next eighth of a cycle, it decreases uniformly from 10 to 8 A; and during the remainder of the half-cycle, it decreases uniformly to zero. Plot a graph of the current wave over half a cycle and determine the r.m.s. and the average values of the current.

18. If the waveform of a voltage has a form factor of 1.15 and a peak factor of 1.5, and if the peak value is 4.5 kV, calculate the average and the r.m.s. values of the voltage.

19. If the waveform of an alternating current is an isosceles triangle having a peak value of 10 A, find the average and the r.m.s. values of the current and the form and peak factors of the wave.

20. An alternating current was measured by a d.c. milliameter in conjunction with a full-wave rectifier (section 18.11). The reading on the milliammeter was 7 mA. Assuming the waveform of the alternating current to be sinusoidal, calculate (a) the r.m.s. value and (b) the maximum value of the alternating current.

CHAPTER 11

■ Resistance and Inductance in
a.c. Circuits

11.1 Alternating current in a circuit possessing resistance only

Consider a circuit having a resistance R ohms connected across the terminals of an alternator A, as in fig. 11.1, and suppose the alternating voltage to be represented by the sine wave of fig. 11.2. If the value of the

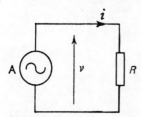

Fig. 11.1 Circuit with resistance only

voltage at any instant B is v volts, the value of the current* at that instant is given by:

$i = v/R$ amperes.

When the voltage is zero, the current is also zero; and since the current is proportional to the voltage, the wave-form of the current is exactly the same as that of the voltage. Also the two quantities are *in phase* with each other; that is, they pass through their zero values at the same instant and attain their maximum values in a given direction at the same instant. Hence the current wave is as shown dotted in fig. 11.2.

---

* An arrow or arrowhead is used to indicate the direction in which the current flows when it is regarded positive. It is immaterial which direction is chosen as positive; but once it has been decided upon for a given circuit or network, the same direction must be adhered to for all the currents and voltages involved in that circuit or network.

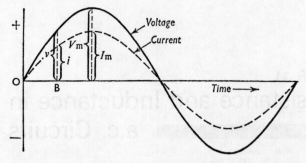

Fig. 11.2 Voltage and current waveforms for a resistive circuit

If V_m and I_m be the maximum values of the voltage and current respectively, it follows that:

$$I_m = V_m/R. \tag{11.1}$$

But the r.m.s. value of a sine wave is 0.707 times the maximum value, so that:

r.m.s. value of voltage $= V = 0.707\,V_m$

and

r.m.s. value of current $= I = 0.707\,I_m$.

Substituting for I_m and V_m in [11.1] we have:

$$\frac{I}{0.707} = \frac{V}{0.707R}$$

$$\therefore \quad I = V/R. \tag{11.2}$$

Hence Ohm's Law can be applied without any modification to an a.c. circuit possessing resistance only.

If the instantaneous value of the applied voltage is represented by:

$$v = V_m \sin \theta,$$

then instantaneous value of current in a resistive circuit

$$= i = \frac{V_m}{R} \sin \theta. \tag{11.3}$$

11.2 Alternating current in a circuit possessing inductance only

Let us consider the effect of a sinusoidal current flowing through a coil having an inductance of L henrys and a negligible resistance, as in fig. 11.3.

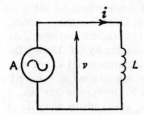

Fig. 11.3 Circuit with inductance only

For instance, let us consider what is happening during the first quarter-cycle of fig. 11.4. This quarter-cycle has been divided into three equal

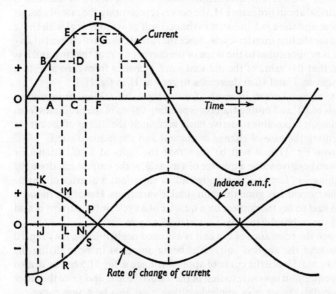

Fig. 11.4 Waveforms of current, rate of change of current and induced e.m.f.

intervals, OA, AC and CF seconds. During interval OA, the current increases from zero to AB; hence the average rate of increase of current is AB/OA amperes/second, and is represented by ordinate JK drawn midway between O and A. From expression [8.1],

the e.m.f., in volts, induced in a coil

$$= -L \times \text{rate of change of current in amperes per second;}$$

consequently, the average value of the induced e.m.f. during interval OA is $-L \times$ AB/OA, namely $-L \times$ JK volts, and is represented by ordinate JQ in fig. 11.4. The negative sign denotes that the induced e.m.f. tends to oppose the growth of the current in the positive direction.

Similarly, during interval AC, the current increases from AB to CE, so that the average rate of increase of current is DE/AC amperes/second, which is represented by ordinate LM in fig. 11.4; and the corresponding induced e.m.f. is $-L \times LM$ volts and is represented by LR. During the third interval CF, the average rate of increase of current is GH/CF, namely NP amperes/second; and the corresponding induced e.m.f. is $-L \times NP$ volts and is represented by NS. At instant F, the current has ceased growing but has not yet begun to decrease; consequently the rate of change of current is then zero. The induced e.m.f. will therefore have decreased from a maximum at O to zero at F. Curves can now be drawn through the derived points, as shown in fig. 11.4.

During the second quarter-cycle, the current decreases, so that the rate of change is negative and the induced e.m.f. becomes positive, tending to prevent the current decreasing. Since the sine wave of current is symmetrical about ordinate FH, the curves representing the rate of change of current and the e.m.f. induced in the coil will be symmetrical with those derived for the first quarter-cycle. Since the rate of change of current at any instant is proportional to the slope of the current wave at that instant, it is evident that the value of the induced e.m.f. increases from zero at F to a maximum at T and then decreases to zero at U in fig. 11.4.

By using shorter intervals, for example by taking ordinates at intervals of 10° and noting the corresponding values of the ordinates from sine values, it is possible to derive fairly accurately the shapes of the curves representing the rate of change of current and the induced e.m.f.

From fig. 11.4, it will be seen that the induced e.m.f. attains its maximum positive value a quarter of a cycle *after* the current has done the same thing—in fact, it goes through all its variations a quarter of a cycle after the current has gone through similar variations. Hence the induced e.m.f. is said to *lag* the current by a quarter of a cycle or the current is said to *lead* the induced e.m.f. by a quarter of a cycle.

Since the resistance of the coil is assumed negligible, we may regard the whole of the applied voltage as being absorbed in neutralizing the induced e.m.f. Hence the curve of applied voltage in fig. 11.5 can be drawn exactly equal and opposite to that of the induced e.m.f.; and since the latter is sinusoidal, the wave of applied voltage must also be a sine curve.

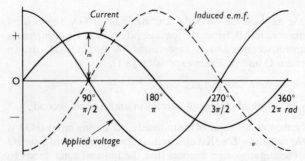

Fig. 11.5   Voltage and current waveforms for a purely inductive circuit

From fig. 11.5 it is seen that the applied voltage attains its maximum positive value a quarter of a cycle earlier than the current; in other words, the voltage applied to a purely inductive circuit leads the current by a quarter of a cycle or 90°, or the current lags the applied voltage by a quarter of a cycle or 90° or $\pi/2$ radians.

The student may quite reasonably ask: If the applied voltage is neutralized by the induced e.m.f., how can there be any current? The answer is that if there were no current there would be no flux, and therefore no induced e.m.f. The current has to vary at such a rate that the e.m.f. induced by the corresponding variation of flux is equal and opposite to the applied voltage. Actually there is a slight difference between the applied voltage and the induced e.m.f., this difference being the voltage required to send the current through the low resistance of the coil.

11.3 Mechanical analogy of an inductive circuit

One of the most puzzling things to a student commencing the study of alternating currents is the behaviour of a current in an inductive circuit. For instance, why should the current in fig. 11.5 be at its maximum value when there is no applied voltage? Why should there be no current when the applied voltage is at its maximum? Why should it be possible to have a voltage applied in one direction and a current flowing in the reverse direction as is the case during the second and fourth quarter-cycles in fig. 11.5?

It may therefore be found helpful to consider a simple mechanical analogy—the simpler the better. In Mechanics, it is found that the *inertia* of a body opposes any change in the *speed* of that body. The effect of inertia is therefore analogous to that of *inductance* in opposing any change in the *current*.

Suppose we take a heavy metal cylinder C (fig. 11.6), such as a pulley

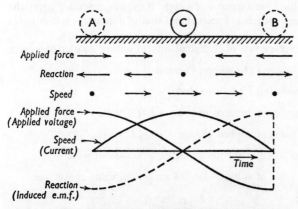

Fig. 11.6 Mechanical analogy of a purely inductive circuit

or an armature, and roll it backwards and forwards on a horizontal surface between two extreme positions A and B. Let us consider the forces and the speed while C is being rolled from A to B. At first the speed is zero, but the force applied to the body is at its maximum causing C to accelerate towards the right. This applied force is reduced—as indicated by the length of the arrows in fig. 11.6—until it is zero when C is midway between A and B. C ceases to accelerate and will therefore have attained its maximum speed from left to right.

Immediately after C has passed the mid-point, the direction of the applied force is reversed and increased until the body is brought to rest at B and then begins its return movement.

The reaction of C, on the other hand, is equal and opposite to the applied force and corresponds to the e.m.f. induced in the inductive circuit.

From an inspection of the arrows in fig. 11.6 it is seen that the speed in a given direction is a maximum a quarter of a complete oscillation after the applied force has been a maximum in the same direction, but a quarter of an oscillation before the reaction reaches its maximum in that direction. This is analogous to the current in a purely inductive circuit lagging the applied voltage by a quarter of a cycle and leading the induced e.m.f. by a quarter of a cycle. Also it is evident that when the speed is a maximum the applied force is zero, and that when the applied force is a maximum the speed is zero; and during the second half of the movement indicated in fig. 11.6, the direction of motion is opposite to that of the applied force. These relationships correspond exactly to those found for an inductive circuit.

11.4 Numerical relationship between current and voltage in a purely inductive circuit

From fig. 11.5 it is seen that the current increases from zero to its maximum value I_m in a quarter of a cycle. If the frequency is f hertz, the duration of one cycle is $1/f$ second and that of a quarter-cycle is $1/(4f)$ second.

Hence, average rate of change of current during quarter of a cycle

$$= I_m \div (1/4f) = 4f I_m \text{ amperes/second.}$$

From expression [8.1] it follows that:

average e.m.f. induced in coil $= -L4f I_m$ volts

∴ average value of applied voltage $= +4f L I_m$ volts.

But in section 10.4 it was shown that for a sinusoidal wave:

average value of voltage $= 0.637 \times$ maximum value of voltage

$$= \frac{2}{\pi} \times V_m.$$

Hence
$$\frac{2}{\pi} V_m = 4fLI_m$$

and
$$\frac{V_m}{I_m} = 2\pi fL.$$

If V and I are the r.m.s. values of the applied voltage and current respectively:

$$\frac{V}{I} = \frac{0.707V_m}{0.707I_m} = 2\pi fL$$

      = *inductive reactance* of the circuit.

Since the reactance is the ratio of the voltage to the current, it is expressed in ohms. It is represented by the symbol X_L. Hence

$$I = \frac{V}{2\pi fL} = \frac{V}{X_L}.$$

[11.4]

It is evident that the inductive reactance is proportional to the frequency and that for a given voltage, the current is inversely proportional to the frequency. These relationships are represented graphically by the straight line and hyperbola respectively in fig. 11.7. If instantaneous

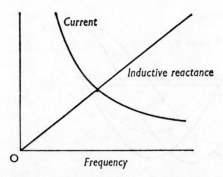

Fig. 11.7 Effect of frequency on inductive resistance and current

value of applied voltage

$$= v = V_m \sin \theta$$

then instantaneous value of current in a purely inductive circuit

$$= i = \frac{V_m}{2\pi fL} \sin (\theta - \pi/2).$$

[11.5]

11.5 Alternating current in a circuit possessing resistance and inductance in series

Suppose the circuit to have resistance R ohms in series with inductance L henrys (fig. 11.8), and suppose the current to be sinusoidal, as shown in fig. 11.9. From section 11.1 it follows that the p.d. across R is in phase with the

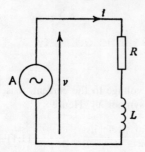

Fig. 11.8 Resistance and inductance in series

current; and in section 11.2 it was shown that the voltage applied to an inductance leads the current by 90°. The curves for these p.d.s are shown dotted in fig. 11.9. At any instant the resultant voltage is the sum of the

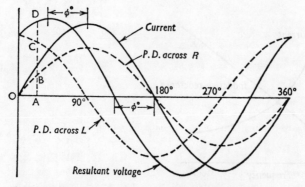

Fig. 11.9 Voltage and current waveforms for fig. 11.8

p.d.s across R and L; thus at instant A, the p.d. across R is AB and that across L is AC, so that the total applied voltage = AB + AC = AD. By adding together the two dotted curves in this way, we can derive the curve representing the resultant voltage across R and L.

It is seen that the resultant voltage attains its maximum positive value ϕ^* degrees before the current does so; similarly the resultant voltage

---

* Small ϕ is used for angles and capital Φ for flux.

passes through zero ϕ degrees before the current passes through zero in the *same* direction.

It will now be evident that the representation of alternating currents and voltages by curves such as those of fig. 11.9 becomes involved when the circuit possesses resistance and reactance in series. Also the calculations become very cumbersome. Both the representation and the calculations can be simplified considerably by the use of phasor diagrams.

11.6 Representation of an alternating quantity by a rotating phasor

In fig. 11.10(a), OA represents to scale the maximum value of the alternating quantity, say, current; i.e. $OA = I_m$. Suppose OA to rotate counter-clockwise about O at a uniform angular velocity. This is purely a conventional direction which has been universally adopted. Also an arrowhead is always drawn at the outer end of the phasor, partly to indicate which end is assumed to move and partly to indicate the precise length of the phasor when two or more phasors happen to coincide.

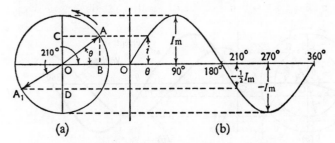

Fig. 11.10 Phasor representation of an alternating quantity

Let us assume that fig. 11.10(a) shows OA when it has rotated through an angle θ from the position occupied when the current was passing through its zero value. If AB and AC be drawn perpendicular to the horizontal and vertical axes respectively:

$$OC = AB = OA \sin \theta$$

$$= I_m \sin \theta$$

$$= i, \text{ namely the value of the current at that instant.}$$

This relationship follows from the method used in fig. 10.1 to construct a sine wave. Hence the projection of OA on the vertical axis represents to scale the instantaneous value of the current. Thus when $\theta = 90°$, the projection is OA itself; when $\theta = 180°$, the projection is zero and corresponds to the current passing through zero from a positive to a negative value; when $\theta = 210°$, the phasor is in position OA_1, and the projection $= OD = \frac{1}{2}OA_1 = -\frac{1}{2}I_m$; and when $\theta = 360°$, the projection is again zero and corresponds to the current passing through zero from a negative to a positive value. It follows that OA rotates through one revolution or 2π radians in one cycle of the current wave.

If f is the frequency in hertz, then OA rotates through f revolutions or $2\pi f$ radians in 1 second. Hence the angular velocity of OA is $2\pi f$ radians/second and is denoted by the symbol ω (omega): i.e.

$\omega = 2\pi f$ radians/second.

If the time taken by OA in fig. 11.10 to rotate through an angle θ radians be t seconds, then:

$\theta =$ angular velocity $\times$ time

$= \omega t = 2\pi f t$ radians.

We can therefore express the instantaneous value of the current thus:

$i = I_m \sin \theta = I_m \sin \omega t = I_m \sin 2\pi f t.$

Let us next consider how two quantities such as voltage and current can be represented by a phasor diagram. Figure 11.11(b) shows the voltage leading the current by an angle ϕ. In fig. 11.11(a), OA represents the

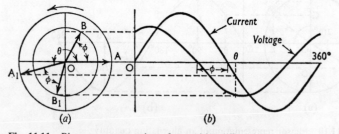

Fig. 11.11   Phasor representation of quantities differeing in phase

maximum value of the current and OB that of the voltage. The angle between OA and OB must be the same angle ϕ as in fig. 11.11(b). Consequently when OA is along the horizontal axis, the current at that instant is zero and the value of the voltage is represented by the projection of OB on the vertical axis. These values correspond to instant O in fig. 11.11(b).

After the phasors have rotated through an angle θ, they occupy positions OA_1 and OB_1 respectively; and the instantaneous values of the current and voltage are again given by the projections of OA_1 and OB_1 on the vertical axis, as shown by the horizontal dotted lines.

If the instantaneous value of the current is represented by

$i = I_m \sin \theta,$

then the instantaneous value of the voltage is represented by

$$v = V_m \sin(\theta + \phi)$$

where $I_m = OA$ and $V_m = OB$ in fig. 11.11(a).

11.7 Addition and subtraction of sinusoidal alternating quantities

Suppose OA and OB in fig. 11.12 to be rotating phasors representing to

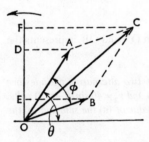

Fig. 11.12   Addition of phasors

scale the maximum values of, say, two alternating voltages having the same frequency but differing in phase by an angle ϕ. Complete the parallelogram OACB and draw the diagonal OC. Project OA, OB and OC on to the vertical axis. Then for the positions shown in fig. 11.12:

> instantaneous value of OA = OD
>
> instantaneous value of OB = OE

and

> instantaneous value of OC = OF.

Since AC is parallel and equal to OB, DF = OE,

∴   OF = OD + DF = OD + OE,

i.e. the instantaneous value of OC

> = sum of the instantaneous values of OA and OB.

Hence OC represents the maximum value of the resultant voltage to the scale that OA and OB represent the maximum values of the separate voltages. OC is therefore termed the *phasor sum* of OA and OB; and it is evident that OC is less than the arithmetic sum of OA and OB except when the latter are in phase with each other. This is the reason why it is seldom correct in a.c. work to add voltages or currents together arithmetically.

If voltage OB is to be subtracted from OA, then OB is produced backwards so that OB_1 is equal and opposite to OB (fig. 11.13). The

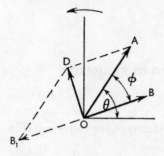

Fig. 11.13   Subtraction of phasors

diagonal OD of the parallelogram drawn on OA and OB_1 represents the *phasor difference* of OA and OB.

Example 11.1   *The instantaneous values of two alternating voltages are represented respectively by $v_1 = 60 \sin\theta$ volts and $v_2 = 40 \sin(\theta - \pi/3)$ volts. Derive an expression for the instantaneous value of* (a) *the sum and* (b) *the difference of these voltages.*

(a) It is usual to draw the phasors in the position corresponding to $\theta = 0$; i.e. OA in fig. 11.14 is drawn to scale along the X axis to represent 60 volts, and OB is drawn $\pi/3$ radians or 60° behind OA to represent 40 volts. The

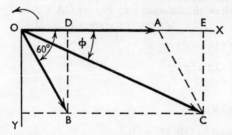

Fig. 11.14   Addition of phasors in example 11.1

diagonal OC of the parallelogram drawn on OA and OB represents the phasor sum of OA and OB. By measurement, OC = 87 volts and angle ϕ between OC and the X axis is 23.5°, namely 0.41 radian; hence:

instantaneous sum of the two voltages = $87 \sin(\theta - 0.41)$ volts.

Alternative, this expression can be found thus:

horizontal component of OA = 60 V,

horizontal component of OB = OD = $40 \cos 60° = 20$ V,

∴   resultant horizontal component $= OA + OD = 60 + 20$

$$= 80\,V = OE \text{ in fig. } 11.14.$$

Vertical component of $OA = 0$,

vertical component of $OB = BD = -40 \sin 60°$

$$= -34.64\,V,$$

∴   resultant vertical component $= -34.64\,V = CE$.

The minus sign merely indicates that the resultant vertical component is *below* the horizontal axis and that the resultant voltage must therefore lag behind OA.

Hence maximum value of resultant voltage

$$= OC = \sqrt{[(80)^2 + (-34.64)^2]}$$

$$= 87.2\,V.$$

If ϕ is the angle of lag of OC behind OA,

$$\tan \phi = EC/OE = -34.64/80 = -0.433$$

∴ $\phi = -23.4° = -0.41$ radian

and instantaneous value of resultant voltage

$$= 87.2 \sin(\theta - 0.41) \text{ volts.}$$

(b) The construction for subtracting OB from OA is obvious from fig. 11.15. By measurement, $OC = 53$ volts and $\phi = 41° = 0.715$ radian.

∴   instantaneous difference of the two voltages

$$= 53 \sin(\theta + 0.715) \text{ volts.}$$

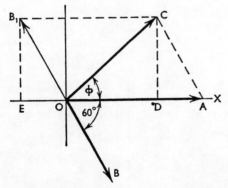

Fig. 11.15 Subtraction of phasors in example 11.1

Alternatively, resultant horizontal component

$$= OA - OE = 60 - 20$$

$$= 40\,V = OD \text{ in fig. 11.15,}$$

and resultant vertical component $= B_1E = 34.64\,V$

$$= DC \text{ in fig. 11.15.}$$

∴   maximum value of resultant voltage

$$= OC = \sqrt{[(40)^2 + (34.64)^2]}$$

$$= 52.9\,V$$

and

$$\tan \phi = DC/OD = 34.64/40$$

$$= 0.866,$$

∴ $\phi = 40.9° = 0.714\,\text{radian}$

and instantaneous value of resultant voltage

$$= 52.9 \sin(\theta + 0.714)\,\text{volts.}$$

11.8 Application of phasors to a circuit possessing resistance and inductance in series

It is evident from fig. 11.8 that the quantity which is common to the resistance and the inductance is the current. Hence we commence the phasor diagram by drawing a phasor OA (fig. 11.16) in any convenient

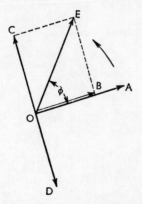

Fig. 11.16   Phasor diagram for fig. 11.8

direction to represent the maximum current, I_m, to scale. From fig. 11.9 it is seen that the p.d. across R is in phase with the current; hence draw OB in phase with OA to represent RI_m to some convenient scale. It is often helpful—especially to a beginner—to draw OA and OB slightly apart so that the identity of each may be easily recognized.

The voltage applied to the inductance L leads the current by 90° and in section 11.4, its peak value was shown to be $2\pi f L I_m$. Hence draw OC 90° in front of OA to represent the maximum value of the p.d. across L. A phasor OD, drawn equal and opposite to OC, represents the peak value of the e.m.f. induced in L.

The maximum value, V_m, of the resultant applied voltage is represented to scale by the diagonal OE of the parallelogram OBEC; and angle ϕ between OA and OE represents the phase difference between the current and the applied voltage.

Since the angle between OB and BE is a right angle:

$$\text{OE} = \text{OB}^2 + \text{BE}^2 = \text{OB}^2 + \text{OC}^2$$

$$\therefore \quad V_m^2 = (RI_m)^2 + (2\pi f L I_m)^2$$

and

$$V_m = I_m\sqrt{[R^2 + (2\pi f L)^2]} = I_m\sqrt{(R^2 + X_L^2)}.$$

If V and I are the r.m.s. values of the voltage and current respectively, then:

$$\frac{V}{I} = \frac{0.707 V_m}{0.707 I_m}$$

$$= \sqrt{(R^2 + X_L^2)} = \text{impedance (in ohms)} \qquad [11.6]$$

or $\quad I = \dfrac{V}{\sqrt{(R^2 + X_L^2)}} = \dfrac{V}{Z}$ $\qquad\qquad\qquad\qquad$ [11.7]

The impedance of any a.c. circuit is always given by the ratio of the voltage to the current, irrespective of the nature of the circuit, and is represented by the symbol Z.

When a circuit is non-reactive, its impedance is the same as the resistance. When the circuit consists of an *inductor* (or choking coil), namely a coil having a very low resistance and used primarily because it possesses inductance, the impedance is practically the same as the reactance; e.g. if $Z = 10\,\Omega$ and $R = 1\,\Omega$, $X_L = \sqrt{(10^2 - 1^2)} = 9.95\,\Omega$.

From fig. 11.16,

$$\tan\phi = \frac{\text{EB}}{\text{OB}} = \frac{\text{OC}}{\text{OB}} = \frac{2\pi f L I_m}{R I_m}$$

$$= \frac{2\pi f L}{R} = \frac{\text{reactance}}{\text{resistance}} \qquad [11.8]$$

Hence ϕ can be obtained. Also,

$$\cos \phi = \frac{OB}{OE} = \frac{RI_m}{ZI_m}$$

$$= \frac{\text{resistance}}{\text{impedance}}$$ [11.9]

Similarly,

$$\sin \phi = \frac{\text{reactance}}{\text{impedance}}$$ [11.10]

11.9 Phasor diagrams drawn with r.m.s. values instead of maximum values

In practice, ammeters and voltmeters usually measure the r.m.s. value of the current and voltage. It is therefore much more convenient to make the phasors represent the r.m.s. rather than the maximum values. Since the r.m.s. value of a sine wave is 0.707 times the peak value, it follows that if the phasors in fig. 11.16, for instance, are drawn to represent to scale the r.m.s. values of the current and voltages, the angles and therefore the phase relationships of the various quantities remain unaffected.

Phasor diagrams are extremely helpful in a.c. calculations, and students should cultivate the habit of introducing a phasor diagram wherever possible.

Example 11.2 *A coil having a resistance of $12\,\Omega$ and an inductance of 0.1 H is connected across a 100-V, 50-Hz supply. Calculate:* (a) *the reactance and the impedance of the coil,* (b) *the current and* (c) *the phase difference between the current and the applied voltage.*

When solving problems of this kind, students should first of all draw a circuit diagram (fig. 11.17) and insert all the known quantities. They

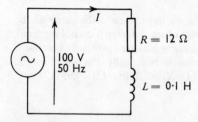

Fig. 11.17 Circuit diagram for example 11.2

should then proceed with the phasor diagram, fig. 11.18. It is not essential to draw the phasor diagram to exact scale, but it is helpful to draw it

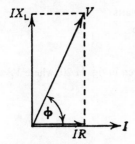

Fig. 11.18 Phasor diagram for example 11.2

approximately correctly since it is then easy to make a rough check of the calculated values.

(a)   Reactance $= X_L = 2\pi f L$

$$= 2\pi \times 50 \times 0.1 = 31.4\,\Omega.$$

   Impedance $= Z = \sqrt{(R^2 + X_L^2)}$

$$= \sqrt{(12^2 + 31.4^2)} = 33.6\,\Omega.$$

(b)   Current $= I = \dfrac{V}{Z} = \dfrac{100}{33.6} = 2.975\,\text{A}.$

(c)   Tan $\phi = \dfrac{X}{R} = \dfrac{41.4}{12} = 2.617.$

Hence $\phi = 69°$.

Example 11.3 *A metal-filament lamp, rated at 1.5 kW, 100 V, is to be operated in series with a choking coil across a 230-V, 50-Hz supply. Calculate: (a) the p.d. across the coil, (b) the inductance of the coil and (c) the phase difference between the current and the supply voltage. Neglect the resistance of the coil.*

(a) The metal filament of the lamp can be regarded as a non-inductive resistor having resistance R in fig. 11.19, where L represents the inductance of the coil to be connected in series.

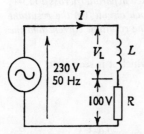

Fig. 11.19 Circuit diagram for example 11.3

$$\text{Rated current of lamp} = \frac{\text{rated power in watts}}{\text{rated voltage}}$$

$$= 1500/100 = 15 \text{ A}.$$

The phasor diagram for this circuit is given in fig. 11.20, where V_L

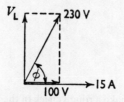

Fig. 11.20   Phasor diagram for example 11.3

represents the voltage across the coil. From this diagram it follows that:

$$230^2 = 100^2 + V_L^2$$

$$\therefore \quad V_L^2 = 42\,900$$

and

$$V_L = 207 \text{ volts} = \text{p.d. across coil.}$$

(b) Since

$$V_L = 2\pi f L I$$

$$\therefore \quad 207 = 2 \times 3.14 \times 50 \times L \times 15$$

and

$$L = 0.044 \text{ H}.$$

(c) From fig. 11.20 it is seen that:

$$\tan \phi = \frac{\text{voltage across coil}}{\text{voltage across lamp}}$$

$$= 207/100 = 2.07.$$

Hence, $\phi = 64.2° = $ angle of lag of current behind the supply voltage.

Example 11.4   *A 40-Ω non-reactive resistor and a coil having an inductance of 0.15 H and a negligible resistance are connected in parallel across a 115-V, 60-Hz supply. Calculate (a) the current in each circuit, (b) the resultant current and (c) the phase difference between the resultant current and the supply voltage.*

(a) The circuit is shown in fig. 11.21, from which it is seen that:

$$\text{current through } R = I_R = 115/40 = 2.875 \text{ A}$$

and

$$\text{current through } L = I_L = \frac{115}{2 \times 3.14 \times 60 \times 0.15} = 2.035 \text{ A}.$$

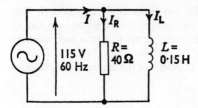

Fig. 11.21 Circuit diagram for example 11.4

When drawing a phasor diagram (fig. 11.22) for parallel circuits it is best to start with the voltage, since this is common to the two circuits. Then I_R is drawn in phase with the supply voltage V and I_L is drawn 90° behind V.

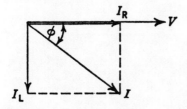

Fig. 11.22 Phasor diagram for example 11.4

(b) The resultant current I is the phasor sum of I_R and I_L; and from fig. 11.22 it is seen that:

$$I^2 = I_R^2 + I_L^2 = (2.875)^2 + (2.035)^2 = 12.4$$

$$\therefore \quad I = 3.52 \, \text{A}.$$

From fig. 11.22,

$$\tan \phi = \frac{I_L}{I_R} = \frac{2.035}{2.875} = 0.708$$

$$\therefore \quad \phi = 35.3°.$$

11.10 The magnetic amplifier

The inductance of an iron-cored coil depends on the permeability of the iron, see section 8.4. However, the permeability of iron depends on the value of the magnetic field strength H, as indicated for example by the graphs in fig. 7.6. The permeability, i.e. the value of B/H, is not constant and as the material reaches saturation then the permeability falls to almost zero. *Saturation* is the term used to describe that part of the *B–H* curve when *B* has an almost constant value, independent of the value of *H*. Thus the inductance of a coil containing an iron core is not constant and, since it is directly proportional to the permeability, drops to almost zero when the core approaches saturation.

Figure 11.23 shows the basic principle of the magnetic amplifier, or

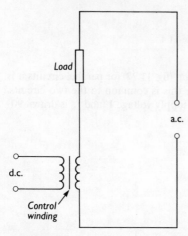

Fig. 11.23

transductor as it is often referred to. An alternating potential difference is applied across a load and the inductor. The circuit current will depend on the reactance of the inductor, reactance X_L being ωL. The larger the value of *L* the greater the reactance and hence the smaller the circuit current. However, the inductance depends on the magnetic field strength *H*. By passing a d.c. current through a control winding the value of *H* in the inductor can be controlled. This direct current establishes an additional flux within the core of the coil which is superimposed on the alternating flux produced by the alternating current in the load circuit. The result of this can be to drive the core into saturation when the directions of the additional flux and that due to the a.c. are in the same direction. This has the effect of producing a drop in the inductance and hence a drop in reactance. This increases the circuit current.

Initially, with no control current, the reactance is made large compared with the load resistance, by using a large inductance, and so the current is small and most of the potential drop is across the inductance. However, with an appropriate control current the core can be driven into saturation and the reactance drop to a very low value. This results in an increase in the circuit current and most of the potential drop now occurring across the load. The average load current, and hence average power dissipated in the load, is thus controlled by the current in the control winding. A small control current can be used to control large values of load current and thus load power.

The circuit shown in fig. 11.23 has a disadvantage that the alternating current in the load circuit will induce a current in the control circuit, like a transformer (see chapter 17). There is also the point that the control current is only having an effect during one half of the cycle of the a.c., that half which gives a flux in the same direction as that given by the control current. A more practical form of magnetic amplifier is thus in the form shown in fig. 11.24. The windings on the core are in opposite directions so

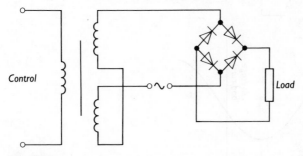

Fig. 11.24

that the induced currents in the control circuit by each winding are in opposite directions and cancel out. The rectifier circuit (see section 23.4) enables both halves of the a.c. to be controlled by the control current. Figure 11.25 shows the load current–control current characteristic for

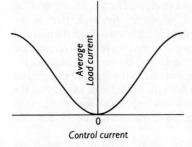

Fig. 11.25

such a circuit. The average load current is effectively zero when there is zero control current due to the very large inductance used and hence large reactance.

With the above circuit and characteristic the magnetic amplifier gives an increase in load current when the control current rises, independent of the direction of flow of the control current, i.e. it is not sensitive to control current polarity. To make the amplifier sensitive to the polarity of the control current an extra winding, referred to as the *bias winding*, is used (fig. 11.26). A steady current is applied to the bias winding and gives a constant flux in the core, regardless of the control current.

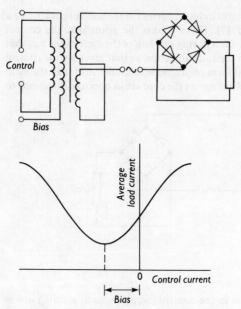

Fig. 11.26 A magnetic amplifier with bias

The greater the change in average load current for a given change in control current the greater the amplification of the system, the amplification being defined as the ratio of output to input. The amplification of the magnetic amplifier can be increased by *positive feedback*. Part of the output is rectified and fed back to the core, in addition to both the bias and control windings. An increase in output current, as a result of an increase in control current, means an increase in the current fed back to the feedback winding and hence a further increase in the flux and so a greater output. The term positive applied to this type of feedback means that the signal fed back adds to the effect which was responsible for producing it. The overall result is a modification in the characteristic and an increased amplification, fig. 11.27.

Compared with semiconductor amplifiers, magnetic amplifiers are bigger and heavier. However they are more robust and have high reliability with good stability.

11.11 Summary of important formulae

For purely resistive circuit,

$$i = \frac{V_m}{R} \sin \theta \qquad\qquad [11.3]$$

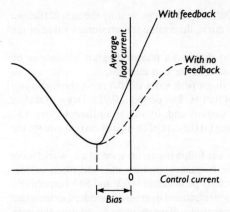

Fig. 11.27 A magnetic amplifier with positive feedback

and

$$I = V/R. \qquad [11.2]$$

For purely inductive circuit,

$$i = \frac{V_m}{2\pi f L} \sin(\theta - \pi/2) \qquad [11.5]$$

and

$$I = \frac{V}{2\pi f L} \qquad [11.4]$$

Inductive reactance $= X_L = 2\pi f L$ ohms.

For R and L in series,

$$\text{impedance} = Z = \sqrt{[R^2 + (2\pi f L)^2]} \text{ ohms} \qquad [11.6]$$

$$I = \frac{V}{\sqrt{[R^2 + (2\pi f L)^2]}} \qquad [11.7]$$

$$\tan \phi = 2\pi f L/R, \qquad [11.8]$$

$$\cos \phi = R/Z, \qquad [11.9]$$

and

$$\sin \phi = 2\pi f L/Z. \qquad [11.10]$$

11.12 Examples

1. Draw sine waves to represent two currents, one having an amplitude of 10 A and the other of 7 A, assuming the smaller current to lag the

larger current by 60°. Derive a curve representing the sum of the two currents, and from this curve determine the maximum value of the resultant current.

Check this result by drawing a phasor diagram to scale or by calculating the phasor sum of the two currents.

2. A sinusoidal current having a peak value of 10 A flows through a coil having an inductance of 0.01 H. The period is 0.02 s. Draw to scale a curve representing the current and, by taking ordinates every 15°, derive curves representing (a) the rate of change of current and (b) the induced e.m.f.

3. Explain the meaning of the following terms used in a.c. work: wave shape, peak value, cycle, frequency, phase, r.m.s. and mean values.

Two sinusoidal e.m.f.s of peak values 100 V and 50 V respectively but differing in phase by 60 electrical degrees are induced in the same circuit. Find *either* by means of a diagram *or* by calculation the peak value of the resultant e.m.f. What is its r.m.s. value?   (U.L.C.I.)

4. What is the justification for the representation of sinusoidally varying quantities by means of phasors?

The currents taken by two parallel circuits are 12 A in phase with the applied voltage and 20 A lagging 30° behind the applied voltage, respectively. Determine the current taken by the combined circuits and its phase with respect to the applied voltage.   (N.C.T.E.C.)

5. Two coils connected in series are fixed together with their planes inclined at an angle of 60° to each other and rotated at a steady speed at right angles to a uniform magnetic field. An e.m.f. having a maximum value of 50 V is induced in each coil. With the aid of a phasor diagram drawn to scale, determine the resultant voltage when (a) the windings of the coils are similarly connected and (b) the connections of one of the coils are reversed.

6. What is meant by the mean value, the r.m.s. value and the peak value of an alternating current? What relationships exist between them when the current is sinusoidal?

Two sinusoidal currents of r.m.s. values 10 A and 20 A and differing in phase by 30° flow in the same conductor. Find either diagrammatically or by calculation the peak value of the resultant current.   (U.L.C.I.)

7. If a current is represented by the expression $i = 10 \sin 628t$ amperes, calculate (a) the frequency and (b) the value of the current 2 ms after it has passed through its zero value.

8. The following expressions represent the instantaneous values of the e.m.f.s induced in two coils which are connected in series:

$$e_1 = 120 \sin(\omega t + \pi/6) \text{ volts,}$$
and
$$e_2 = 80 \sin(\omega t - \pi/4) \text{ volts.}$$

Determine an expression representing the instantaneous value of the resultant e.m.f. when the coils are connected to give (a) the sum and (b) the difference of the two e.m.f.s.

9. Three circuits in parallel take currents which can be represented by $i_1 = 10\sin 314t$; $i_2 = 7\sin(314t - \pi/3)$; $i_3 = 12\sin(314t + \pi/4)$. Sketch a phasor diagram to represent the three currents and their resultant. Calculate the resultant, and express it in the same form as the individual currents. Give the r.m.s. value and the frequency of the resultant current.

10. The following e.m.f.s are being generated in three coils, A, B and C:

$$e_A = 100\sin \omega t$$

$$e_B = 70\sin(\omega t - \pi/3)$$

and

$$e_C = 120\cos \omega t.$$

If the three coils are connected in series, calculate (a) the maximum value of the resultant e.m.f. and (b) the phase of the resultant e.m.f. relative to the e.m.f. in coil A. Check your result by means of a phasor diagram drawn to scale.

11. An alternating voltage, represented by $v = 60\sin 800t$ volts, is maintained across a coil having a resistance of $20\,\Omega$ and an inductance of $40\,mH$. Determine (a) the frequency, (b) an expression for the instantaneous value of the current and (c) the r.m.s. value of the current.

12. A coil having an inductance of $0.2\,H$ and negligible resistance is connected across a 100-V a.c. supply. Calculate the current if the frequency is (a) 30 Hz and (b) 500 Hz.

13. A coil of wire when connected to a d.c. supply of 100 V takes a current of 10 A. When the coil is connected to an a.c. supply of 100 V at a frequency of 50 Hz, the current is 5 A. Explain the difference and calculate the inductance of the coil. (N.C.T.E.C.)

14. A coil having a resistance of $10\,\Omega$ and an inductance of $0.2\,H$ is connected to a 100-V, 50-Hz supply. Calculate (a) the impedance of the coil, (b) the reactance of the coil, (c) the current taken, (d) the phase difference between the current and the applied voltage. (N.C.T.E.C.)

15. A coil having a resistance of $20\,\Omega$ takes a current of 4 A when connected to a 230-V, 50-Hz supply. Calculate (a) the inductance of the coil and (b) the phase difference between the current and the applied voltage.

16. A 50-Ω non-reactive resistor and a coil having an inductance of $0.1\,H$ and negligible resistance are connected in parallel across a 200-V 60-Hz supply. Calculate (a) the current in each circuit, (b) the resultant current and (c) the phase angle between the resultant current and the applied voltage. Sketch the phasor diagram.

17. A coil having an inductance of $0.08\,H$ and negligible resistance is connected in parallel with a non-reactive resistor across a 240-V, 50-Hz supply. The resultant current is 15 A. Sketch the phasor diagram and calculate (a) the current through the resistance of the resistor and (b) the phase difference between the resultant current and the supply voltage.

CHAPTER 12

▰▰▰▰▰ Power in an a.c. Circuit

12.1 Power in a non-reactive circuit

In section 10.5 it was explained that when an alternating current flows through a circuit having a resistance R ohms, the average heating effect over a complete cycle is I^2R watts, where I is the r.m.s. value of the current in amperes.

If V be the r.m.s. value of the applied voltage, in volts, then for a non-reactive circuit, $V = IR$, so that:

average value of the power $= I^2R = I \times IR$

$= IV$ watts.

Hence the power in a non-reactive circuit is given by the product of the ammeter and voltmeter readings, exactly as in a d.c. circuit.

12.2 Power in a purely inductive circuit

Consider a coil wound with such thick wire that the resistance is negligible in comparison with the inductive reactance X_L ohms. When such a coil is connected across a supply voltage V, the current is given by $I = V/X_L$ amperes. Since the resistance is very small, the heating effect and therefore the power are also very small, even though the voltage and the current be large. Such a curious conclusion—so different from anything we have experienced in d.c. circuits—requires fuller explanation if its significance is to be properly understood. Let us therefore consider fig. 12.1, which shows the applied voltage and the current for a purely inductive circuit, the current being a quarter of a cycle behind the voltage.

The power at any instant is given by the product of the voltage and the current at that instant; thus at instant L, the p.d. is LN volts and the current is LM amperes, so that the power at that instant is LN × LM watts and is represented to scale by LP.

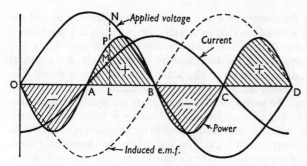

Fig. 12.1 Power waveform for purely inductive circuit

By repeating this calculation at various instants we can deduce the curve representing the variation of power over one cycle. It is seen that during interval OA the applied voltage is positive, but the current is negative, so that the power is negative; and that during interval AB, both the current and the voltage are positive, so that the power is positive.

The power curve is found to be symmetrical about the horizontal axis OD. Consequently the shaded areas marked '−' are exactly equal to those marked '+', so that the mean value of the power over the complete cycle OD is zero.

It is necessary, however, to consider the significance of the positive and negative areas if we are to understand what is really taking place. So let us consider an alternator P (fig. 12.2) connected to a coil Q whose resistance is negligible, and let us assume that the voltage and current are as represented in fig. 12.1. At instant A, there is no current and therefore no magnetic field through and around Q. During interval AB, the growth of the current is accompanied by a growth of flux as shown by the dotted

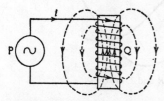

Fig. 12.2 Magnetic field of an inductive circuit

lines in fig. 12.2. But the existence of a magnetic field involves some kind of a strain in the space occupied by the field and the storing of energy in that field, as already dealt with in section 8.7. The current, and therefore the magnetic energy associated with it, reach their maximum values at instant B; and since the loss in the coil is assumed negligible, it follows that at that instant the whole of the energy supplied to the coil during interval AB, and represented by the shaded area marked '+', is stored in the magnetic field.

During interval BC, the current and its magnetic field are decreasing; and the e.m.f. induced by the collapse of the magnetic flux is in the same direction as the current. But any circuit in which the current and the induced or generated e.m.f. are in the same direction acts as a generator of electrical energy (section 8.2). Consequently the coil is now acting as a generator transforming the energy of its magnetic field into electrical energy, the latter being sent to alternator P to drive it as a motor. The energy thus returned is represented by the shaded area marked '−' in fig. 12.1; and since the positive and negative areas are equal, it follows that during alternate quarter-cycles electrical energy is being sent from the alternator to the coil, and during the other quarter-cycles the same amount of energy is sent back from the coil to the alternator. Consequently the net energy absorbed by the coil during a complete cycle is zero; in other words, the average power over a complete cycle is zero.

12.3 Power in a circuit having resistance and inductance in series

It has been shown in section 11.8 that in a circuit having a resistance R ohms in series with an inductance L henrys, the current lags the applied voltage by an angle ϕ such that $\tan \phi = 2\pi f L/R$. In fig. 12.3 the current and voltage are shown with a phase difference of 60°. The power wave is again

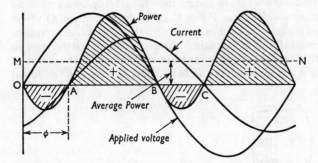

Fig. 12.3   Power waveform for circuit with R and L in series

derived from the product of the instantaneous values of current and voltage. It is seen that the positive area representing the energy absorbed by the circuit during interval AB is greater than the negative area representing the energy returned from the circuit to the alternator during interval BC. Also it is found that the power curve is symmetrical about a dotted line MN drawn midway between the positive and negative peaks of the power curve. Hence the height of this dotted line above the zero axis PABC represents the average value of the power over one cycle.

By drawing power curves for voltages and currents having the same amplitudes but various phase differences, it can be seen that the larger the phase difference, the smaller is the difference between the area representing the positive and negative energies, and the smaller, therefore, is the mean value of the power.

In section 10.5 it was explained that in any circuit having a resistance R ohms, the average heating effect due to a current having an r.m.s. value of I amperes is I^2R watt. On the other hand, the product of the r.m.s. values of the current and applied voltage is IV *voltamperes*, the latter term being used to distinguish this quantity from the power, expressed in watts.

Since the number of watts is very often less than the number of voltamperes, the latter has to be multiplied by some factor, equal to or less than unity, to give the power in watts. This factor is therefore termed the *power factor*, i.e.

power in watts = number of voltamperes × power factor

or

$$\text{power factor} = \frac{\text{power in watts}}{\text{product of r.m.s. values of voltage and current}}$$

[12.1]

$$= \frac{I^2R\,(\text{watts})}{IV\,(\text{amperes})} = \frac{IR}{V}.$$

The phasor diagram for a circuit having resistance and inductance in series has already been given in fig. 11.18, from which it is seen that $IR/V = \cos\phi$, where ϕ is the phase difference between the current and the applied voltage. Hence

power factor $= \cos\phi$ [12.2]

It should be pointed out that expression [12.1] for power factor is always correct, whereas expression [12.1] is correct only when both the current and the voltage waveforms are sinusoidal, since it has been derived from a phasor diagram, and phasor diagrams are based upon sine waves (see section 11.6).

If Z be the impedance of the circuit in ohms, then $V = IZ$,

$\therefore$ power factor $= \dfrac{IR}{V} = \dfrac{IR}{IZ} = \dfrac{R}{Z}.$ [12.3]

Example 12.1   *A coil having a resistance of 6 Ω and an inductance of 0.03 H is connected across a 50-V, 60-Hz supply. Calculate (a) the current, (b) the phase angle between the current and the applied voltage, (c) the power factor, (d) the voltamperes and (e) the power.*

(a) The phasor diagram for such a circuit is given in fig. 11.18.

Reactance of circuit $= 2\pi f L = 2 \times 3.14 \times 60 \times 0.03$

$$= 11.31\,\Omega.$$

From [11.6],

$$\text{impedance} = \sqrt{[6^2 + (11.31)^2]} = 12.8\,\Omega$$

and

$$\text{current} = 50/12.8 = 3.91\text{ A}.$$

(b) From [11.8],

$$\tan\phi = \frac{X}{R} = \frac{11.31}{6} = 1.885.$$

Hence,

$$\phi = 62°\,3'.$$

(c) From [12.2],

$$\text{power factor} = \cos 62°\,3' = 0.469$$

or from [12.3],

$$\text{power factor} = 6/12.8 = 0.469.$$

(d) Voltamperes $= 50 \times 3.91 = 195.5$ VA.

(e)   Power = voltamperes × power factor

$$= 195.5 \times 0.469 = 91.7\text{ W}.$$

Or, alternatively:

$$\text{power} = I^2 R = (3.91)^2 \times 6 = 91.7\text{ W}.$$

12.4 Active and reactive (or wattless) components of the current

Let us again consider a circuit consisting of resistance and inductance in series, and suppose the applied voltage and the current to be represented by the phasors OV and OI respectively in fig. 12.4.

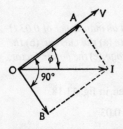

Fig. 12.4 Active and reactive components of current

Let us resolve the current I into two components, one component being in phase with the voltage V and the other lagging V by 90° (or quarter-cycle); for instance, in fig. 12.4, draw IA perpendicular to OV, and IB perpendicular to OB, which is at rightangles to OV. Consequently OAIB is a rectangle and the diagonal OI represents the phasor sum of two currents represented by OA and OB; i.e. a current OI may be replaced by two currents represented in magnitude and phase by OA and OB respectively.

Since OI represents the actual current, I amperes, and since OA = OI cos ϕ:

$$\therefore \quad \text{power} = IV\cos\phi = V \times \text{OI} \cos\phi$$

$$= V \times \text{OA watts.}$$

Hence component OA of the current is such that the product of OA and the voltage gives the power. Therefore OA is termed the *active* or *power component* of the current, i.e.

active or power component of current = $I \cos \phi$ [12.4]

Since OB lags the voltage by 90°,

power due to a component OB = OB $\times V \times \cos 90° = 0$.

Hence OB is referred to as the *reactive* or *wattless component* of the current. From fig. 12.4 it is seen that:

reactive or wattless component of current = OB = AI

$$= I \sin \phi \quad\quad [12.5]$$

and

reactive voltamperes (or vars) = $IV \sin \phi$ [12.6]

Example 12.2 *A single-phase motor is taking a current of 45 A from a 400-V supply at a power factor of 0.8 lagging. Calculate (a) the active and reactive comonents of the current, (b) the kilooltmperes, (c) the power taken from the supply, (d) the reactive kilooltmperes and (e) the output power, assuming the efficiency to be 86 per cent.*

(a) From [12.4],

active component of current = $45 \times 0.8 = 36$ A.

Since cos ϕ = power factor = 0.8

$$\therefore \quad \sin \phi = \sqrt{(1 - \cos^2 \phi)} = \sqrt{[1 - (0.8)^2]} = 0.6.$$

From [12.5],

reactive component of current = $45 \times 0.6 = 27$ A.

(b) Number of kilovoltamperes = number of voltamperes/1000

$$= 400 \times 45/1000 = 18 \text{ kVa.}$$

(c)   Number of kilowatts = number of kilovoltamperes × power factor

$$= 18 \times 0.8 = 14.4 \, \text{kW}.$$

(d)   Reactive kilovoltamperes (or kilovars)

$$= \text{kVA} \times \sin \phi = 18 \times 0.6 \, \text{kVAr}.$$

(e)   Output power = input power × efficiency

$$= 14.4 \, [\text{kW}] \times 0.86$$

$$= 12.38 \, \text{kW}.$$

Example 12.3   *An alternator is aupplying a load of 300 kW at a power factor of 0.6. If the power factor is raised to unity, how many more kilowatts can the alternator supply for the same kVA loading?*

Since the power in kilowatts

= number of kilovoltamperes × power factor.

∴   number of kilovoltamperes = 300/0.6 = 500 kVA.

When the power factor is raised to unity:

number of kilowatts = number of kilovoltamperes

$$= 500 \, \text{kW}.$$

Hence increased power supplied by alternator = 500 − 300 = 200 kW.

12.5 The practical importance of power factor

If an alternator is rated to give, say, 2000 A at a voltage of 400 V, it means that these are the highest current and voltage values the machine can give without the temperature exceeding a safe value. Consequently the rating of the alternator is given as 400 × 2000/1000 = 800 kVA. The phase difference between the voltage and the current depends upon the nature of the load and not upon the generator. Thus if the power factor of the load is unity, the 800 kVA are also 800 kW; and the engine driving the generator has to be capable of developing this power together with the losses in the generator. But if the power factor of the load is, say, 0.5, the power is only 400 kW; so that the engine is only developing about one-half of the power of which it is capable, though the alternator is supplying its rated output of 800 kVA.

Similarly the conductors connecting the alternator to the load have to be capable of carrying 2000 A without excessive temperature rise. Consequently they can transmit 800 kW if the power factor is unity, but only 400 kW at 0.5 power factor, for the same rise of temperature.

It is therefore evident that the higher the power factor of the load, the greater is the *power* that can be generated by a given alternator and transmitted by a given conductor.

The matter may be put another way by saying that, for a *given power*, the lower the power factor the larger must be the size of the alternator to generate that power, and the greater must be the cross-sectional area of the conductor to transmit it; in other words, the greater is the cost of generation and transmission of the electrical energy. This is the reason why supply authorities do all they can to improve the power factor of their loads either by the installation of capacitors (Chapter 13) or special machines or by the use of tariffs which encourage consumers to do so.

12.6 Summary of important formulae

$$\text{Power factor} = \frac{\text{power in watts}}{\text{r.m.s. volts} \times \text{r.m.s. amperes}} \qquad [12.1]$$

or,

power in watts $= IV \times$ power factor.

For voltage and current having sinusoidal waveforms and phase difference ϕ,

power factor $= \cos \phi$ $\qquad$ [12.2]

For circuit having resistance R and impedance Z,

power factor $= R/Z$ $\qquad$ [12.3]

Active or power component of current $= I \cos \phi$ $\qquad$ [12.4]

Reactive or wattless component of current $= I \sin \phi$ $\qquad$ [12.5]

Reactive voltamperes (or vars) $= IV \sin \phi.$ $\qquad$ [12.6]

12.7 Examples

1. A circuit having a power factor of 0.866 lagging takes a current of 10 A from a 240-V 50-Hz supply. Write down an expression for the instantaneous values of voltage and current, and find the value of the current when the voltage is (a) passing through its maximum value, (b) 0.005 s later and (c) after a further 0.005 s.
2. An alternating e.m.f. of 100 V, r.m.s. value, at 50 Hz, is applied to a 20-Ω non-inductive resistor. Plot to scale the voltage and current over one complete cycle and deduce the curve of power and state its mean value. What energy is used per cycle in watt seconds? (N.C.T.E.C.)

3. A coil having a resistance of $12\,\Omega$ and an inductance of $0.0159\,H$ is connected to a 260-V, 50-Hz supply. Calculate the energy in joules expended in the coil in 5 min. Draw a phasor diagram showing the voltage and current in their correct relative phase positions, to scales of $1\,cm = 20\,V$, $1\,cm = 2\,A$.

(U.L.C.I.)

4. The power factor of a load is 0.866 lagging. The voltage is 200 V and the current is 5 A. Calculate the equivalent series reactance and resistance of the load.

5. A coil having a resistance of $10\,\Omega$ and an inductance of $0.05\,H$ is connected across a 200-V, 50-Hz supply. Calculate (a) the current, (b) the phase difference between the current and the applied voltage and (c) the mean power absorbed.

6. Three 100-watt, 115-V lamps, connected in parallel, are to be supplied at their rated voltage from 230-V, 50-Hz mains by the use of a suitable choking coil of negligible resistance in series with the bank of lamps. Calculate (a) the voltage across the coil, (b) the inductance of the coil and (c) the power factor of the whole circuit.

7. When a d.c. supply of 5 V is applied across a certain circuit, the current is 0.1 A. When the d.c. supply is replaced by a sinusoidal supply at a frequency of 80 Hz, a p.d. of 10 V is required to cause a current of 0.1 A to flow in the same circuit. Calculate (a) the d.c. resistance of the circuit, (b) the power absorbed under the d.c. conditions and (c) the impedance of the circuit at 80 Hz.

   If the power absorbed by the circuit in the a.c. case is 0.6 W, calculate: (d) the power factor of the circuit and (e) its resistance to alternating current of the value stated.

8. Two coils are connected in series. With a direct current of 1 A, the voltages across the coils are 60 V and 80 V respectively. With an alternating current of 1 A, the voltages are 70 V and 100 V respectively. Calculate the power factor of each coil and that of the whole circuit.

9. The coil of an electromagnet has a resistance of $8\,\Omega$. When the coil is connected across a 250-V, 50-Hz supply, the current is 10 A and the power absorbed is 1 kW. Calculate (a) the inductance of the circuit at this value of the current and (b) the iron losses.

10. If a bank of lamps takes 5 kW at 250 V, calculate the impedance of a reactor of negligible resistance required to reduce the voltage across the lamps to 150 V, assuming that the resistance of the lamps does not change.

11. A coil having a resistance of $20\,\Omega$ takes 950 W from a 230-V, 50-Hz supply. Calculate (a) the inductance and the power factor of the coil, (b) the active and reactive components of the current and (c) the reactive voltamperes.

12. A 100-Ω resistor and a coil having an inductance of 0.3 H and negligible resistance are connected in parallel across a 400-V, 60-Hz supply. Calculate (a) the current in each circuit, (b) the resultant current, (c) the phase angle between the resultant current and the

supply voltage, (d) the power and (e) the power factor of the combined circuits.

13. An inductor having an inductance of 0.5 H and an impedance of 200 Ω at 50 Hz is connected in parallel with a 200-Ω non-reactive resistor. This parallel circuit is then connected across a 100-V, 50-Hz supply. Calculate (a) the resultant current and (b) the total power dissipated.

14. A load takes 100 A at power factor 0.7 lagging, from a 230-V, 50-Hz supply. A second load takes 44 A at power factor 0.9 leading. Calculate the values of the power in kilowatts, the kilovoltamperes and the power factor of the total load.

15. A coil A takes 6 A and dissipates 200 W when connected to a 100-V, 50-Hz supply, while another coil B takes 8 A and dissipates 600 W when connected to a similar supply. Calculate (a) the current and (b) the total power when the coils are connected in series across a 200-V, 50-Hz supply.

16. A single-phase induction motor runs off a 400-V supply with an efficiency of 85 per cent and a power factor of 0.8 when the output power is 5 kW. Calculate (a) the current, (b) the active and reactive components of the current and (c) the reactive kilovoltamperes taken by the motor.

17. A motor developing 1.5 kW is taking 12 A at power factor 0.75 from a 200-V supply. Calculate its efficiency.

18. Explain the operation of a magnetic amplifier and the purpose and application of both bias and positive feedback.

CHAPTER 13
▉ Capacitance in an a.c. Circuit

13.1 Alternating current in a circuit possessing capacitance only

Figure 13.1 shows a capacitor C connected in series with an ammeter A

Fig. 13.1 Circuit with capacitance only

across the terminals of an alternator; and the alternating voltage applied to C is represented in fig. 13.2. Suppose this voltage to be positive when it makes D positive relative to E.

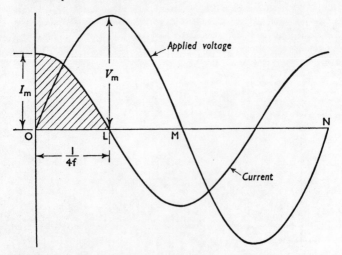

Fig. 13.2 Voltage and current waveforms for capacitive circuit

If the p.d. across the capacitor increases by v volts in t seconds and if the average charging current is i amperes, the increase of charge is $i \times t$ coulombs. If the capacitance is C farads, the increase of charge is also Cv coulombs,

$$\therefore \quad it = Cv$$

and

$$i = C \times v/t = C \times \text{rate of change of p.d.}$$

In fig. 13.2, the p.d. is increasing at the maximum rate at instant O; consequently the charging current is also at its maximum value I_m at that instant.

At instant L, the applied voltage has reached its maximum value V_m; and for a very brief interval of time, the p.d. is neither increasing nor decreasing, so that there is no current. During interval LM, the applied voltage is decreasing. Consequently the capacitor discharges, and the discharge current is in the negative direction.

At instant M, the slope of the voltage curve is at its maximum, i.e. the p.d. is varying at the maximum rate; consequently the current is also a maximum at that instant.

From a comparison of the voltage and current curves in fig. 13.2 it is seen that the current is leading the voltage by a quarter of a cycle. A simple mechanical analogy may again assist in understanding this relationship.

13.2 Analogies of capacitance in an a.c. circuit

If the piston P in fig. 9.9 be moved backwards and forwards, the to-and-fro movement of the water causes the diaphragm to be distended in alternate directions. This hydraulic analogy, when applied to capacitance in an a.c. circuit, becomes rather complicated owing to the inertia of the water and of the piston; and as we do not want to take the effect of inertia into account at this stage, it is more convenient to consider a very light flexible strip L (fig. 13.3), such as a metre rule, having one end rigidly clamped. Let

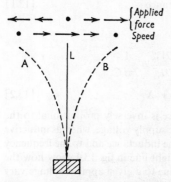

Fig. 13.3 Mechanical analogy of a purely capacitive circuit

us apply an alternating force comparatively slowly by hand so as to oscillate L between poisitions A and B.

When L is in position A, the applied force is at its maximum towards the *left*. As the force is relaxed, L moves towards the *right*. Immediately L has passed the centre position the applied force has to be increased towards the right, while the speed in this direction is decreasing. These variations are indicated by the lengths of the arrows in fig. 13.3. From the latter it is seen that the speed towards the right is a maximum a quarter of a cycle before the applied force is a maximum in the same direction. The speed is therefore the analogue of the alternating current, and the applied force is that of the applied voltage.

13.3 Numerical relationship between current and voltage in a purely capacitive circuit

At instant L in fig. 13.2, the p.d. across C is V_m so that the charge on C is CV_m coulombs. *But the charge is also equal to the product of the average current during interval OL and the* duration of that interval; hence, if I_m is the maximum value of the current, the average current is $(2/\pi)I_m$ (section 10.4)

and

$$CV_m = \frac{2}{\pi} \times I_m \times \frac{1}{4f}$$

where f = frequency in hertz;

$$\therefore \quad \frac{V_m}{I_m} = \frac{1}{2\pi fC} = X_C = \text{capacitive reactance} \qquad [13.1]$$

Hence,

$$\frac{\text{r.m.s. value of p.d.}}{\text{r.m.s. value of current}} = \frac{V}{I} = \frac{0.707V_m}{0.707I_m} = \frac{1}{2\pi fC}$$

and

$$I = 2\pi fCV = V/X_C. \qquad [13.2]$$

It is seen that capacitive reactance is inversely proportional to the capacitance and to the frequency of the supply voltage, whereas inductive reactance is directly proportional to the inductance and to the frequency (section 11.4). The hyperbola and straight line in fig. 13.4 show how the capacitive reactance and the current due to a given applied voltage vary with the frequency.

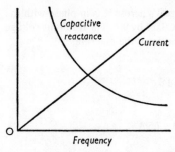

Fig. 13.4 Effect of frequency on capacitive reactance and current

Example 13.1 *A 30-μF capacitor is connected across a 400-V, 50-Hz supply. Calculate* (a) *the reactance of the capacitor and* (b) *the current.*

From [13.1],

$$\text{reactance} = \frac{1}{2 \times 3.14 \times 50 \times 30 \times 10^{-6}}$$

$$= 106.2 \, \Omega.$$

From [13.2]:

current $= 400/106.2 = 3.77$ A.

Figure 13.5 represents the voltage and current phasors for this circuit.

Fig. 13.5 Phasor diagram for a purely capacitive circuit

Example 13.2 *A metal-filament lamp, rated at 750 W, 100 V, is to be connected in series with a capacitor across a 230-V, 60-Hz supply. Calculate* (a) *the capacitance required and* (b) *the power factor of the combination.*

(a) The circuit is given in fig. 13.6, where R represents the lamp. In the

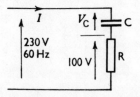

Fig. 13.6 Circuit diagram for example 13.2

phasor diagram of fig. 13.7, the voltage V_R across R is in phase with the current I, while the voltage V_C across C lags by 90°. The resultant voltage V is the phasor sum of V_R and V_C and from the diagram:

$$V = V_R^2 + V_C^2$$

$$\therefore \quad (230)^2 = (100)^2 + V_C^2$$

$$\therefore \quad V_C = 207\,\text{V}.$$

Rated current of lamp $= 750/100 = 7.5\,\text{A}$.

From [13.2]

$$7.5 = 2 \times 3.14 \times 60 \times C \times 207$$

$$\therefore \quad C = 96 \times 10^{-6}\,\text{F}$$

$$= 96\,\mu\text{F}.$$

(b)   Power factor $= \cos\phi = V_R/V$ (from fig. 13.7)

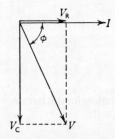

Fig. 13.7   Phasor diagram for example 13.2

$$= 100/230 = 0.435\text{ leading}.$$

13.4 Alternating current in a circuit possessing resistance, inductance and capacitance in series (fig. 13.8)

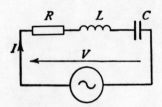

Fig. 13.8   Circuit with R, L and C in series

Let V and I be the r.m.s. values of the supply voltage and current respectively. The p.d. across R is RI volts in phase with the current and is represented by OA in fig. 13.9. From expression [11.4], the p.d. across L is $2\pi f LI$, leading the current by 90°, and is represented by phasor OB. From expression [13.2], the p.d. across C is $1/(2\pi f C)$, lagging the current by 90°, and is represented by phasor OC. Since OB and OC are in direct opposition, their resultant is $OD = OB - OC$, OB being assumed greater than OC in fig. 13.9; and the supply voltage is the phasor sum of OA and

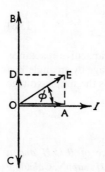

Fig. 13.9 Phasor diagram for fig. 13.8

OD, namely OE. From the phasor diagram:

$$OE^2 = OA^2 + OD^2 = OA^2 + (OB - OC)^2$$

$$\therefore \quad V^2 = (RI)^2 + \left(2\pi f LI - \frac{1}{2\pi f C}\right)^2$$

so that

$$V = I \sqrt{\left\{R^2 + \left(2\pi f L - \frac{1}{2\pi f C}\right)^2\right\}}$$

and

$$I = \frac{V}{\sqrt{\left\{R^2 + \left(2\pi f L - \frac{1}{2\pi f C}\right)^2\right\}}} = \frac{V}{Z} \qquad [13.3]$$

where

$Z = impedance$ of circuit in ohms.

From expression [13.3] it is seen that:

$$\text{resultant reactance} = 2\pi f L - \frac{1}{2\pi f C}$$

$$= \text{inductive reactance} - \text{capacitive reactance}.$$

If ϕ = phase difference between the current and the supply voltage,

$$\tan \phi = \frac{AE}{OA} = \frac{OD}{OA} = \frac{OB - OC}{OA} = \frac{2\pi f L I - I/(2\pi f C)}{RI}$$

$$= \frac{\text{inductive reactance} - \text{capacitive reactance}}{\text{resistance}} \qquad [13.4]$$

$$\cos \phi = \frac{OA}{OE} = \frac{RI}{ZI} = \frac{\text{resistance}}{\text{impedance}} \qquad [13.5]$$

$$\sin \phi = \frac{AE}{OE} = \frac{\text{resultant reactance}}{\text{impedance}} \qquad [13.6]$$

If the inductive reactance is greater than the capacitive reactance, $\tan \phi$ is positive and the current lags the applied voltage by an angle ϕ; if less, $\tan \phi$ is negative, signifying that the current leads the supply voltage by an angle ϕ.

Example 13.3 *A resistance of* 12 Ω, *an inductance of* 0.15 H *and a capacitance of* 100 μF *are in series across a* 100-V, 50-Hz *supply. Calculate* (a) *the impedance,* (b) *the current,* (c) *the voltages across resistance, inductance and capacitance respectively,* (d) *the power factor, and* (e) *the power.*

The circuit diagram is the same as that of fig. 13.8.

(a) From [13.3]:

$$Z = \sqrt{\left\{ 12^2 + \left(2 \times 3.14 \times 50 \times 0.15 - \frac{10^6}{2 \times 3.14 \times 50 \times 100} \right)^2 \right\}}$$

$$= \sqrt{\{144 + (47.1 - 31.85)^2\}} = 19.4\,\Omega.$$

(b) $\text{Current} = \dfrac{V}{Z} = \dfrac{100}{19.4} = 51.5\,\text{A}.$

(c) Voltage across $R = V_R$

$$= 12 \times 5.15 = 61.8\,\text{V}.$$

Voltage across $L = V_L$

$$= 47.1 \times 5.15 = 242.5\,\text{V}.$$

Voltage across $C = V_C$

$$= 31.85 \times 5.15 = 164\,\text{V}.$$

These voltages and current are represented by the phasors of fig. 13.10, and the significance of the voltages across the inductance and the capacitance being greater than the applied voltage is explained in section 13.5.

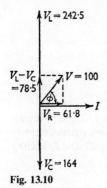

Fig. 13.10

(d) Power factor $= \cos \phi$

$$= \frac{V_R}{V} = \frac{61.8}{100} = 0.618.$$

Since the current is lagging the supply voltage, the power factor is said to be 0.618 lagging.

(e) Power $= IV \cos \phi$

$$= 5.15 \times 100 \times 0.168 = 318 \text{ W}.$$

Or, alternatively:

$$\text{power} = I^2 R = (5.15)^2 \times 12 = 318 \text{ W}.$$

Example 13.4 *Circuits having a resistance of 50 Ω, an inductance of 0.15 H and a capacitance of 100 μF respectively are in parallel across a 100-V, 50-Hz supply. Calculate* (a) *the current in each circuit,* (b) *the resultant current,* (c) *the power factor and* (d) *the power.*

(a) The circuit diagram is given in fig. 13.11, where I_R, I_L and I_C represent the currents through the resistor, inductor and capacitor respectively.

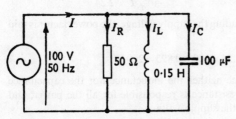

Fig. 13.11 Circuit diagram for example 13.4

$$I_R = 100/50 = 2\,\text{A}$$

$$I_L = \frac{100}{2 \times 3.14 \times 50 \times 0.15} = 2.125\,\text{A}$$

and

$$I_C = 2 \times 3.14 \times 50 \times 100 \times 10^{-6} \times 100 = 3.14\,\text{A}.$$

In the case of parallel circuits, the first phasor (fig. 13.12) to be drawn is that representing the quantity that is common to those circuits, namely the voltage. I_R is then drawn in phase with V, I_L lagging $90°$ and I_C leading $90°$.

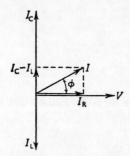

Fig. 13.12   Phasor diagram for example 13.4

(b)   The resultant of I_C and $I_L = I_C - I_L$

$$= 3.14 - 2.125$$

$$= 1.015\,\text{A, leading } V \text{ by } 90°.$$

The current I taken from the alternator is the resultant of I_R and $(I_C - I_L)$, and from fig. 13.12:

$$I^2 = I_R^2 + (I_C - I_L)^2 = 2^2 + (1.015)^2 = 5.03$$

$$\therefore \quad I = 2.24\,\text{A}.$$

(c)   Power factor of the combined circuits $= \cos \phi$

$$= \frac{I_R}{I} = \frac{2}{2.24} = 0.893.$$

Since the current is leading the supply voltage, the power factor is said to be 0.893 leading.

(d)   Power $= IV \cos \phi = 2.24 \times 100 \times 0.893 = 200\,\text{W}.$

Or alternatively, since neither the inductance nor the capacitance absorbs any power, the resistance is responsible for all the power; and since I_R is in phase with the supply voltage,

$$\text{power} = I_R V = 2 \times 100 = 200\,\text{W}.$$

13.5 Electrical resonance in a series circuit

In the preceding section it was shown that for a circuit having resistance, inductance and capacitance in series, the resultant reactance is $[2\pi f L - (1/2\pi f C)]$. Consequently, if we arrange the values of f, L and C so that the inductive reactance and the capacitive reactance neutralize each other, the resultant reactance is zero and the impedance is the same as the resistance. Under these circumstances it is possible for each of the voltages across the inductance and the capacitance to be many times the supply voltage, and the circuit is then said to be in *resonance*, i.e. in a state of electrical vibration. This effect is best illustrated by an example.

Example 13.5 *A circuit has a resistance of 4 Ω, an inductance of 0.5 H and a variable capacitance in series and is connected across a 100-V, 50-Hz supply. Calculate* (a) *the capacitance to give resonance and* (b) *the voltages across the inductance and the capacitance.*

(a) For resonance:

inductive reactance = capacitive reactance

i.e.
$$2\pi f L = 1(2\pi f C)$$

$$\therefore \quad C = \frac{1}{4 \times (3.14)^2 \times (50)^2 \times 0.5}$$

$$= 20.3 \times 10^{-6} \text{ F}$$

$$= 20.3 \; \mu\text{F}.$$

(b) At resonance:

impedance = resistance = 4 Ω

$$\therefore \quad I = V/R = 100/4 = 25 \text{ A}.$$
From [11.4]:

p.d. across inductance = $V_L = 2 \times 3.14 \times 50 \times 0.5 \times 25$

$$= 3925 \text{ V}$$

p.d. across capacitance = V_C = p.d. across inductance

$$= 3925 \text{ V}.$$

Or, alternatively, from [13.2]:

$$V_C = \frac{25 \times 10^6}{2 \times 3.14 \times 50 \times 20.3} = 3925 \text{ V}.$$

The voltages and current are represented by the respective phasors of fig. 13.13; and fig. 13.14 shows how the current taken by this circuit varies with the frequency, the applied voltage being assumed constant at 100 V.

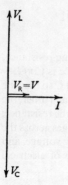

Fig. 13.13 Phasor diagram for series circuit at resonance

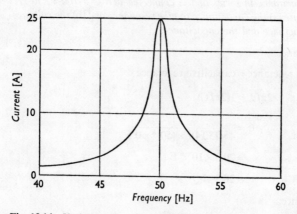

Fig. 13.14 Variation of current with frequency for circuit of example 13.5

In the above example, the voltages across the inductance and the capacitance at resonance are each nearly forty times the supply voltage. This voltage ratio is termed the *Q-factor* of a series circuit, i.e.

$$Q\text{-factor of a series circuit} = \frac{\text{voltage across } L \text{ (or } C)}{\text{supply voltage}}$$

$$= \frac{2\pi f L I}{R I} = \frac{2\pi f L}{R} \qquad [13.7]$$

= voltage magnification obtained at resonance.

A fuller explanation, however, is necessary to understand the physical significance of this voltage magnification.

13.6 Natural frequency of oscillation of a circuit possessing inductance and capacitance in series

Suppose C in fig. 13.15 to be a capacitor whose capacitance can be varied from, say, 1 to 20 microfarads, and suppose L to be an inductor whose inductance is variable between, say, 0.1 and 0.01 henry. A loudspeaker P is

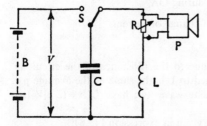

Fig. 13.15 Capacitor discharged through an inductor

connected across a variable resistor R. A two-way switch S enables C to be charged from a battery B and discharged through L, and R is adjusted to give a convenient volume of sound in P. It is found that each time C is discharged through L, a pizzicato note is emitted by P, similar to the sound produced by plucking the string of a violin or 'cello. Further, the pitch of the note can be varied by varying either C or L—the larger the capacitance and the inductance, the lower the pitch. But it is well known that a musical sound requires the vibration of some medium for its production and that the pitch is dependent upon the number of vibrations per second. Hence it follows that the discharge current through P is an alternating current of diminishing amplitude as shown by the full line in fig. 13.16, the p.d. across the capacitor being represented by the dotted line.

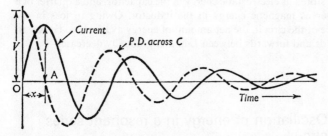

Fig. 13.16 Voltage and current waveforms for fig. 13.15

From expression [9.20], the energy stored in C at instant O is $\frac{1}{2}CV^2$ joules. At instant A there is no p.d. across and therefore no energy in C; but the current is I amperes, so that the energy stored in L is $\frac{1}{2}LI^2$ joules (section 8.7). If we neglect the energy wasted in the resistance of the circuit during interval OA, then:

$$\tfrac{1}{2}LI^2 = \tfrac{1}{2}CV^2$$

and

$$I = V\sqrt{(C/L)} \qquad\qquad\qquad [13.8]$$

If x seconds be the duration of the quarter-cycle OA:

average e.m.f. induced in L during OA

$$= -L \times \text{average rate of change of current}$$

$$= -L \times I/x \text{ volts.}$$

If the small voltage drop due to the resistance of the circuit be neglected, the average p.d. applied to L is the same as the average p.d. across C. Hence—assuming a sine wave—we have $LI/x = (2/\pi)V$ (see footnote, p. 194).

Substituting for I the value given in expression [13.8]:

$$\frac{LV}{x} \times \sqrt{\left(\frac{C}{L}\right)} = \frac{2}{\pi}V,$$

$$\therefore \quad x = \frac{\pi}{2}\sqrt{(LC)} \text{ seconds,}$$

so that the duration of one cycle $= 4x = 2\pi\sqrt{(LC)}$ seconds, and

frequency = number of cycles/second

$$= \frac{1}{2\pi\sqrt{(LC)}} \text{ hertz.} \qquad\qquad [13.9]$$

This quantity is termed the *natural frequency* of the circuit and represents the frequency with which energy is oscillating backwards and forwards between the capacitor and the inductor, the energy being at one moment stored as electrostatic energy in the capacitor, and a quarter of a cycle later as magnetic energy in the inductor. Owing to loss in the resistance of the circuit, the net amount of energy available to be passed backwards and forwards between L and C gradually decreases.

13.7 Oscillation of energy in a resonant series circuit

In section 13.5 it was explained that resonance occurs in an a.c. circuit when

$$2\pi fL - 1/(2\pi fC) = 0,$$

i.e. when

$$f = \frac{1}{2\pi\sqrt{(LC)}} \qquad\qquad [13.10]$$

From expressions [13.9] and [13.10] it follows that the condition for resonance is that the frequency of the applied alternating voltage is the same as the natural frequency of oscillation of the circuit. This condition enables a large amount of energy to be maintained in oscillation between L and C; thus, if the current and the p.d. across the capacitor in a resonant circuit be represented by the curves of fig. 13.17, the magnetic energy stored in L at instant A is $\frac{1}{2}LI_m^2$ joules, and the electrostatic energy in C at instant B is $\frac{1}{2}CV_{cm}^2$ joules.

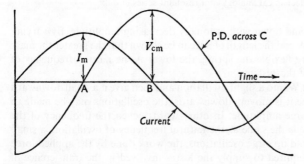

Fig. 13.17 Waveforms of current and p.d. for a capacitor

Since $I_m = 2\pi f C\ V_{cm}$, and from [13.10], $L = 1/(2\pi f)^2 C$,

$$\therefore \quad \tfrac{1}{2}LI_m^2 = \frac{1}{2} \times \frac{1}{(2\pi f)^2 C} \times (2\pi f C V_{cm})^2$$

$$= \tfrac{1}{2}CV_{cm}^2,$$

i.e. the magnetic energy in L at instant A is exactly equal to the electrostatic energy in C at instant B, and the power taken from the supply is simply that required to move this energy backwards and forwards between L and C through the resistance of the circuit.

A series resonant circuit of this type is frequently termed an *acceptor*, since the current is a maximum at resonance. It is much used in radio work.

13.8 Mechanical analogy of a resonant series circuit

It was pointed out in section 11.3 that inertia in mechanics is analogous to inductance in the electric circuit, and in section 9.5 that elasticity is analogous to capacitance. A very simple mechanical analogy of an electrical circuit possessing inductance, capacitance and a very small resistance can therefore be obtained by attaching a mass W (fig. 13.18) to the lower end of a helical spring S, the upper end of which is rigidly supported. If W is pulled down a short distance and then released, it will

Fig. 13.18   Mechanical analogy of a resonant series circuit

oscillate up and down with gradually decreasing amplitude. By varying the mass of W and the length of S it can be shown that the greater the mass and the more flexible the spring, the lower is the natural frequency of oscillation of the system.

If we set W into a slight oscillation and then give it a small downward tap each time it is moving downwards, the oscillations may be made to grow to a large amplitude. In other words, when the frequency of the applied force is the same as the natural frequency of oscillation, a small force can build up large oscillations, the work done by the applied force being that required to supply the losses involved in the transference of energy backwards and forwards between the kinetic and potential forms of energy.

Examples of resonance are very common; for instance, the rattling of some loose member of a vehicle at a particular speed or of a loudspeaker diaphragm when reproducing a sound of a certain pitch, and oscillations of the pendulum of a clock and of the balance wheel of a watch due to the small impulse given regularly through the escapement mechanism from the mainspring.

13.9 Resonance in parallel circuits

Suppose a coil having an inductance L and a very low resistance R to be connected in parallel with a capacitor of capacitance C, as in fig. 13.19.

Current through coil $= I_1 = V/[\sqrt{\{R^2 + (2\pi f L)^2\}}]$; and from [11.8], the phase angle ϕ between I_1 and the applied voltage is such that tan $\phi = 2\pi f L/R$. Since R is very small compared with $2\pi f L$, tan ϕ is very large, so that ϕ is nearly 90°, as shown in fig. 13.20.

Current taken by capacitor $= I_C = 2\pi f CV$, leading V by 90°.

If I_1 and I_C are such that the resultant current I is in phase with the supply voltage, as shown in fig. 13.20, the circuit is said to be in resonance.

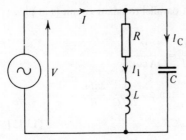

Fig. 13.19 Parallel circuit

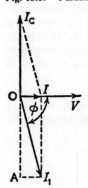

Fig. 13.20 Phasor diagram for fig. 13.19

From fig. 13.20,

$$I_C = OA = I_1 \sin \phi \qquad\qquad [13.11]$$

But

$$\sin \phi = \frac{\text{reactance of coil}}{\text{impedance of coil}} = \frac{2\pi f L}{\sqrt{\{R^2 + (2\pi f L)^2\}}}.$$

Substituting for I_C, I_1 and $\sin \phi$ ([13.11], we have:

$$2\pi f C V = \frac{2\pi f L V}{R^2 + (2\pi f L)^2}$$

$$\therefore \quad f = \frac{1}{2\pi L} \sqrt{(L/C - R^2)}.$$

If R^2 is very small compared with L/C, as in radio circuits,

$$f = \frac{1}{2\pi \sqrt{(LC)}},$$

which is the same as the resonance* frequency of a series circuit (section 13.5).

---

* It is the *circuit* and not the *frequency* that resonates; hence we speak of a *resonant circuit* and of the *resonance frequency* (i.e. the frequency at which the circuit resonates).

Since the resultant current in a parallel resonant circuit is in phase with the supply voltage,

$$\text{impedance of such a circuit} = \frac{V}{I} = \frac{V}{\text{OA} \cot \phi}$$

$$= \frac{V}{I_{\mathrm{C}}} \tan \phi = \frac{1}{2\pi f C} \cdot \frac{2\pi f L}{R}$$

$$= L/CR. \qquad\qquad [13.12]$$

This means that a parallel resonant circuit is equivalent to a non-reactive resistor of $L/(CR)$ ohms. This quantity is often termed the *dynamic impedance* of the circuit; and it is obvious that the lower the resistance of the coil, the higher is the dynamic impedance of the parallel circuit. This type of circuit, when used in radio work, is referred to as a *rejector* since its impedance is a maximum and the resultant current a minimum at resonance.

If the alternator voltage in fig. 13.19 is kept constant but its frequency varied, the current decreases to a minimum at the resonance frequency; thus fig. 13.21 represents the variation of the resultant current for an

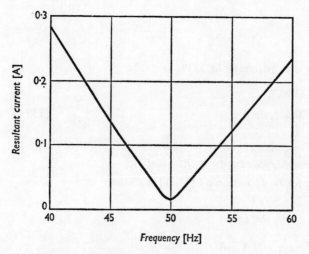

Fig. 13.21   Resonance curve for a rejector

inductor having a resistance of $4\,\Omega$ and an inductance of 0.5 H connected in parallel with a 20.3-μF capacitor across a constant voltage of 100 V, the values being the same as those of the series circuit of example 13.5. Since figs. 13.14 and 13.21 refer to circuits having the same values of R, L and C, it is of interest to note that a change of 10 per cent in the frequency from the resonance value reduces the current to about an eighth in the series circuit and increases the resultant current about eight times when the circuits are in parallel.

It is evident from fig. 13.20 that when the coil has a low resistance, the current circulating in L and C, at resonance, is much larger than the resultant current taken from the supply. The current magnification at resonance is termed the Q-factor of a parallel circuit, i.e.

$$Q\text{-factor of a parallel circuit} = \frac{\text{current in capacitor}}{\text{resultant current}}$$

$$= \frac{I_C}{I} = \frac{I_1 \sin \phi}{I} = \frac{\sin \phi}{\cos \phi}$$

$$= \tan \phi = 2\pi f L/R,$$

namely the same relationship as that given in expression [13.7] for the Q-factor of a series circuit. Hence the Q-factor is a measure of voltabe magnification in a series circuit and of current magnification in a parallel circuit.

13.10 Bandwidth

Both the series and the parallel LCR circuits have responses which change when the frequency changes with the current being a maximum or a minimum at a particular frequency, the resonant frequency. Thus, for example, fig. 13.14 shows how the current varies with frequency for a series LCR circuit. The curve is symmetrical about the resonance frequency. A term used to describe the frequency response of such a circuit is the bandwidth.

The *bandwidth* **B** is the separation between the frequencies at which the power developed by the circuit falls to one half of the maximum value it has at resonance. The points at which this occurs are called the *half-power points*.

The power developed at resonance is proportional to I_{max}^2, where I_{max} is the current at resonance—a maximum in the case of a series circuit. Thus at the half-power points the power is proporttional to $\frac{1}{2}I_{max}^2$. The current at which this occurs is given by

$$I^2 = \frac{1}{2}I_{max}^2$$

$$I = \sqrt{(\frac{1}{2}I_{max}^2)}$$

$$I = \frac{I_{max}}{\sqrt{2}}.$$

Thus the half-power points are when the current has fallen to $1/\sqrt{2}$ of its maximum value. It can also be shown that the bandwidth is dependent on the Q-factor,

$$B = \frac{f_r}{Q}$$

where f_r is the resonant frequency. As the Q-factor increases then the bandwidth decreases and the circuit becomes more selective. Since $Q = 2\pi f_r L/R$ or $\frac{1}{2}\pi f_r CR$ (see equation [13.7]), then high values of inductance and low values of capacitance produce a high Q and low bandwidth, hence high selectivity. A low value of resistance also produces a high Q-factor and so low bandwidth and high selectivity.

13.11 Coupled circuits

A transformer at its simplest consists of two coils with a changing current in one coil, the primary coil, inducing an e.m.f. in the other coil, the secondary. Thus without physical connections between the two coils there is an interaction. Such coils are said to be coupled.

Consider two LCR circuits with the same resonant frequency, as in fig. 13.22, with the two inductance coils coupled. If the frequency of the

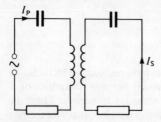

Fig. 13.22

applied voltage in the primary circuit is varied and the current in that circuit I_p plotted as a function of frequency, a graph of the form shown in fig. 13.23 is obtained. Instead of just the single resonant frequency that would have been obtained by either circuit alone there are now two resonant frequencies.

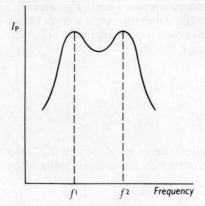

Fig. 13.23

The effect of the coupling is to increase the bandwidth. The increase in the bandwidth can be controlled by varying the degree of coupling. The terms *loose* and *tight* are used to describe the coupling, a loose coupling being when there is little interaction between the circuits and the bandwidth is not increased very much. With tight coupling there is considerable interaction between the circuits and a significant increase in the bandwidth. Thus by changing the coupling the bandwidth can be controlled.

13.12 Filter circuits

A filter is a network that is designed to transmit signals having a certain range or ranges of frequencies, referred to as the *pass band*, and suppress signals of other frequencies, the suppressed bands being referred to as the *attenuation band* or bands. The term *active filter* is used for a network of resistors and capacitors built around a device such as an operational amplifier, the term *passive filter* is used for a network of resistors, capacitors and inductors without such amplification. In this book the discussion is restricted to passive filters.

Standard types of filter are:

1. *Low pass.* This filter allows low frequencies to pass and absorbs high frequencies, i.e. pass band 0 to f_c, attenuation band f_c to infinity (f_c is the cut-off frequency).
2. *Band pass.* This filter allows a selected band of frequencies to pass and absorbs frequencies both below and above this band, i.e. pass band f_1 to f_2, attenuation bands 0 to f_1 and f_2 to infinity.
3. *Band stop* (or band reject). This filter absorbs a selected band of frequencies and allows frequencies both below and above this band to pass, i.e. pass bands 0 to f_1 and f_2 to infinity, attenuation band f_1 to f_2.
4. *High pass.* This filter allows high frequencies to pass and absorbs low frequencies, i.e. pass band f_c to infinity and attenuation band 0 to f_c.

Figure 13.24 shows the types of circuits used to give the above filters and their characteristics.

13.13 Summary of important formulae

$$\text{Capacitive reactance} = X_C = 1/2\pi f C \qquad [13.1]$$

For circuit with R, L and C in series,

$$\text{impedance} = Z = \sqrt{\left\{R^2 + \left(2\pi f L - \frac{1}{2\pi f C}\right)^2\right\}} \qquad [13.3]$$

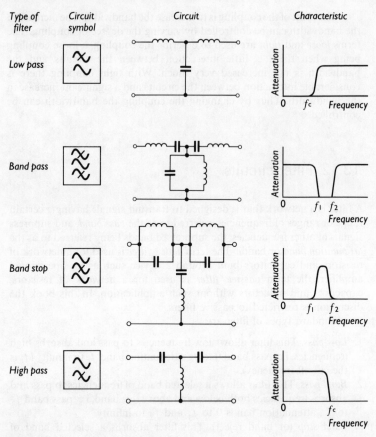

Type of filter	Circuit symbol	Circuit	Characteristic

Fig. 13.24

$$\tan \phi = \frac{2\pi f L - 1/(2\pi f C)}{R} \qquad [13.4]$$

$$\cos \phi = R/Z \qquad [13.5]$$

$$\sin \phi = \frac{2\pi f L - 1/(2\pi f C)}{Z}. \qquad [13.6]$$

$$Q\text{-factor} = 2\pi f L/R. \qquad [13.7]$$

When a capacitor, charged to a p.d. of V volts, is discharged through an inductor of low resistance,

$$\text{peak value of initial current} = V\sqrt{(C/L)} \qquad [13.8]$$

and

$$\text{frequency of oscillations} = \frac{1}{2\pi\sqrt{(LC)}}. \qquad [13.9]$$

For resonance in a series circuit,

$$f = \frac{1}{2\pi\sqrt{(LC)}}.$$ [13.10]

For resonance in a parallel circuit,

$$f = \frac{1}{2\pi L} \sqrt{(L/C - R^2)}$$

$$= \frac{1}{2\pi\sqrt{(LC)}} \text{ when } R^2 \ll L/C.$$

Dynamic impedance of parallel resonant circuit $= L/CR$ [13.12]

13.14 Examples

1. A 50-μF capacitor is connected across a 110-V, 30-Hz supply. Calculate (a) the reactance and (b) the current.

2. A 20-Ω resistor and a 100-μF capacitor are connected in series across a 200-V, 50-Hz supply. Calculate (a) the current, (b) the p.d.s across the resistor and the capacitor, (c) the phase difference between the current and the applied voltage, (d) the power and (e) the power factor. Sketch the phasor diagram.

3. A current of 4.8 A is taken when a 30-Ω non-reactive resistor and a loss-free capacitor are connected in series across a 240-V, 50-Hz supply. Calculate the impedance of the circuit, the capacitance of the capacitor, the power and the circuit power factor.

4. It is desired to operate a 100-W, 120-V lamp at its rated voltage from a 240-V, 50-Hz supply. Calculate the capacitance of a capacitor connected in series with the lamp to enable this to be accomplished. What would be the circuit power factor?

5. A 130-Ω resistor and a 30-μF capacitor are connected in parallel across a 200-V, 50-Hz supply. Calculate (a) the current in each circuit, (b) the resultant current, (c) the phase difference between the resultant current and the applied voltage, (d) the power and (e) the power factor.

6. A loss-free capacitor is connected in parallel with a 200-Ω non-reactive resistor. The combination takes a current of 2 A when connected across a 250-V, 50-Hz supply. Calculate the capacitance and the power factor of the combination.

7. A circuit has a resistance of 10 Ω, an inductance of 0.1 H and a capacitance of 50 μF in series across a 100-V, 50-Hz supply. Calculate (a) the current, (b) the p.d.s across R, L and C respectively, (c) the phase angle between the current and the applied voltage, (d) the power factor and (e) the power.

8. A coil having a resistance of 10 Ω and an inductance of 0.05 H is

connected in series with a 150-μF capacitor across a 250-V, 50-Hz supply. Calculate (a) the voltage across the coil and (b) the power factor of the whole circuit, stating whether it is lagging or leading.

9. A coil is connected in series with a 60-μF capacitance across a 200-V 50-Hz supply. The current is 3 A and the power absorbed is 144 W. Calculate (a) the p.d. across the capacitor, (b) the resistance and inductance of the coil, (c) the power factor of the coil and (d) the power factor of the whole circuit.

10. A coil having a resistance of 5 Ω and an inductance of 0.2 H is connected in series with a variable capacitor across a 30-V, 50-Hz supply. Calculate the capacitance required to produce resonance and the corresponding values of (a) the current, (b) the voltages across the coil and the capacitor, (c) the power factor and (c) the Q-factor.

11. A circuit consists of a coil having a resistance of 20 Ω and an inductance of 0.1 H connected in series with a 0.1 μF capacitor. Calculate (a) the frequency at which the circuit takes the maximum current, (b) the value of the current and of the power assuming the supply voltage to be 1 V, and (c) the Q-factor.

12. A circuit consists of a 10-Ω resistor, a 300-μF capacitor and an inductor having an inductance of 0.06 H and a negligible resistance connected in parallel across a 100-V, 25-Hz supply. Calculate (a) the current in each branch, (b) the resultant current, (c) the power factor of the whole circuit and (d) the power.

13. An inductor takes 10 A and dissipates 1920 W when connected across a 240-V, 50-Hz supply. When a loss-free capacitor is connected in parallel with the inductor, the magnitude of the supply current remains unaltered. Draw the phasor diagram and calculate the capacitance of the capacitor.

14. A coil has a resistance of 2 Ω and an inductance of 0.07 H. Calculate the capacitance of a capacitor required to produce resonance when connected in parallel with the coil across a 50-Hz supply. What is the value of the Q-factor?

15. A coil having a resistance of 10 Ω and an inductance of 0.1 H is connected in parallel with a 50-μF capacitor across a variable-frequency source. Calculate the frequency at which the resultant current is in phase with the supply voltage and the corresponding value of the Q-factor.

    If the corresponding value of the supply voltage is 100 V, calculate (a) the current in each circuit, (b) the resultant current and (c) the power factor of the whole circuit.

16. A 2-μF capacitor is charged to a p.d. of 200 V and then connected across a coil having an inductance of 0.05 H and negligible resistance. Calculate (a) the maximum value of the current in the coil and (b) the frequency of the oscillations.

17. A single-phase motor takes 50 A at a power factor of 0.6 lagging from a 250-V, 50-Hz supply. Calculate the capacitance of a shunting capacitor required to raise the power factor to 0.9 lagging.

18. A single-phase motor takes 15 A from a 230-V, 50-Hz supply at a

power factor of 0.7 lagging. An 80-μF capacitor is connected in parallel with the motor. Determine (a) the resultant current and (b) the power factor of the combination.

CHAPTER 14
Direct-Current Machines

14.1 General arrangement of a d.c. machine

Figure 14.1 shows the general arrangement of a four-pole d.c. generator or motor. The fixed part consists of four iron cores C, referred to as *pole cores*, attached to an iron or steel ring R, called the *yoke*. The pole cores are

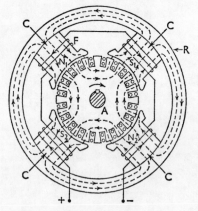

Fig. 14.1   General arrangement of a four-pole d.c. machine

usually made of steel plates riveted together and bolted to the yoke, which may be of cast steel or fabricated rolled steel. Each pole core has pole tips, partly to support the field winding and partly to increase the cross-sectional area of the airgap and thus reduce its reluctance. Each pole core carries a winding F so connected as to excite the poles alternately N and S.

14.2 Armature

The rotating part of a d.c. machine is referred to as the *armature*. The armature core, labelled A in fig. 14.1, consists of iron laminations, about 0.4 to 0.6 mm thick, insulated from one another and assembled on the shaft in the case of small machines and on a cast-iron spider in the case of large machines. If the core were made of solid iron, as shown in fig. 14.2(a) for a two-pole machine, then if the armature were rotated clockwise when

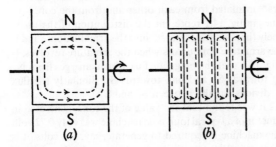

Fig. 14.2   Eddy currents

viewed from the right-handside of the machine, e.m.f.s would be generated in the core in exactly the same way as they are generated in conductors placed on the armature, and these e.m.f.s would circulate currents—known as *eddy currents*—in the core, as shown dotted in fig. 14.2(a). Owing to the very low resistance of the core, these eddy currents would be considerable and would cause a large loss of power in and excessive heating of the armature.

If the core is made of laminations insulated from one another, the eddy currents are confined to their respective sheets, as shown in fig. 14.2(b), and the eddy-current loss is thereby reduced. Thus if the core is split up into five laminations, the e.m.f. per lamination is only a fifth of that generated in the solid core, and the cross-sectional area per path is also reduced to about a fifth, so that the resistance per path is roughly five times that of the solid core. Consequently the current per path is about $\frac{1}{25}$th of that in the solid core. Hence:

$$\frac{I^2R \text{ loss per lamination}}{I^2 \text{ loss in solid core}} = \left(\frac{1}{25}\right)^2 \times 5 = \frac{1}{125}\text{(approx.)}.$$

Since there are five laminations, then:

$$\frac{\text{total eddy current loss in laminated core}}{\text{total eddy current loss in solid core}} = \frac{5}{125} = \left(\frac{1}{5}\right)^2.$$

It follows that the eddy-current loss is approximately proportional to the square of the thickness of the laminations. Hence the eddy-current loss can be reduced to any desired value, but if the thickness of the laminations is made less than about 0.4 mm, the reduction in the loss does not justify the extra cost of construction. Eddy-current loss can also be reduced considerably by the use of silicon–iron alloy—usually about 4 per cent of silicon—due to the resistivity of this alloy being much higher than that of ordinary iron.

Slots are stamped on the periphery of the laminations, partly to accommodate and provide mechanical security to the armature winding and partly to give a shorter airgap for the magnetic flux to cross between the pole-face and the armature 'teeth'. In fig. 14.1, each slot has two circular conductors*, insulated from each other and from the core.

The dotted lines in fig. 14.1 represent the distribution of the *useful* magnetic flux, namely that flux which passes into the armature core and is therefore cut by the armature conductors when the armature revolves. It will be seen from fig. 14.1 that the magnetic flux which emerges from N_1 divides, half going towards S_1 and half towards S_2. Similarly the flux emerging from N_2 divides equally between S_1 and S_2.

In example 6.2 it was found that the value of the e.m.f. generated in one armature conductor of a typical four-pole machine was only 0.72 volt. Consequently, if the machine is required to generate, say, 500 volts, it is necessary to use a large number of conductors and to connect them in such a way that their e.m.f.s help one another, and at the same time to arrange for the magnetic field to be as intense and the speed as high as is practicable.

In fig. 10.5 we considered a conductor A situated on a revolving armature and passing successively under N and S poles. The alternating e.m.f. generated in the conductor varied through one cycle every time the conductor passed a pair of poles. A d.c. generator, however, has to give a voltage that remains constant in direction and as constant as possible in magnitude. It is therefore necessary to use a *commutator* to enable a steady or direct voltage to be obtained from the alternating e.m.f. generated in the rotating conductors.

Figure 14.3 shows a longitudinal or axial section and an end elevation of half of a relatively small commutator. It consists of a large number of wedge-shaped copper segments or bars C, assembled side by side to form a ring, the segments being insulated from one another by thin mica sheets P.

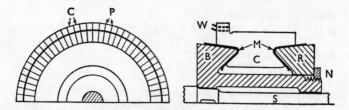

Fig. 14.3   Commutator of a d.c. machine

---

* The term *conductor*, when applied to armature windings, refers to the active portion of the winding, namely that part which cuts the flux, thereby generating an e.m.f.; for example, if an armature has 40 slots and if each slot contains 8 wires, the armature is said to have 8 conductors per slot and a total of 320 conductors.

The segments are shaped as shown so that they can be clamped securely between a V-ring B, which is part of a cast-iron bush or sleeve, and another V-ring R which is tightened and kept in place by a nut N. The bush is keyed to shaft S.

The copper segments are insulated from the V-rings by collars of micanite M, namely thin mica flakes cemented together with shellac varnish and moulded to the exact shape of the rings. These collars project well beyond the segments so as to reduce surface leakage of current from the commutator to the shaft. At the end adjacent to the winding, each segment has a milled slot to accommodate two armature wires W which are soldered to the segment.

14.3 Ring-wound armature

The action of the commutator is most easily understood if we consider the earliest form of armature winding, namely the ring-wound armature shown in fig. 14.4. In this diagram, C represents a core built of sheet-iron

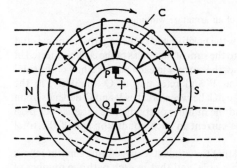

Fig. 14.4 Ring-wound armature

rings or laminations, insulated from one another. The core is wound with eight coils, each consisting of two turns; and the two ends of any one coil are connected to adjacent segments of the commutator. P and Q represent two carbon brushes, namely two blocks of specially treated carbon, bearing on the commutator. Actually these brushes are pressed by springs against the outer surface of the segments, but to avoid confusion in fig. 14.4 they are shown on the inside. The term 'brush' is a relic of the time when current was collected by a bundle of copper wires arranged somewhat like a brush.

The dotted lines in fig. 14.4 represent the distribution of the magnetic flux; and it will be seen that this magnetic flux is cut only by that part of the winding that lies on the external surface of the core. It is also evident from fig. 14.4 that between brushes Q and P, there are two paths in parallel through the armature winding and that all the conductors are divided equally between these two paths. Furthermore, as the armature rotates, this state of affairs remains unaltered.

Suppose the armature to be driven clockwise. Then, by applying the Right-hand Rule of section 6.4, we find that the direction of the e.m.f.s generated in conductors under the N pole is towards the paper, and that of the e.m.f.s in the conductors under the S pole is outwards from the paper, as indicated by the arrowheads. The result is that in each of the two *parallel* paths, the same number of conductors is generating e.m.f. acting from Q towards P. The effect is similar to that obtained if equal numbers of cells (in series) were connected in two parallel circuits, as shown in fig. 14.5.

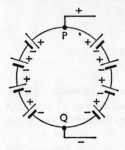

Fig. 14.5　Equivalent circuit of an armature winding

In this case it is clear that P is the positive and Q the negative terminal, and that the total e.m.f. is equal to the sum of the e.m.f.s of the cells connected in series between Q and P. Also, the current in each path is a half of the total current flowing outwards at P and returning at Q. Similarly, in the armature winding, the total e.m.f. between brushes P and Q depends upon the e.m.f. per conductor and the number of conductors in series per path between Q and P; and the current in each conductor is half the total current at each brush.

It follows from fig. 14.4 that the conductors generating e.m.f. are those which are moving opposite a pole, and that in each path the number of conductors simultaneously generating e.m.f. remains practically constant from instant to instant and is unaffected by the rotation of the armature. Hence, the p.d. between P and Q must also remain practically constant from instant to instant—a result made possible by the commutator.

The ring winding of fig. 14.4 has the disadvantages: (1) it is very expensive to wind, since each turn has to be taken round the core by hand; (2) only a small portion of each turn is effective in cutting magnetic flux. Hence, modern armatures are 'drum' wound, as shown in fig. 14.6.

14.4　Double-layer drum windings

Let us consider a four-pole armature with, say, 11 slots, as in fig. 14.6. In order that all the coils may be similar in shape and therefore may be

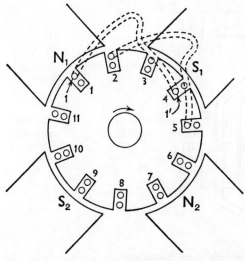

Fig. 14.6 Arrangement of a double-layer winding

wound to the correct shape before being assembled on the core, they have to be made such that if side 1 of a coil occupies the outer half of one slot, the other side 1′ occupies the inner half of another slot. This necessitates a kink in the end-connections in order that the coils may overlap one another as they are being assembled. Figure 14.7 shows the shape of the end-

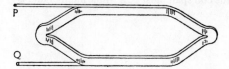

Fig. 14.7 An armature coil

connections of a single coil consisting of a number of turns, and fig. 14.8(b) shows how three coils, 1–1′, 2–2′ and 3–3′, are arranged in the slots so that their end-connections overlap one another, the end elevation of the end-connections of coils 1–1′ and 3–3′ being as shown in fig. 14.8(a). The end-connection of coil 2–2′ has been omitted from fig. 14.8(a) to enable the shape of the other end-connections to be shown more clearly. In fig. 14.7, the two ends of the coil are brought out to P and Q; and as far as the connections to the commutator segments are concerned, the number of turns on each coil is of no consequence. Hence, in winding diagrams it is customary to show only one turn between any pair of segments.

From fig. 14.6 it is evident that if the e.m.f.s generated in conductors 1 and 1′ are to assist each other, 1′ must be moving under a S pole when 1 is moving under a N pole; thus, by applying the Right-hand Rule (section 6.4) to fig. 14.6 and assuming the armature to be rotated clockwise, we find

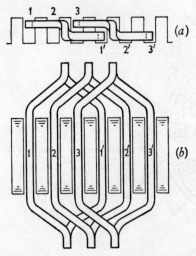

Fig. 14.8   Arrangement of overlap of end-connections

that the direction of the e.m.f. generated in conductor 1 is towards the paper whereas that generated in conductor 1′ is outwards from the paper. Hence, the distance between coil-sides 1 and 1′ must be approximately a pole pitch. With 11 slots it is impossible to make the distance between 1 and 1′ exactly a pole pitch, and in fig. 14.6 one side of coil 1–1′ is shown in slot 1 and the other side is in slot 4. The coil is then said to have a *coil span* of 4–1, namely 3. In practice, the coil span must be a whole number and is approximately equal to

$$\frac{\text{total number of slots}}{\text{total number of poles}}.$$

In the example shown in fig. 14.6 a very small number of slots has for simplicity been chosen. In actual machines the number of slots per pole usually lies between 10 and 15 and the coil span is slightly less than the value given by the above expression. For instance, if a four-pole armature has 47 slots, the number of slots per pole $= 11\frac{3}{4}$, and the coil span would be 11; consequently, if one side of a coil lies in the outer half of slot 6, the other side lies in the inner half of slot $(6+11)$, namely slot 17.

Let us now return to the consideration of the 11-slot armature. The 11 coils are assembled in the slots with a coil span of 3, and we are now faced with the problem of connecting to the commutator segments the 22 ends that are projecting from the winding.

Apart from a few special windings, armature windings can be divided into two groups, depending upon the manner in which the wires are joined to the commutator, namely:

(a) lap windings,
(b) wave windings.

In lap windings the two ends of any one coil are taken to adjacent segments as in fig. 14.9(a), where a coil of two turns is shown; whereas in wave windings the two ends of each coil are bent in opposite directions and taken to segments some distance apart, as in fig. 14.9(b).

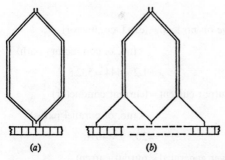

(a)   (b)

Fig. 14.9   (a) Coil of a lap winding; (b) Coil of a wave winding

A lap winding has as many paths in parallel between the negative and positive brushes as there are of poles; for instance, with an 8-pole lap winding, the armature conductors form eight parallel paths between the negative and positive brushes. A wave winding, on the other hand, has only two paths in parallel, irrespective of the number of poles. Hence, if a machine has p pairs of poles,

no. of parallel paths with a lap winding $= 2p$

and

no. of parallel paths with a wave winding $= 2$.

For a given cross-sectional area of armature conductor and a given current density in the conductor, it follows that the total current from a lap winding is $2p$ times the current per conductor whereas the total current from a wave winding is only twice the current per conductor. Hence the total current from a lap winding is p times that from a wave winding. On the other hand, for a given number of armature conductors, the number of conductors in series per path in a wave winding is p times that in a lap winding. Consequently, for a given generated e.m.f. conductor, the voltage between the negative and positive brushes with a wave winding is p times that with a lap winding. It may therefore be said that, in general, lap windings are used for low-voltage, heavy-current machines.

Example 14.1   *A 4-pole armature is wound with 888 conductors. The magnetic flux and speed are such that the average e.m.f. generated in each conductor is 1.2 V. The machine is being driven as a generator. Calculate the terminal voltage on no load and the total power generated when the current per conductor is 20 A if the armature winding is (a) wave-connected and (b) lap-connected.*

(a) With the armature winding wave-connected,

$$\text{no. of parallel paths} = 2$$

$\therefore$ no. of conductors in series per path $= 888/2 = 444$.

Hence,

$$\text{terminal voltage on no load} = (\text{e.m.f./conductor})$$
$$\times (\text{no. of conductors/path})$$
$$= 1.2 \times 444 = 532.8 \text{ V}.$$

$$\text{Output current} = (\text{current/conductor})$$
$$\times (\text{no. of parallel paths})$$
$$= 20 \times 2 = 40 \text{ A}.$$

$\therefore$ electrical power generated $=$ output current
$$\times \text{generated e.m.f.}$$
$$= 40 \times 532.8 = 21\,312 \text{ W}$$
$$= 21.31 \text{ kW}.$$

(b) With the armature winding lap-connected,

$$\text{no. of parallel paths} = \text{no. of poles}$$
$$= 4,$$

$\therefore$ no. of conductors in series per path $= 888/4 = 222$.

Hence,

$$\text{terminal voltage on no load} = 1.2 \times 222 = 266.4 \text{ V}.$$

$$\text{Output current} = 20 \times 4 = 80 \text{ A}.$$

$\therefore$ electrical power generated $= 80 \times 266.4 = 21\,312 \text{ W}$
$$= 21.31 \text{ kW}.$$

It will be seen from the above example that the total power generated for a given e.m.f. per conductor and a given current per conductor is the same whether the armature winding is wave-connected or lap-connected.

14.5 Calculation of e.m.f. generated in an armature winding

When an armature is rotated through one revolution, each conductor cuts the magnetic flux emanating from all the N poles and also that entering all the S poles. Consequently, if

Φ = useful flux per pole, in webers, entering or leaving the armature

p = number of *pairs* of poles

and

N = speed in revolutions/minute, time of 1 revolution

$= 60/N$ seconds

and time taken by a conductor to move one pole pitch

$$= \frac{60}{N} \cdot \frac{1}{2p} \text{ seconds}$$

$\therefore$ average rate at which conductor cuts the flux

$$= \Phi \div \left(\frac{60}{N} \cdot \frac{1}{2p} \right) = \frac{2\Phi N p}{60} \text{ webers/second}$$

and average e.m.f. generated in each conductor

$$= \frac{2\Phi N p}{60} \text{ volts.}$$

If

Z = total number of armature conductors,

c = number of parallel paths through winding between positive and negative brushes

$= 2$ for a wave winding

and

$= 2p$ for a lap winding,

$\therefore$ Z/c = number of conductors in series in each path.

The brushes are assumed to be in contact with segments connected to conductors in which no e.m.f. is being generated, and the e.m.f. generated in each conductor, while it is moving between positions of zero e.m.f., varies as shown by curve OCDE in fig. 10.6. Also, it has been shown that the number of conductors in series in each of the parallel paths between the brushes remains practically constant, hence

total e.m.f. between brushes

= average e.m.f. per conductor

× no. of conductors in series per path

$$= \frac{2\Phi N p}{60} \times \frac{Z}{c}$$

i.e.

$$E = 2\frac{Z}{c} \times \frac{Np}{60} \times \Phi \text{ volts.} \qquad [14.1]$$

Example 14.2 *A four-pole wave-connected armature has 51 slots with 12 conductors per slot and is rotated at 900 rev/min. If the useful flux per pole is 25 mWb, calculate the value of the generated e.m.f.*

Total number of conductors $= Z = 51 \times 12 = 612$; $c = 2$; $p = 2$; $N = 900$ rev/min; $\Phi = 0.025$ weber.

Using expression [14.1], we have:

$$E = 2 \times \frac{612}{2} \times \frac{900 \times 2}{60} \times 0.025$$

$$= 459 \text{ volts.}$$

Example 14.3 *An eight-pole lap-connected armature rotated at 350 rev/min is required to generate 260 V. The useful flux per pole is about 0.05 Wb. If the armature has 120 slots, calculate a suitable number of conductors per slot.*

For an eight-pole lap winding, $c = 8$.

Hence, $$260 = 2 \times \frac{Z}{8} \times \frac{350 \times 4}{60} \times 0.05$$

$\therefore$ $\qquad\qquad\qquad Z = 890$ (approximately)

and number of conductors per slot $= 890/120 = 7.4$ (approx.).

This value must be an even number; hence 8 conductors per slot would be suitable.

Since this arrangement involves a total of $8 \times 120 = 960$ conductors, and since a flux of 0.05 weber per pole with 890 conductors gave 260 V, then with 960 conductors the same voltage is generated with a flux of $0.05 \times (890/960) = 0.0464$ weber/pole.

14.6 Production of torque in d.c. machines

Let us, for simplicity, consider a two-pole machine having an armature with eight slots and two conductors per slot, as shown in fig. 14.10. The curved lines between the conductors and the commutator segments represent the front end-corrections of the armature winding and those on the outside of the armature represent the back end-connections. The armature winding—like all modern d.c. windings—is of the double-layer type, the end connections of the outer layer being represented by full lines and those of the inner layer by dotted lines.

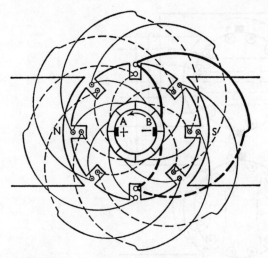

Fig. 14.10 A two-pole armature winding

Brushes A and B are placed so that they are making contact with conductors which are moving midway between the poles and have therefore no e.m.f. generated in them. If the armature is driven anticlockwise, the direction of the e.m.f.s generated in the various conductors is represented by the dots and crosses. By following the directions of these e.m.f.s, it is found that brush A is positive and B is negative. In diagrams where the end-connections are omitted, it is usual to show the brushes midway between the poles, as in fig. 14.11.

In general, an armature has ten to fifteen slots per pole, so that the conductors are more uniformly distributed around the armature core than is suggested by fig. 14.10; and for simplicity we may omit the slots and consider the conductors uniformly distributed as in fig. 14.11(a). The latter shows the distribution of the magnetic flux due to the field current I_f *when there is no armature current*, the flux in the gap being practically radial and uniformly distributed.

When the machine is operating as a d.c. *generator*, the direction of the current in the armature conductors is the same as that of the e.m.f. generated in the conductors. If the armature is being driven anticlockwise, it follows from Fleming's Right-hand Rule (section 6.4) that the direction of the e.m.f.s generated in the conductors moving in the magnetic flux emerging from the N pole is outwards from the paper, whereas the direction of the e.m.f.s in the conductors moving in the magnetic flux entering the S pole is towards the paper, as indicated by the dots and crosses in fig. 14.11(a). Hence, when the machine is operating as a *generator*, the direction of the current in all the conductors on the left-hand side of the vertical axis through brushes A and B is outwards from the paper, and that of the current in all the conductors on the right-hand side of that axis is towards the paper, as shown in fig. 14.11(b).

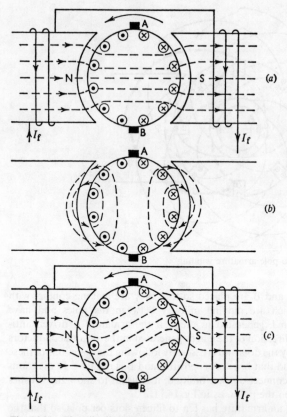

Fig. 14.11   Flux-distribution due to (a) field current alone, (b) armature current alone, (c) field and armature currents of a d.c. generator

Let us next consider the case when there is no field current and the armature is *stationary*. Current is sent through the armature winding from an external source, the direction of this current being arranged to be the same as that of the armature current when the machine is operating as a generator. This armature current sets up a magnetic flux, the distribution of which is indicated by the dotted lines in fig. 14.11(b). At the centre of the armature core and in the pole shoes, the direction of this flux is at right angles to the flux produced by the field current I_f in fig. 14.11(a). Hence the reason why the flux due to the armature current is referred to as *cross flux*.

A comparison of figs. 14.11(a) and (b) shows that over the leading* halves of the pole-faces the direction of the cross flux in the airgap is opposite to that of the main flux, thereby reducing the flux density in these regions. On the other hand, over the trailing halves of the pole faces, the

---

* The pole tip which is first met during rotation by a point on the armature surface is known as the *leading pole tip* and the other as the *trailing pole tip*.

two fluxes are in the same direction, so that the flux density in these regions is strengthened. Consequently the effect is to distort or twist the magnetic flux in the direction of rotation, as shown in fig. 14.11(c). It will be noticed that the flux in the airgaps is skewed and the length of the flux paths in the gaps thereby increased.

It has been pointed out in section 5.3 that magnetic flux behaves like stretched elastic cords. It follows that in fig. 14.11(c) the distorted flux exerts a clockwise pull on the armature. To overcome the tangential component of this pull, the engine has to exert a torque to drive the generator in an anticlockwise direction. This torque is *additional* to that required to drive the generator on *no* load.

If a current is sent through the armature winding in the reverse direction, as shown in fig. 14.12, the distribution of the resultant flux due to the field and armature currents is distorted in a clockwise direction. This flux is now so skewed as to exert a torque driving the armature in an anticlockwise direction. The machine is now operating as a d.c. motor. The magnitude of the torque is discussed in section 16.3.

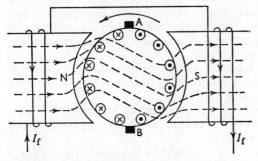

Fig. 14.12   Flux distribution in a d.c. motor

It may be mentioned at this point that in *all* d.c. and a.c. machines, the torque on the rotating element is due to the resultant flux in the *airgap* being skewed as indicated in figs. 14.11(c) and 14.12. There is, of course, an equal and opposite torque exerted on the fixed element of the machine. When the flux in the airgap is radial, as shown in fig. 14.11(a), there is *no* torque.

14.7 Summary of important formulae

$$E = 2\frac{Z}{c} \times \frac{Np}{60} \times \Phi \text{ volts} \qquad [14.1]$$

where

$c = 2$ for wave winding

$= 2p$ for lap winding

and

$p = $ no of *pairs* of poles.

14.8 Examples

1. A six-pole armature is wound with 498 conductors. The flux and the speed are such that the average e.m.f. generated in each conductor is 2 V. The current in each conductor is 120 A. Calculate the total current and the generated e.m.f. of the armature if the winding is connected (a) wave, (b) lap. Also calculate the total power generated in each case.

2. A four-pole armature is wound with 564 conductors and driven at 800 rev/min, the flux per pole being 20 mWb. The current in each conductor is 60 A. Calculate the total current, the e.m.f. and the electrical power generated in the armature if the armature conductors are connected (a) wave, (b) lap.

3. An eight-pole lap-connected armature has 96 slots with 6 conductors per slot and is driven at 500 rev/min. The useful flux per pole is 0.09 Wb. Calculate the generated e.m.f.

4. A four-pole armature is lap-connected with 624 conductors and is driven at 1200 rev/min. Calculate the useful flux per pole required to generate an e.m.f. of 250 V.

5. A six-pole armature is wave-connected with 410 conductors. The flux per pole is 0.025 Wb. Find the speed at which the armature must be driven to generate an e.m.f. of 485 V.

CHAPTER 15

Direct-Current Generators

15.1 Armature and field connections

The general arrangement of the brush and field connections of a four-pole machine is shown in fig. 15.1. The four brushes B make contact with the

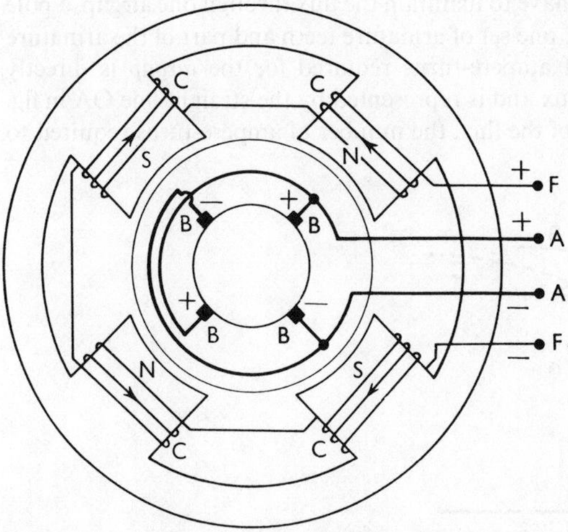

Fig. 15.1 Armature and field connections

commutator; and it has been explained in sections 14.5 and 14.6 that these brushes are consecutively positive and negative. Hence, the two positive brushes are connected to the positive terminal A. Similarly the negative brushes are connected to the negative terminal A_1. From fig. 14.10 it will be seen that the brushes are situated approximately in line with the centres of the poles. This position enables them to make contact with conductors in which little or no e.m.f. is being generated since these conductors are then moving between the poles.

The four exciting or field coils C are usually joined in series and the ends are brought out to terminals F and F_1. These coils must be so connected as to produce N and S poles alternately. The arrowheads in fig. 15.1 indicate the direction of the field current when F is positive.

In general, we may divide the methods used for connecting the field and armature windings into the following groups:

(a) *Separately-excited generators*—the field winding being connected to a source of supply other than the armature of its own machine.

(b) *Self-excited generators*, which may be subdivided into:

   (i) *Shunt-wound generators*—the field winding being connected across the armature terminals.
   (ii) *Series-wound generators*—the field winding being connected in series with the armature winding.
   (iii) *Compound-wound generators*—a combination of shunt and series windings.

Before we discuss the above systems in greater detail, let us consider the relationship between the magnetic flux and the exciting ampere-turns of a generator on no load. From fig. 14.1 it will be seen that the ampere-turns of one field coil have to maintain the flux through one airgap, a pole core, part of the yoke, one set of armature teeth and part of the armature core. The number of ampere-turns required for the *airgap* is directly proportional to the flux and is represented by the straight line OA in fig. 15.2. For low values of the flux, the number of ampere-turns required to

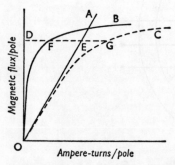

Fig. 15.2   Magnetization curve of a machine

send the flux through the *iron* portion of the magnetic circuit is very small; but when the flux exceeds a certain value, some parts—especially the teeth—begin to get saturated and the number of ampere-turns increases far more rapidly than the flux, as shown by curve B. Hence, if DE represents the number of ampere-turns/pole required to maintain flux OD across the airgap and if DF represents the number of ampere-turns/pole to send this flux through the iron portion of the magnetic circuit, then:

total ampere-turns/pole to produce flux $OD = DE + DF = DG$.

By repeating this procedure for various values of the flux, we can derive the *magnetization curve* C representing the relationship between the useful magnetic flux per pole and the total ampere-turns/pole.

15.2 Separately-excited generator

Figure 15.3 shows a simple method of representing the armature and field windings, A and A_1 being the armature terminals and F and F_1 the field terminals already referred to in fig. 15.1. The field winding is connected in series with a resistor R and an ammeter A to a battery or another gnerator.

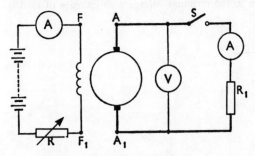

Fig. 15.3 Separately-excited generator

Suppose the armature to be driven at a *constant* speed or no load (i.e. with switch S open) and the magnetizing current to be increased from zero up to the maximum permissible value and then reduced to zero, the generated e.m.f. being noted for various values of the magnetizing current. It is found that the relationship between the two quantities is represented by curves P and Q in fig. 15.4, P being for increasing values of the excitation and Q for decreasing values. The difference between the curves is due to hysteresis (section 7.10), and OR represents the e.m.f. generated by the residual magnetism in the poles. If the test is repeated, the e.m.f. curve is found to follow the dotted line and then merge into curve P. These curves are termed the *internal* or *open-circuit characteristics* of the machine.

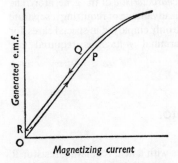

Fig. 15.4 Open-circuit characteristic

From expression [14.1], it is evident that for a given speed, the e.m.f. generated in a given machine is proportional to the flux per pole; and for a given field winding, the number of ampere-turns/pole is proportional to the magnetizing current. Hence the shape of the open-circuit characteristic P in fig. 15.4 is the same as that of the magnetization curve C in fig. 15.2; and curves P and Q in fig. 15.4 indicate how the flux varies with increasing and decreasing values of the magnetizing current.

Let us next consider the effect of load upon the terminal voltage. This can be determined experimentally by closing switch S (fig. 15.3), varying the value of the load resistance and noting the terminal voltage for each load current. Curve M in fig. 15.5 is typical of the relationship for such a machine. The decrease in the terminal voltage with increase of load is

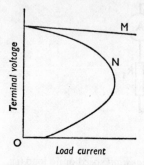

Fig. 15.5 Load characteristics of separately-excited and shunt generators

mainly due to the resistance drop in the armature circuit; thus, if the load current is 100 A and the resistance of the armature circuit is 0.08 Ω, the voltage drop in the armature circuit is $100 \times 0.08 = 8$ V. Consequently, if the generated e.m.f. be 235 V, the terminal p.d. $= 235 - 8 = 227$ V. In general, if $E =$ generated e.m.f., $I_a =$ armature current, $R_a =$ resistance of armature circuit and $V =$ terminal p.d., then:

$$V = E - I_a R_a.$$ [15.1]

The curve representing the variation of terminal voltage with load current is termed the *load* or *external characteristic* of the generator. The separately-excited generator has the disadvantage of requiring a separate source of direct current and is therefore only employed in special cases, for instance when a wide range of terminal voltage is required (see section 16.8).

15.3 Shunt-wound generator

The field winding is connected in series with a field regulating resistor R across the armature terminals as shown in fig. 15.6, and is therefore in parallel or 'shunt' with the load. The power absorbed by the shunt circuit is limited to about 2 to 3 per cent of the rated output of the generator by winding the field coils with a large number of turns of comparatively thin wire.

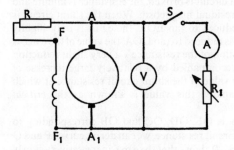

Fig. 15.6   Shunt-wound generator

A shunt generator will excite only if the poles have some residual magnetism and the resistance of the shunt circuit is less than some critical value, the actual value depending upon the machine and upon the speed at which the armature is driven. Suppose curve P in fig. 15.7 to represent the

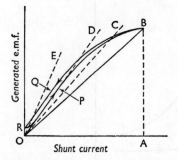

Fig. 15.7   Variation of e.m.f. with shunt circuit resistance

open-circuit characteristic of a shunt generator with *increasing* excitation; then, for a shunt current OA, the e.m.f. is AB and

$$\text{corresponding resistance of shunt circuit} = \frac{\text{terminal voltage}}{\text{shunt current}}$$

$$= \text{AB/OA} = \tan \text{BOA}$$

$$= \text{slope of OB.}$$

When the machine is running at its rated speed with the shunt circuit disconnected, the residual magnetism generates an e.m.f. OR. Immediately the shunt circuit is closed, this e.m.f., OR, circulates a small current through the shunt winding; but the magnetic flux produced by this current generates a larger e.m.f., which in turn circulates a still larger shunt current, etc., until the e.m.f. reaches the value AB for a shunt circuit resistance equal to AB/OA.

Figure 15.7 shows the effect of varying the resistance of the shunt circuit; thus, when the shunt circuit is broken, the resistance is infinite and the e.m.f. OR is due to the residual magnetism. When the shunt circuit is closed and its resistance reduced to tan EOA, the e.m.f. is only slightly larger than OR; but when it is reduced to tan DOA, the value of the e.m.f. is extremely sensitive to variation of the resistance—a very slight reduction of the shunt resistance is accompanied by a relatively large increase of e.m.f. Since tan DOA is the value of the shunt circuit resistance at which the excitation increases rapidly, this value is known as the *critical resistance of the shunt circuit*.

By drawing lines such as OE, OD, OC and OB corresponding to various values of the shunt circuit resistance, we can derive curves P and Q of fig. 15.8. Portion *ab* of curve P shows that the e.m.f. increases very slowly with decrease of resistance so long as the latter is greater than the critical value; but when the resistance is about the critical value, the e.m.f. is very unstable—a slight decrease of resistance or a slight increase of speed is accompanied by a large increase of e.m.f.

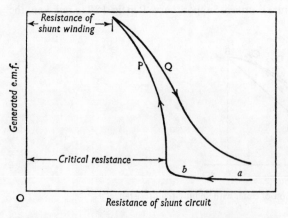

Fig. 15.8   Variation of e.m.f. with shunt circuit resistance

It will be seen from curve Q in fig. 15.8 that hysteresis causes the variation of voltage with increasing shunt resistance to be comparatively gradual.

The variation of terminal voltage with load current for a shunt generator is greater than that for the corresponding separately-excited generator and is represented by curve N in fig. 15.5. This greater variation is due to the decrease in terminal voltage accompanying an increase of load causing a reduction of the shunt current and therefore of flux and of generated e.m.f. It is seen that as the load increases, these variations become more marked. Ultimately, the load reaches an unstable value when the effect of reducing the load resistance still further is to cause the terminal voltage to fall so much that the load current decreases.

The shunt machine is the type of d.c. generator most frequently employed, but the load current must be limited to a value that is well below the maximum value, thereby avoiding excessive variation of the terminal voltage.

15.4 Series-wound generator

Since the whole of the armature current passes through the field winding (fig. 15.9), the latter consists of comparatively few turns of thick wire or strip.

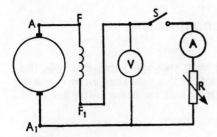

Fig. 15.9 Series-wound generator

When the machine is on open circuit, i.e. when S is open, the terminal voltage OP (fig. 15.10) is very small, being due to the residual flux in the poles. If S is closed with the load resistance R comparatively large, the

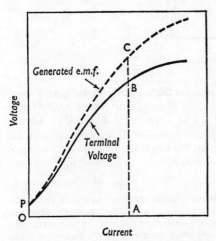

Fig. 15.10 Characteristics of a series generator

machine does not excite; but as R is reduced, a value is reached when a slight reduction of R is accompanied by a relatively large increase of terminal voltage. The full line in fig. 15.10 shows the variation of terminal voltage with load current for a certain series generator driven at a constant speed; and the dotted line represents the generated e.m.f. of the same machine when driven at the same speed with the armature on open circuit and the field separately excited. Thus, for a current OA, AC is the generated e.m.f. and AB is the corresponding terminal voltage on load. The difference, BC, is mainly due to the *IR* drop in the field and armature windings.

It is obvious from fig. 15.10 that a series-wound generator is quite unsuitable when the voltage is to be maintained constant or even approximately constant over a wide range of load current.

15.5 Compound-wound generator

In fig. 15.11, C represents the shunt coils and S the series coils, these windings being usually connected* so that their ampere-turns assist one another. Consequently the larger the load, the greater are the flux and the generated e.m.f.

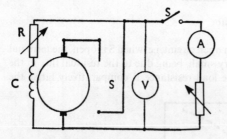

Fig. 15.11 Compound-wound generator

If curve S in fig. 15.12 represents the load characteristic with shunt winding alone, then, by the addition of a small series winding, the fall of terminal voltage with increase of load is reduced as indicated by curve P. Such a machine is said to be *under-compounded*. By increasing the number of series turns we can arrange for the machine to maintain its terminal voltage (curve Q) practically constant between no load and full load,

---

* In practice, it is of little consequence whether the shunt winding is connected 'long-shunt' as in fig. 15.11 or 'short-shunt', i.e. directly across the armature terminals, since the shunt current is very small compared with the full-load current and the number of series turns is very small compared with the number of shunt turns.

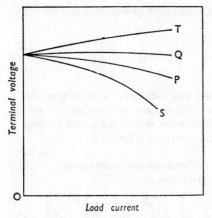

Fig. 15.12 Load characteristics of shunt-wound and compound-wound generators

in which case the machine is said to be *level-compounded*. If the number of series turns be increased still further, the terminal voltage increases with increase of load—as represented by curve T. The machine is then said to be *over-compounded*.

15.6 Examples

1. A six-pole d.c. generator having a lap-wound armature is required to give a terminal voltage of 240 V when supplying an armature current of 400 A. The armature has 84 slots and is driven at 700 rev/min. The resistance of the armature circuit is 0.03 Ω and the useful flux per pole is about 0.03 Wb. Calculate the number of conductors per slot and the actual value of the useful flux per pole.

2. The following table gives the open-circuit voltages for different field currents of a shunt generator driven at a constant speed:

O.C. voltage (volts)	120	240	334	400	444	470
Field current (amperes)	0.5	1.0	1.5	2.0	2.5	3.0

   Plot a graph showing the variation of generated e.m.f. with exciting current and from this graph derive the value of the generated e.m.f. when the shunt circuit has a resistance of (a) 160 Ω, (b) 210 Ω and (c) 300 Ω. Also find the value of the critical resistance of the shunt circuit.

3. Sketch and explain the curves connecting (a) e.m.f. and speed at constant excitation and (b) e.m.f. and excitation at constant speed for a separately-excited generator. The curve of induced e.m.f. for a

separately-excited generator when run at 1300 rev/min on open circuit is given by:

E.m.f. (volts)	12	44	73	98	113	122	127
Exciting current (amperes)	0	0.2	0.4	0.6	0.8	1.0	1.2

Deduce the curve of e.m.f. and excitation when the generator is running separately excited at 1000 rev/min. To what voltage will the generator build up on no load when shunt excited and running at 1000 rev/min, if the total field resistance is 100 Ω?

(N.C.T.E.C.)

4. The following figures give the open-circuit characteristics of a d.c. shunt generator driven at 600 rev/min:

Field current (amperes)	0	1	1.5	2	2.5	3	3.5
Armature e.m.f. (volts)	7.5	93	125	165	186	202	215

Plot the open-circuit characteristic for a speed of 750 rev/min and determine the voltage to which the machine will excite if the field-circuit resistance is 80 Ω. What additional resistance would have to be inserted in the field circuit to reduce the voltage to 200 V at 750 rev/min?

5. The open-circuit characteristic of a certain d.c. generator, driven at 1000 rev/min, is given below:

Field current (amperes)	2	4	6	8	10
Generated e.m.f. (volts)	142	270	358	404	428

The generator is now shunt-connected and driven at 1200 rev/min, the resistance of the field circuit being 55 Ω. What e.m.f. will it generate? What is the approximate value of the critical resistance of the shunt circuit when the speed is 1200 rev/min?

CHAPTER 16

Direct-Current Motors

16.1 A d.c. machine as generator or motor

There is no difference of construction between a d.c. generator and a d.c. motor. In fact, the only difference is that in a generator the generated e.m.f. is greater than the terminal voltage, whereas in a motor the generated e.m.f. is less than the terminal voltage. For instance, suppose a shunt generator D (fig. 16.1) to be driven by an engine and connected through a

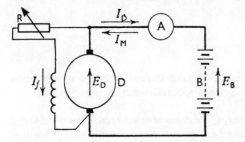

Fig. 16.1   Shunt-wound machine as a generator or motor

centre-zero ammeter A to a battery B. If the field regulator R is adjusted until the reading on A is zero, the e.m.f., E_D, generated in D is then exactly equal to the e.m.f., E_B, of the battery. If R is now reduced, the e.m.f. generated in D exceeds that of B, and the excess e.m.f. is available to circulate a current I_D through the resistance of the armature circuit, the battery and the connecting leads. Since I_D is in the same direction as E_D, machine D is a generator of electrical energy.

Next, suppose the supply of steam or oil to the engine driving D to be cut off. The speed of the set falls, and as E_D decreases, I_D becomes less, until when $E_D = E_B$ there is no circulating current. But E_D continues to decrease and becomes less than E_B, so that a current I_M flows in the reverse direction. Hence B is now supplying electrical energy to drive D as an electric motor.

The speed of D continues to fall until the difference between E_D and

E_B is sufficient to circulate the current necessary to maintain the rotation of D.

It will be noticed that the direction of the field current I_f is the same whether D is running as a generator or a motor.

The relationship between the current, the e.m.f., etc., for machine D may be expressed thus: if

E = e.m.f. generated in armature,

V = terminal voltage,

R_a = resistance of armature circuit

and

I_a = armature current,

then, when D is operating as a generator, it follows from expression [15.1] that:

$$E = V + I_a R_a.$$ [16.1]

When the machine is operating as a motor, the e.m.f., E, is less than the applied voltage V, and the direction of the current I_a is the reverse of that when the machine is acting as a generator; hence

$$E = V - I_a R_a$$

or

$$V = E + I_a R_a.$$ [16.2]

Since the e.m.f. generated in the armature of a motor is in opposition to the applied voltage, it is sometimes referred to as a *back e.m.f.*

Example 16.1   *The armature of a d.c. machine has a resistance of 0.1 Ω and is connected to a 230-V supply. Calculate the generated e.m.f. when it is running* (a) *as a generator giving 80 A and* (b) *as a motor taking 60 A.*

(a)   Voltage drop due to armature resistance = $80 \times 0.1 = 8$ V.

From [16.1],   generated e.m.f. = $230 + 8 = 238$ V.

(b)   Voltage drop due to armature resistance = $60 \times 0.1 = 6$ V.

From [16.2],   generated e.m.f. = $230 - 6 = 224$ V.

16.2 Speed of a motor

In section 14.5 it was shown that the relationship between the generated e.m.f., speed, flux, etc., is represented by:

$$E = 2\frac{Z}{c} \times \frac{Np}{60} \times \Phi.$$ [14.1]

For a given machine, Z, c and p are fixed; and in such a case we can write:

$$E = kN\Phi$$

where

$$k = 2\frac{Z}{c} \cdot \frac{p}{60}.$$

Substituting for E in expression [16.2] we have:

$$V = kN\Phi + I_aR_a$$

$$\therefore \quad N = \frac{V - I_aR_a}{k\Phi} \qquad [16.3]$$

The value of I_aR_a is usually less than 5 per cent of the terminal voltage V, so that:

$$N \simeq \frac{V}{k\Phi}. \qquad [16.4]$$

In words, this expression means that the speed of an electric motor is approximately proportional to the voltage applied to the armature and inversely proportional to the flux; and all methods of controlling the speed involve the use of either or both of these relationships.

Example 16.2 *A four-pole motor is fed at 440 V and takes an armature current of 50 A. The resistance of the armature circuit is 0.28 Ω. The armature is wave-connected with 888 conductors and the useful flux per pole is 0.023 Wb. Calculate the speed.*

From expression [16.2] we have:

$$440 = \text{generated e.m.f.} + 50 \times 0.28$$

$$\therefore \quad \text{generated e.m.f.} = 440 - 14 = 426 \text{ V}.$$

Substituting in the e.m.f. equation [14.1], we have:

$$426 = 2 \times \frac{888}{2} \times \frac{N \times 2}{60} \times 0.023$$

$$\therefore \quad N = 626 \text{ rev/min}.$$

16.3 Torque developed by a d.c. motor

It was explained in section 14.6 that the combined effect of the field and armature currents in a d.c. machine is to distort the magnetic flux in the airgaps in such a way as to exert a torque on the armature. In the case of a generator, this torque is in opposition to that extended by the engine driving the generator; whereas in a motor, initially at standstill, the torque is responsible for accelerating the machine and its load. After the motor has attained a steady speed, the torque available at the shaft is responsible for driving the load.

Let us start with expression [16.2] and multiply each term by I_a, namely the total armature current, thus:

$$VI_a = EI_a + I_a^2 R_a.$$

VI_a represents the total electrical power supplied to the armature, and $I_a^2 R_a$ represents the loss due to the resistance of the armature circuit. The difference between these two quantities, namely EI_a, therefore represents the mechanical power developed by the armature. All of this mechanical power is not available externally since some of it is absorbed as friction loss at the bearings and at the brushes and some is wasted as hysteresis loss (section 7.10) and in circulating eddy currents in the iron core (section 14.2).

If T is the torque*, in newton metres, exerted on the armature to develop the mechanical power just referred to, and if N is the speed in revolutions/minute, then:

$$\text{work done in 1 revolution} = 2\pi \times T \text{ joules}$$

$\therefore$ mechanical power developed = work done per second

$$= 2\pi TN/60 \text{ watts}$$

Hence

$$2\pi TN/60 = EI_a \qquad [16.5]$$

$$= 2\frac{Z}{c} \cdot \frac{Np}{60} \Phi \cdot I_a$$

so that

$$T = 0.318 \frac{I_a}{c} \cdot Zp\Phi \text{ newton metres}\dagger$$

$$[16.6]$$

For a given machine, Z, c and p are fixed; in which case:

$$T \propto I_a \times \Phi \qquad [16.7]$$

In words, the torque of a given d.c. motor is proportional to the product of the armature current and the flux per pole.

---

* In many textbooks, the value of the torque is derived from expression [5.1]. This method gives the correct result; but it should be realized that with a slotted armature, flux density in the slots is extremely low so that the force on the conductors is very small. Practically the whole of the torque is exerted on the teeth.
† Expression [16.6] also represents the torque to be applied to the armature of a d.c. *generator* to generate the output of power and the I^2R loss in the *armature*. Additional torque has to be applied to supply the iron loss in the armature core and the friction losses at the commutator and the bearings of the generator.

Example 16.3 *Calculate the gross torque developed by the motor referred to in Example 16.2.*

From expression [16.2], $I_a = 50$ A, $E = 426$ V and $N = 626$ rev/min.

∴ mechanical power developed by armature

$$= 50 \times 426 = 21\,300 \text{ W}.$$

Substituting in expression [16.5], we have:

$$2\pi T \times 626/60 = 21\,300$$

$$T = 325 \text{ N m}.$$

16.4 Starting resistor

If the armature referred to in example 16.2 were stationary and then switched across a 440-V supply, there would be no generated e.m.f. and the current would tend to grow to $440/0.28 = 1572$ A. Such a current, in addition to subjecting the armature to a severe mechanical shock, would blow the fuses, thereby disconnecting the supply from the motor. It is therefore necessary (except with very small motors) to connect a variable resistor in series with the armature, the resistance being reduced as the armature accelerates. Such an arrangement is termed a starter. If the starting current in the above example is to be limited to, say, 80 A, the total resistance of the starter and armature must be $440/80 = 5.5\,\Omega$, so that the resistance of the starter alone must be $(5.5 - 0.28) = 5.22\,\Omega$.

Figure 16.2 shows a starting resistor R subdivided between four contact-studs S and connected to a shunt-wound motor. One end of the shunt winding is joined to study 1; consequently when arm A is moved

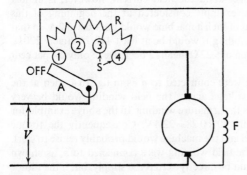

Fig. 16.2 Shunt-wound motor with starter

from 'Off' to that stud, the full voltage is applied to the shunt winding and the whole of R is in series with the armature. The armature current instantly grows to a value I_1 (fig. 16.3) where:

$$I_1 = \frac{\text{supply of voltage } V}{\text{resistance of (armature + starter)}}.$$

Since the torque is proportional to (armature current × flux), it follows that the maximum torque is immediately available to accelerate the armature.

As the armature accelerates, its e.m.f. grows and the armature current decreases as indicated by curve *ab*. When the current has fallen to some pre-arranged value I_2, arm A is moved over to stud 2, thereby cutting out sufficient resistance to allow the current to rise once more to I_1. This operation is repeated until A is on study 4 and the whole of the starting resistance is cut out of the armature circuit. The motor continues to accelerate and the current to decrease until it settles down at some value I

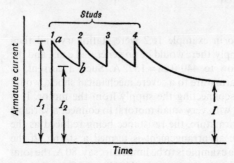

Fig. 16.3 Variation of starting current

(fig. 16.3) such that the torque due to this current is just sufficient to enable the motor to cope with its load.

It is evident from fig. 16.2 that when A is on stud 4 the whole of R is in the field circuit. The effect upon the field current, however, is almost negligible. Thus, for the example considered at the beginning of this section, the shunt current of such a machine would not exceed 2 A, so that the resistance of shunt winding F would be at least 440/2, namely 220 Ω. Consequently the addition of 5.22 Ω means a decrease of only 2.4 per cent in the field current.

If the field winding were connected to *a* as in fig. 16.4, then at the instant of starting, the voltage across the field winding would be very small, namely that across the armature winding. In the above example, for instance, this p.d. is only $80 \times 0.28 = 22.4$ V. Consequently the torque would be extremely small and the machine would probably refuse to start. On the other hand, if the field winding were connected to *b*, as shown dotted in fig. 16.4, it would be directly across the supply; but if the motor were stopped by moving the starter arm back to 'Off', the field would still remain excited. If such a field current were switched off by opening switch S quickly, the sudden collapse of the flux would induce a very high e.m.f. in F and might result in a breakdown of the insulation.

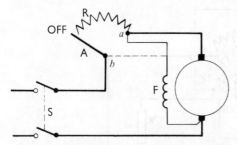

Fig. 16.4 Wrong methods of connecting the shunt winding

With the connections shown in fig. 16.2, it is obvious that the armature winding, the starting resistor and the field winding form a closed circuit. Consequently when A returns to 'Off', the kinetic energy of the machine and its load maintains rotation for an appreciable period, and during most of that time the machine continues to excite as a shunt generator. The field current therefore decreases comparatively slowly and there is no risk of an excessive e.m.f. being induced.

16.5 Protective devices on starters

It is very desirable to provide the starter with protective devices to enable the starter arm to return to 'Off',

(a) when the supply fails, thus preventing the armature being directly across the mains when this voltage is restored, and
(b) when the motor becomes overloaded or develops a fault causing the machine to take an excessive current.

Figure 16.5 shows a starter fitted with such features. The no-volt release NVR consists of a coil wound on a U-shaped iron core. The starter arm carries an iron plate B, so that when it is in the 'On' position, as shown, the core of NVR is magnetized by the field current and holds B against the tendency of a spiral spring G to return A to 'Off'. Should the supply fail, the motor stops, NVR is de-energized, and the residual magnetism in it is not sufficient to hold A against the counterclockwise torque of G.

If a connection C be made between the core of NVR and the supply side of the coil, the shunt current flows from A via B and C, and the starting resistor is thereby short-circuited.

The coil of the *overload release* OLR is wound on an iron core and connected in series with the motor. An iron plate L, pivoted at one end, carries an extension which, when lifted, connects together two pins *p, p*. When the current taken by the motor exceeds a certain value, the magnetic pull on L is sufficient to lift it, thereby short-circuiting the coil of NVR and releasing the starter arm A. The critical value of the overload current is

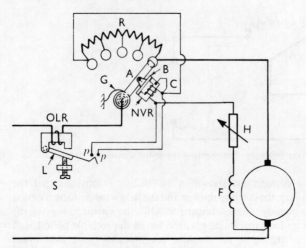

Fig. 16.5   Starter with no-volt and overload releases

dependent upon the length of air-gap between L and the core of OLR and is controlled by a screw adjustment S.

16.6 Speed characteristics of electric motors

With very few exceptions, d.c. motors are shunt-, series- or compound-wound. The connections of a shunt motor have already been given in figs. 16.2 and 16.5; and figs 16.6 and 16.7 show the connections for series and compound motors respectively, the starter in each case being shown in the

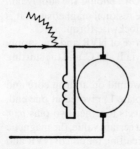

Fig. 16.6   Series-wound motor

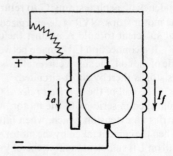

Fig. 16.7   Compound-wound motor

'On' position. (Students should always make a practice of including starters in diagrams of motor connections.) In compound motors, the series and shunt windings almost invariably assist each other, as indicated in fig. 16.7.

The speed characteristic of a motor usually represents the variation of speed with input current or input power, and its shape can be easily derived from expression [16.3], namely:

$$N = \frac{V - I_a R_a}{k\Phi}.$$

In shunt motors, the flux Φ is only slightly affected by the armature current and the value of $I_a R_a$ at full load rarely exceeds 5 per cent of V, so that the variation of speed with input current may be represented by curve A in fig. 16.8. Hence shunt motors are suitable where the speed has to remain approximately constant over a wide range of load.

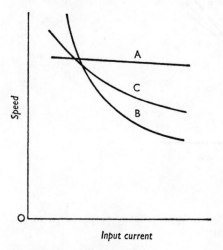

Fig. 16.8 Speed characteristics

In series motors, the flux increases at first in proportion to the current and then less rapidly owing to magnetic saturation (fig. 15.2). Also R_a in the above expression now includes the resistance of the field winding. Hence the speed is roughly inversely proportional to the current, as indicated by curve B in fig. 16.8. It will be seen that if the load falls to a very small value, the speed may become dangerously high. A series motor should therefore not be employed when there is any such risk; for instance, it should never be belt-coupled to its load except in very small machines such as vacuum cleaners.

Since the compound motor has a combination of shunt and series excitations, its characteristic (curve C in fig. 16.8) is intermediate between those of the shunt and series motors, the exact shape depending upon the values of the shunt and series ampere-turns.

16.7 Torque characteristics of electric motors

In section 16.3 it was shown that for a given motor:

torque ∝ armature current × flux per pole.

Since the flux in a shunt motor is practically independent of the armature current:

∴   torque of a shunt motor ∝ armature current

and is represented by the straight line A in fig. 16.9.

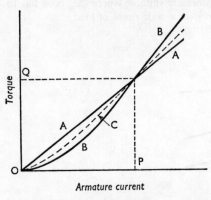

Fig. 16.9   Torque characteristics

In a series motor the flux is approximately proportional to the current up to full load, so that:

torque of a series motor ∝ (armature current)^2, approx.

Above full load, magnetic saturation becomes more marked and the torque does not increase so rapidly.

Curves A, B and C in fig. 16.9 show the relative shapes of torque curves for shunt, series and compound motors having the same full-load torque OQ with the same full-load armature current OP, the exact shape of curve C depending upon the relative value of the shunt and series ampere-turns at full load.

From fig. 16.9 it is evident that for a given current below the full-load value the shunt motor exerts the largest torque; but for a given current above that value the series motor exerts the largest torque.

The maximum permissible current at starting is usually about 1.5 times the full-load current. Consequently where a large starting torque is required, such as for hoists, cranes, electric trains, etc., the series motor is the most suitable machine.

Example 16.4 *A series motor runs at 600 rev/min when taking 110 A from a 230-V supply. The resistance of the armature circuit is 0.12 Ω and that of the series winding is 0.03 Ω. Calculate the speed when the current has fallen to 50 A, assuming the useful flux per pole for 110 A to be 0.024 Wb and that for 50 A to be 0.0155 Wb.*

Total resistance of armature and series windings

$$= 0.12 + 0.03 = 0.15\,\Omega,$$

∴   e.m.f. generated when current is 110 A

$$= 230 - 110 \times 0.15 = 213.5\,\text{V}.$$

In section 16.2 it was shown that for a given machine:

generated e.m.f. = a constant (say k) × speed × flux.

Hence with 110 A,

$$213.5 = k \times 600 \times 0.024$$

$$\therefore \quad k = 14.82.$$

With 50 A,

generated e.m.f. $= 230 - 50 \times 0.15 = 222.5\,\text{V}.$

But

the new e.m.f. = k × new speed × new flux

$$\therefore \quad 222.5 = 14.82 \times \text{new speed} \times 0.0155$$

∴   speed for 50 A = 969 rev/min.

16.8 Speed control of d.c. motors

It has already been explained in section 16.2 that the speed of a d.c. motor can be altered by varying either the flux or the armature voltage or both; and the methods most commonly employed are:

(a) A variable resistor, termed a *field regulator*, in series with the shunt winding—only applicable to shunt and compound motors. Such a field regulator is indicated by H in fig. 16.5. When the resistance is increased, the field current, the flux and the back e.m.f. are reduced. Consequently more current flows through the armature and the increased torque enables the armature to accelerate until the back e.m.f. is again nearly equal to the applied voltage (see example 16.5).

With this method it is possible to increase the speed to three or four times that at full excitation, but it is not possible to reduce the speed below that value. Also, with any given setting of the regulator, the speed remains approximately constant between no load and full load.

(b) A resistor, termed a *diverter*, in parallel with the field winding—only applicable to series-wound motors. In fig. 16.10, R represents the variable portion of the diverter and r the fixed portion, the function of r being to prevent the series winding being short-circuited. The smaller the value of R, the greater is the current diverted from the series winding and the higher, in consequence, is the speed of the motor. The minimum speed for a given input current is obtained by moving the slider on to the off-stud S, thereby breaking the circuit through the diverter.

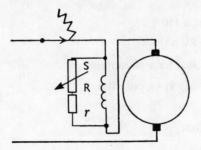

Fig. 16.10 Speed control of series-wound motor by a diverter

(c) A resistor, termed a *controller*, in series with the armature. The electrical connections for a controller are exactly the same as for a starter, the only difference being that in a controller the resistor elements are designed to carry the armature current indefinitely, whereas in a starter they can only do so for a comparatively short time without getting excessively hot.

For a given armature current, the larger the controller resistance in circuit, the smaller is the p.d. across the armature and the lower, in consequence, is the speed.

This system has several disadvantages: (i) the relatively high cost of the controller; (ii) much of the input energy may be dissipated in the controller and the overall efficiency of the motor considerably reduced thereby; (iii) the speed may vary greatly with variation of load due to the change in the p.d. across the controller causing a corresponding change in the p.d. across the motor; thus, if the supply voltage is 240 V, and if the current decreases so that the p.d. across the controller falls from, say, 100 V to 40 V, then the p.d. across the motor increases from 140 V to 200 V.

The principal advantage of the system is that speeds from zero upwards are easily obtainable, and the method is chiefly used for controlling the speed of cranes, hoists, trains, etc., where the motors are frequently started and stopped and where efficiency is of secondary importance.

(d) Exciting the field winding off a constant-voltage system and supplying the armature from a separate generator, as shown in fig. 16.11. M represents the motor whose speed is to be controlled; M_1G is a motor-generator set consisting of a shunt motor coupled to a separately-excited generator. The voltage applied to the armature of M can be varied between zero and a maximum by means of R. If provision is made for

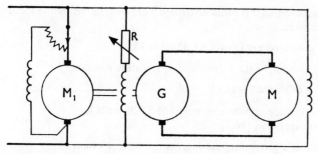

Fig. 16.11 Ward–Leonard system of speed control

reversing the excitation of G, the speed of M can then be varied from a maximum in one direction to the same maximum in the reverse direction.

This method is often referred to as the *Ward–Leonard* system and is used for controlling the speed of motors driving colliery winders., rolling mills, etc. The advantages of the system are:

(i) Starting and speed control are easily effected with a single regulator.
(ii) There is no resistor in series with the armature circuit to absorb any power; consequently the efficiency at low speeds is much higher than that obtainable with the controller system.
(iii) For a given value of R and therefore a given e.m.f. generated in G, the speed of M is almost independent of its load.
(iv) Retardation of M may be effected by increasing R so that the e.m.f. generated in G is less than that of M. Consequently much of the kinetic energy of M and its load is transferred electrically to G, which then runs as a motor driving M_1 as a generator, thereby sending electrical energy back into the mains. Where M has to be started and stopped frequently, the total energy *regenerated* may amount to a considerable percentage of the total energy absorbed from the mains.

The main disadvantage of the system is the initial cost of the motor-generator set.

Example 16.5 *A shunt motor is running at 626 rev/min (example 16.2) when taking an armature current of 50 A from a 440-V supply. The armature circuit has a resistance of 0.28 Ω. If the flux is suddenly reduced by 5 per cent, find: (a) the maximum value to which the current increases momentarily and the ratio of the corresponding torque to the initial torque, (b) the ultimate steady value of the armature current, assuming the torque due to the load to remain unaltered.*

(a) From example 16.2:

initial generated e.m.f. $= 440 - 50 \times 0.28 = 426$ V.

Immediately after the flux is reduced 5 per cent, i.e. before the speed has begun to increase:

new generated e.m.f. $= 426 \times 0.95 = 404.7$ V

$\therefore$ corresponding voltage drop due to armature resistance

$$= 440 - 404.7 = 35.3 \text{ V}$$

and corresponding armature current $= 35.5/0.28 = 126$ A.
From expression [16.7]:

torque of a given machine $\propto$ armature current $\times$ flux

$$\therefore \quad \frac{\text{new torque}}{\text{initial torque}} = \frac{\text{new current}}{\text{initial current}} \times \frac{\text{new flux}}{\text{initial flux}}$$

$$= \frac{126}{50} \times 0.95 = 2.394.$$

Hence the sudden reduction of 5 per cent in the flux is accompanied by more than a twofold increase of torque; this is the reason why the motor accelerates.

(b) After the speed and current have attained steady values, the torque will have decreased to the original value, so that:

new current $\times$ new flux $=$ original current $\times$ original flux

$\therefore$ new armature current $= 50 \times 1/0.95 = 52.6$ A.

Example 16.6   *A series motor is taking 30 A from a 230-V supply. The armature and field circuits have a total resistance of 0.6 Ω. Calculate the resistance required in series with the armature to reduce the speed by 40 per cent if the current at the new speed is 18 A. Assume the flux to be proportional to the current.*

When the motor is taking 30 A,

generated e.m.f. $= 230 - 30 \times 0.6 = 212$ V.

But the generated e.m.f. is proportional to (speed $\times$ flux) and the flux is proportional to the current:

$\therefore$ e.m.f. generated at the new speed $= 212 \times \dfrac{60}{100} \times \dfrac{18}{30}$

$$= 76.3 \text{ V}.$$

Corresponding voltage across motor terminals

$$= 76.3 + 18 \times 0.6 = 87.1 \text{ V}$$

$\therefore$ p.d. across the series resistor $= 230 - 87.1 = 142.9$ V

and

value of the series resistance $= 142.9/18 = 7.94$ Ω.

16.9 Applications of shunt, series and compound motors

In the preceding sections the characteristics of the three types of motors have been dealt with, and we may summarize the discussion thus:

Shunt motors are used—

(a) When the speed has to be maintained approximately constant between no load and full load; e.g. for driving a line of shafting.

(b) When it is required to drive the load at various speeds, any one speed being maintained constant for a relatively long period; e.g. for individual driving of such machines as lathes, etc. The shunt regulator enables the required speed control to be obtained easily and economically.

Series motors are used—

(a) When a large starting torque is required, e.g. for driving hoists, cranes, trains, etc.

(b) When the motor can be direct-coupled to a load, such as a fan, whose torque increases with speed.

Where constancy of speed is not essential, the decrease of speed with increase of load has the advantage that the power absorbed by the motor does not increase as rapidly as the torque; for instance, when the torque is doubled, the power usually increases by only about 50 or 60 per cent.

A series motor should not be used when the load may decrease to a very small value; thus it should not be used for driving a centrifugal pump or for a belt-drive except in very small machines.

Compound motors are used—

(a) When a large starting torque is required but where the load may fall to such a small value that a series motor would reach a dangerously high speed.

(b) When the load is of a fluctuating nature; e.g. for driving stamping presses, etc. The shunt excitation prevents the speed becoming excessive on light load and the decrease of speed with increase of load enables the flywheel usually fitted to such a machine to give up some of its kinetic energy, thereby assisting the motor in dealing with the peak load.

(c) When the supply voltage is subject to fluctuations, for instance, on a traction system. The series winding reduces the fluctuation of armature current, partly by its inductance and partly by its influence on the value of the flux and therefore on that of the generated e.m.f.

16.10 Summary of important formulae

For a generator,

$$E = V + I_a R_a. \qquad [16.1]$$

For a motor,

$$V = E + I_a R_a. \qquad [16.2]$$

For a given motor,

$$N = \frac{V - I_a R_a}{k\Phi} \qquad [16.3]$$

$$\simeq \frac{V}{k\Phi}. \qquad [16.4]$$

$$T = 0.132 \frac{I_a}{c} \cdot Zp\Phi \text{ newton metres.} \qquad [16.6]$$

For a given motor,

$$T \propto I_a \Phi. \qquad [16.7]$$

16.11 Examples

1. A shunt-wound machine runs as a motor at 1000 rev/min when connected across a 460-V supply and taking an armature current of 30 A. Calculate the speed at which the machine has to be driven as a generator giving a terminal voltage of 460 V when the armature current is 30 A. The armature circuit has a resistance of 1 Ω. Assume the flux to remain constant.

2. A shunt machine driven as a generator at 700 rev/min is supplying an armature current of 60 A at a terminal voltage of 220 V. If the same machine runs as a motor with the same terminal voltage and an armature current of 40 A, calculate the speed. Resistance of armature circuit = 0.2 Ω.

3. A shunt motor is running off a 220-V supply taking an armature current of 15 A, the resistance of the armature circuit being 0.8 Ω. Calculate the value of the e.m.f. generated in the armature winding.

   If the flux is suddenly reduced by 10 per cent, to what value will the armature current increase momentarily?

4. A shunt machine is running as a motor off a 500-V system, taking an armature current of 50 A. If the field current is suddenly increased so as to increase the flux by 20 per cent, calculate the current which will momentarily be fed back into the mains. Neglect the shunt current and assume the resistance of the armature circuit to be 0.5 Ω.

5. A four-pole motor has its armature lap-connected with 1040 conductors and runs at 1200 rev/min when taking an armature current of 60 A from a 230-V supply. The resistance of the armature circuit is 0.2 Ω. Calculate the useful flux per pole and the torque in newton metres.

6. A series motor runs at 700 rev/min off a 500-V supply when taking 70 A. What will be the speed when the load is such that the motor takes only 30 A? Assume the resistance of the armature and field circuits to be 0.6 Ω, the useful flux per pole with 70 A to be 0.03 Wb and that with 30 A to be 0.022 Wb. Also calculate the torque, in newton metres, when the current is 70 A.

7. A six-pole shunt motor has its armature wave-connected with 936 conductors. The useful flux per pole is 0.02 Wb, and the resistance of the armature circuit is 0.7 Ω. Calculate (a) the speed and (b) the torque, in newton metres, when the armature current is 30 A and the terminal voltage is 460 V.

8. A shunt motor runs at 800 rev/min off a 460-V supply when taking an armature current of 50 A. Calculate the speed at which it will run off a 230-V supply when the armature current is 30 A. The resistance of the armature circuit is 0.4 Ω. Assume the flux to have decreased to 75 per cent of the original value.

9. Explain briefly why a shunt-wound motor needs a starter instead of a plain switch.

A shunt-wound motor has a field resistance of 400 Ω and an armature resistance of 0.1 Ω, and runs off a 240-V supply. The armature current is 50 A and the motor speed is 900 rev/min. Assuming a straight-line magnetization curve, calculate (a) the additional resistance in the field circuit to increase the speed to 1000 rev/min for the same armature current, and (b) the speed with the original field current and an armature current of 200 A.

(E.M.E.U.)

10. A shunt motor having an armature resistance of 1 Ω takes an armature current of 20 A when running on load at 800 rev/min off a 250-V supply. If the load torque remains unchanged, calculate the value of the armature current and of the speed if the flux due to the field winding is reduced by 20 per cent.

11. A shunt motor has an armature resistance of 0.5 Ω and runs at 1000 rev/min when the armature current is 40 A and the supply voltage is 250 V. If the torque remains constant, calculate the speed and the armature current if the flux per pole is strengthened by 30 per cent.

12. A series motor runs at 500 rev/min when taking a current of 30 A from a 230-V supply. Calculate the speed when the current is 20 A, assuming the resistance of the armature and field circuits to be 0.8 Ω and the flux to be proportional to the current.

13. A shunt motor is running at 900 rev/min off a 440-V supply with no resistance in series with the armature, the armature current being 80 A. Calculate the resistance required in series with the armature to

reduce the speed to 500 rev/min, if the armature current has then decreased to 45 A. Resistance of the armature circuit is 0.25 Ω. Assume the flux to remain constant.

14. A separately-excited motor runs at 600 rev/min when the voltage across the armature is 160 V and the armature current is 40 A. Calculate the speed when the armature voltage is 250 V and the armature current is 60 A. The resistance of the armature circuit is 0.3 Ω. Assume the flux per pole to remain unaltered.

15. Calculate the percentage change of speed of a shunt motor having an armature resistance of 0.2 Ω when the armature current changes from 100 A to 50 A. The supply voltage is constant at 250 V. Assume the flux per pole to remain unaltered.

16. Calculate the torque, in newton metres, developed by a d.c. motor having an armature resistance of 0.25 Ω and running at 750 rev/min when taking an armature current of 60 A. The supply voltage is 440 V.

17. Explain the principle of speed control of a d.c. shunt motor by variation of the field current.

   A shunt motor runs at 900 rev/min from a 460-V supply when taking an armature current of 25 A. Calculate the speed at which it will run from a 230-V supply when taking an armature current of 15 A. The resistance of the armature circuit is 0.8 Ω. Assume the flux per pole at 230 V to have decreased to 75 per cent of its value at 460 V.
   (E.M.E.U.)

18. The full-load input power of a certain shunt motor is 20 kW. The machine is to be switched on to a 230-V supply. Calculate the value of the starting resistance necessary to limit the starting current to 1.5 times the full-load current. The resistance of the armature circuit is 0.2 Ω. Neglect the shunt current.

   Also calculate the value of the generated e.m.f. of the motor when the current has fallen to the full-load value, assuming that the whole of the starting resistance is still in circuit.

19. Explain why a d.c. motor requires a 'starter'.

   A shunt motor has an armature resistance of 0.3 Ω and is to be connected across a 230-V supply. Calculate (a) the value of the resistance required in series with the armature to limit the armature current to 75 A at starting, and (b) the value of the generated e.m.f. when the armature current has fallen to 50 A with this value of resistance still in circuit.
   (E.M.E.U.)

20. The starter for a certain shunt motor has a resistance of 2.75 Ω. The armature and shunt circuits have resistances of 0.25 Ω and 200 Ω respectively. Calculate the armature and shunt currents when the armature is at standstill with the starter arm on the first stud and the field winding connected (a) correctly and (b) to the junction of the starting resistor and the armature. The supply voltage is 230 V.

CHAPTER 17

Transformers

17.1 The function of transformers

The function of a transformer is to transform an alternating voltage at one level to an alternating voltage at another level, such a function being required in both power systems and in electronics. For instance, the voltage produced at the generator may be of the order of 20 kV. This is stepped up, to perhaps 275 kV, for transmission by the overhead line system and then stepped down to values which it can be used at in, say, factories, perhaps 11 kV, and homes, 240 V. Transformers are used to do this stepping up and stepping down of the voltages. They are also used in electronic circuits to change voltage levels, e.g. to step down the mains supply voltage of 240 V to a level suitable for semiconductor circuits, perhaps 5 V. Transformers are also used to change the effective impedance of a circuit so that maximum power can be transferred between circuits.

17.2 Principle of action of a transformer

The principle of operation of the transformer is mutual inductance between two coils which are linked by a common magnetic field (see section 8.8). Figure 17.1 shows the general arrangement. C is an iron core

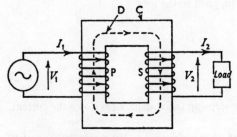

Fig. 17.1 A transformer

on which two coils are wound, the vertical parts of the core being referred to as the limbs and the top and bottom parts as yokes. Coil P is connected to the supply and is called the primary coil, coil S which is connected to the load being the secondary coil.

The alternating voltage V_1 applied to the primary produces an alternating current through the primary coil. This then produces an alternating magnetic flux in the core, the line D indicating the general path of the flux. In an ideal transformer there are no flux losses and all this flux links both the primary and the secondary coils. Because we have the same flux linking each turn of the primary and secondary coils we must have the same induced e.m.f. produced for each turn, the induced e.m.f. being proportional to the rate of change of flux (see section 6.2 and Faraday's law). Thus if the primary coil has N_1 turns and the secondary coil N_2 turns:

induced e.m.f. in primary $= N_1 \times$ e.m.f. induced per turn

induced e.m.f. in secondary $= N_2 \times$ e.m.f. induced per turn

Hence

$$\frac{\text{induced e.m.f. in primary}}{\text{induced e.m.f. in secondary}} = \frac{N_1}{N_2}$$

When the secondary coil is on open circuit, i.e. no load, its terminal voltage V_2 is the same as the induced e.m.f. If no current is taken from the secondary coil then no energy is taken. The consequence of this is, for an ideal transformer, that no energy is taken from the primary coil and so there must be no, or little net current in the primary. This means that the net e.m.f. in the primary coil is zero, i.e. the induced e.m.f. is equal to but opposing the input voltage. Thus

$$\frac{V_1}{V_2} = \frac{N_1}{N_2} \qquad [17.1]$$

Since the ideal transformer is considered to be 100 per cent efficient there is no power loss and thus the

volt-amperes supplied to the primary

$\qquad$ = volt-amperes supplied to the load

$$I_1 V_1 = I_2 V_2.$$

Thus

$$\frac{V_1}{V_2} = \frac{I_2}{I_1} = \frac{N_1}{N_2}. \qquad [17.2]$$

Thus a transformer which steps up the voltage, steps down the current.

Example 17.1 *A single-phase transformer has 250 primary turns and 50 secondary turns. What will be the secondary voltage and current if the* *transformer has an a.c. input of 240 V at 5 A (r.m.s. values)?*

Using equation [17.1],

$$\frac{V_1}{V_2} = \frac{N_1}{N_2}$$

Hence

$$V_2 = \frac{50 \times 240}{250}$$

$$= 48 \text{ V}.$$

The current can be obtained using equation [17.2].

$$\frac{I_2}{I_1} = \frac{N_1}{N_2}$$

$$I_2 = \frac{250 \times 5}{50}$$

$$= 25 \text{ A}.$$

17.3 The ideal transformer

With the ideal transformer there are:

1. No flux losses, the same flux linking each turn of both the primary and secondary coils.
2. The volts per turn is the same for each winding, whether it be primary or secondary, i.e. V/N is the same for all windings (see equation [17.1]).
3. The primary and secondary ampere-turns are equal, i.e. $NI = a$ constant (see equation [17.2]).

Example 17.2 *The primary of a transformer has 1200 turns and takes a current of 0.5 A when supplied with an alternating input of 100 V. What is (a) the volts per turn, (b) the secondary voltage if it has 300 turns, (c) the ampere-turns? The transformer may be considered to be an ideal transformer.*

(a) Volts per turn $= \dfrac{100}{1200}$

$$= 0.8333 \text{ V per turn.}$$

(b) Secondary voltage = volts per turn × turns

$$= 0.0833 \times 300$$

$$= 25.0 \text{ V}.$$

(c) Ampere-turns $= 0.5 \times 1200$

$$= 600 \text{ ampere-turns.}$$

17.4 The e.m.f. equation of a transformer

An alternating current in the primary gives rise to alternating flux within the transformer core. If the current is sinusoidal and we assume that the flux is always directly proportional to the current, then

$$\Phi = \Phi_m \sin 2\pi f t$$

describes the variation with time of the flux. Φ is the flux at time t, Φ_m is the maximum value of the flux and f the frequency with which the flux, and the primary current, alternates. The induced e.m.f. per turn is given by (see chapter 6) using calculus notation for rate of change:

$$E = -\frac{d\Phi}{dt}$$

$$= -\frac{d(\Phi_m \sin 2\pi f t)}{dt}$$

$$= -2\pi f \Phi_m \cos 2\pi f t. \tag{17.3}$$

The maximum value of this induced e.m.f. per turn occurs when the cosine term has its maximum value of 1. Thus

$$\text{Maximum induced e.m.f. per turn} = 2\pi f \Phi_m \tag{17.4}$$

The r.m.s. value of this e.m.f. is given by

$$\text{r.m.s. value of induced e.m.f. per turn} = \frac{\text{maximum value}}{\sqrt{2}}$$

$$= \frac{2\pi f \Phi_m}{\sqrt{2}}$$

$$= 4.44 f \Phi_m. \tag{17.5}$$

Hence if the primary has N_1 turns then the r.m.s. value of the primary e.m.f. is $4.44 N_1 f \Phi_m$. With a secondary having N_2 turns the r.m.s. value of the secondary e.m.f. is $4.44 N_2 f \Phi_m$.

Example 17.3 *A single-phase transformer, rated at 100 VA, 240/12 V, 50 Hz, has 120 turns on the secondary. Calculate (a) the values of the primary and secondary currents, (b) the number of primary turns, (c) the maximum value of the flux. Assume the transformer is ideal.*

(a) Volt-amperes supplied to the primary = 100 VA.

Since the primary voltage is 240 V the primary current I_1 must be given by

$$240 I_1 = 100$$

$$I_1 = 0.42 \text{ A}.$$

Since the secondary voltage is 12 V the secondary current I_2 must be given by

$$12I_2 = 100$$

$$I_2 = 8.33 \text{ A}.$$

(b) Using equation [17.1]

$$\frac{V_1}{V_2} = \frac{N_1}{N_2}$$

$$N_1 = \frac{120 \times 240}{12}$$

$$= 2400.$$

(c) Using equation [17.5]

r.m.s. value of induced e.m.f. per turn $= 4.44 f \Phi_m$

Hence
$$\Phi_m = \frac{240}{2400 \times 4.44 \times 50}$$

$$= 4.5 \times 10^{-4} \text{ Wb}.$$

17.5 Phasor diagram for a transformer on no load

As indicated in equation [17.3], the e.m.f. induced by a sinusoidal flux is out of phase with that flux.

Flux equation

$$\Phi = \Phi_m \sin 2\pi ft$$

E.m.f. equation

$$E = -2\pi f \Phi_m \cos 2\pi ft$$

$$E = -E_m \cos 2\pi ft$$

where the maximum induced e.m.f. per turn is $E_m = 2\pi f \Phi_m$.

Since $-\cos \theta = \sin(\theta - 90°)$, the equation for the e.m.f. can be written as

$$E = E_m \sin(2\pi ft - \pi/2).$$

The e.m.f. thus lags the flux by a quarter of a cycle, $\pi/2$ or $90°$. The phasors for the e.m.f.s induced in the primary and the secondary are thus drawn $90°$ behind the phasor for Φ. The value of these e.m.f.s is proportional to the number of windings in each coil.

When a transformer is on no load the net e.m.f. in the primary is virtually zero, the phasor for V_1, the input voltage, can thus be drawn equal and opposite to that representing the induced e.m.f. in the primary.

Figure 17.2 shows the phasor diagram for a transformer on no load. For the example given the induced e.m.f.s in the primary and the secondary have been drawn of equal length, this would imply equal numbers of turns in the primary and secondary. Current I_0 is the no load current flowing in the primary.

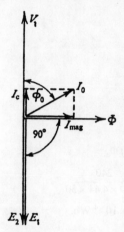

Fig. 17.2　Phasor diagram for a transformer on no load

The no load current can be resolved into two components. The component I_{mag} in the direction of the flux phasor is the current responsible for producing the flux. The component I_c is in the direction of V_1 and is the current supplying the hysteresis and eddy current losses in the iron core and the negligible $I^2 R$ loss in the primary winding. Neglecting this $I^2 R$ loss, the losses in the iron of the core are thus

$$\text{Iron loss} = I_c V_1 \qquad \qquad [17.6]$$

The relationship between the components of I_0 can be obtained by applying the Pythagorus theorem to the part of the figure involving the current phasor and its components. Thus

$$I_0^2 = I_c^2 + I_{mag}^2$$
$$I_0 = \sqrt{(I_c^2 + I_{mag}^2)}. \qquad \qquad [17.7]$$

The no load power factor is $\cos \Phi_0$. Thus

$$\cos \Phi_0 = \frac{I_c}{I_0}. \qquad \qquad [17.8]$$

Figure 17.3 shows the waveforms of the voltages, magnetizing current and the flux for the transformer on no load.

Example 17.4　*A 240/12-V single-phase transformer when connected to a 50-Hz supply takes 0.50 A and dissipates 20 W. If the secondary is on open*

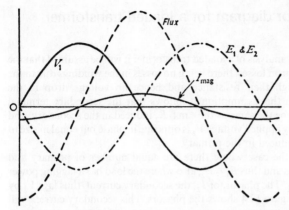

Fig. 17.3 Waveforms of voltages, magnetizing current and flux for a transformer on no load

circuit what is (a) *the iron loss,* (b) *the power factor and* (c) *the magnetizing current.*

(a) The power dissipated is due to the iron loss. Thus iron loss is 20 W.

(b) Using equation [17.6]

$$\text{iron loss} = I_c V_1.$$

But

$$I_c = I_0 \cos \Phi_0,$$

hence

$$\text{iron loss} = V_1 I_0 \cos \Phi_0.$$

Hence

$$\cos \Phi_0 = \frac{20}{240 \times 0.50}$$

$$= 0.167 \text{ lagging.}$$

(c) Consideration of the phasor diagram in fig. 17.2 shows that

$$\frac{I_{\text{mag}}}{I_0} = \sin \Phi_0.$$

Hence since $\Phi_0 = 80.4°$,

$$I_{\text{mag}} = 0.50 \times \sin 80.4°$$

$$= 0.49 \text{ A.}$$

17.6 Phasor diagram for a loaded transformer

To simplify the analysis of a loaded transformer it will be assumed that the only losses are iron losses, there being no losses in the windings due to the currents through their resistance and hence no voltage drop in the windings. With this assumption it follows that the secondary terminal voltage V_2 will be the same as the e.m.f. E_2 induced in the secondary, and that the primary applied voltage V_2 is opposite in phase but equal in size to the e.m.f. E_1 induced in the primary.

Consider the case where there are equal number of primary and secondary turns and thus $E_1 = E_2$, also where the load has a lagging power factor of $\cos \Phi_2$. The phasor for I_2, the secondary current, thus lags V_2 by an angle Φ_2. Figure 17.4 shows the phasors. This secondary current will

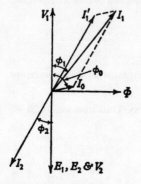

Fig. 17.4   Phasor diagram for a loaded transformer

attempt to create flux in the core. However, the flux in the core depends only on the applied primary voltage V_1. This means that the effect of the secondary current must be nullified by a primary current. Thus in the figure I_1' represents the component of the primary current which is equal in size to I_2 but opposite in phase, thus cancelling out the magnetizing effect of the secondary current. Phasor I_0 is the no-load current in the primary, as discussed in the previous section. The phasor sum of I_1' and I_0 gives the total primary current I_1 and the power factor on the primary side of the transformer is $\cos \Phi_1$.

Example 17.5   *A single-phase transformer having 1000 turns on the primary and 200 turns on the secondary has a no load current of 4 A at a power factor of 0.2 lagging. Calculate the primary current and power factor when the secondary current is 220 A at a power factor of 0.8 lagging. Assume that the voltage drop in the windings is negligible.*

When the voltage drop is negligible there must be negligible resistance and hence all the losses are effectively just iron losses.

I_1' is the primary current to neutralize the magnetizing effect of the secondary current. Thus the ampere turns due to I_1' must be equal and opposite to those due to I_2. Hence

$$I_1' \times 1000 = 220 \times 200$$

$$I_1' = 44 \, \text{A}.$$

From the phasor diagram in fig. 17.4,

$$I_1 \cos \Phi_1 = I_1' \cos \Phi_2 + I_0 \cos \Phi_0$$

$$= 44 \times 0.8 + 4 \times 0.2$$

$$= 36 \, \text{A}.$$

Also

$$I_1 \sin \Phi_1 = I_1' \sin \Phi_2 + I_0 \sin \Phi_0$$

Since $\cos \Phi_2 = 0.8$ then $\sin \Phi_2 = 0.6$, also since $\Phi_0 = 0.2$ then $\sin \Phi_0 = 0.98$. Thus

$$I_1 \sin \Phi_1 = 44 \times 0.6 + 4 \times 0.98$$

$$= 30.3 \, \text{A}.$$

Since $\sin^2 \Phi_1 + \cos^2 \Phi_1 = 1$, then

$$I_1^2 = (I_1 \sin \Phi_1)^2 + (I_1 \cos \Phi_1)^2$$

$$= 36^2 + 30.3^2$$

$$I_1 = 47.1 \, \text{A}.$$

The primary power factor $= \cos \Phi_1$. But

$$\tan \Phi_1 = \frac{I_1 \sin \Phi_1}{I_1 \cos \Phi_1}.$$

Hence

$$\Phi_1 = 40.1°.$$

Thus primary power factor $= 0.77$ lagging.

17.7 Iron loss

The iron loss is the power dissipated in the core of the transformer. Two types of loss are responsible for the iron loss, the *hysteresis loss* and the *eddy current loss*.

$$\text{Iron loss} = \text{hysteresis loss} + \text{eddy current loss.} \qquad [17.9]$$

When an alternating current passes through the primary the magnetic state of the transformer core is taken through a complete magnetization cycle, a hysteresis loop (see section 7.10). Energy is consumed in this magnetizing, demagnetizing, magnetizing in the reverse direction, demagnetizing cycle. The energy consumed is proportional to the area enclosed by the hysteresis loop. From a consideration of such areas the following equation has been derived

$$\text{hysteresis loss} = kvf B_m^n$$

where k is a constant for a particular specimen, known as the hysteresis coefficient, v is the volume of the iron, f the frequency of the alternating supply responsible for the magnetization, B_m is the maximum value of the flux density in the core and n is the number known as the Steinmetz coefficient and has a value within the range 1.6 to 2.2.

The hysteresis loss can be reduced by choosing a material for the transformer core that has a small area enclosed by its hysteresis loop.

The eddy current loss arises from the fact that currents are produced in the core due to the changing magnetic flux, such currents being referred to as eddy currents (see section 14.2 for another example of eddy currents). The core is an electrical conductor and when the magnetic flux is linked with a conductor changes then an e.m.f. is induced in it. This induced e.m.f. sets up the eddy currents.

The electrical resistance of a solid iron core can be very small and so the induced e.m.f. can set up very large eddy currents. Such large currents mean large power losses. By using a core made up of thin laminations which are insulated from each other, by perhaps layers of varnish, the loss can be considerably reduced. One of the effects of using laminations is to increase the resistance and so reduce the size of the eddy currents. Eddy current losses can also be reduced by using a steel for the core which has a high resistivity.

The eddy current loss is proportional to the square of the frequency and the square of the maximum flux density for a given core.

17.8 The transformer core

The transformer core is a magnetic circuit, having as its function the linking of the windings of the primary and secondary coils as efficiently as possible. The hysteresis loop for a magnetic material indicates the suitability of a material for use as a transformer core. The various factors that determine such a choice are:

1. The greater the value of the flux density B for a given magnetic field strength H the better. The magnetic field strength is proportional to the current and thus as the flux density is proportional to the flux, this statement is really stating that the bigger the flux produced for a given current the better. Thus in fig. 17.5(a) material 1 is preferable to material 2.

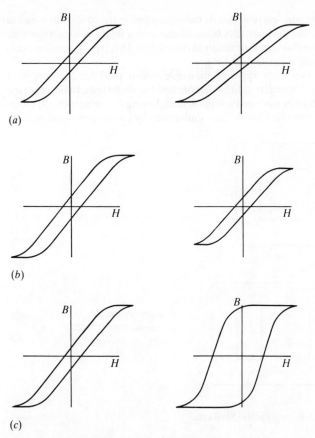

Fig. 17.5 (a) Material (1) is preferable, B/H is greater; (b) Material (1) is preferable, B_{max} is greater; (c) Material (1) is preferable, the loop area is smaller

2. The level of the flux density B before saturation should be as high as possible. At saturation an increase in magnetic field strength, and hence current, has no effect on the flux density. Distortion thus occurs when the current used is near to, or at, the saturation value. Thus in fig. 17.5(b) material 1 is preferable to material 2.

3. The bigger the area enclosed by the hysteresis loop the greater the hysteresis loss (see section 17.7), thus a low area loop is desirable. Thus in fig. 17.5(c) material 1 is preferable to material 2.

The form of the core is determined by the need to reduce eddy current losses (see section 17.7). To reduce the size of the eddy currents the electrical resistance of the core has to be made as large as possible. At low frequencies, like that of the mains supply, and at audio frequencies, this increased resistance is obtained by making the core from thin laminations, each being insulated from its neighbours. The thinner the laminations the smaller the eddy current loss. At radio frequencies the core is made from a steel dust, the dust particles being mixed with a high electrical resistance paste and moulded to the required core shape. This type of transformer is said to have a ferrite dust core.

The two basic types of magnetic circuit used for the core of a transformer are referred to as the core and the shell-types. In the core type the transformer has just two limbs, each having the same uniform cross-sectional area (fig. 17.6(a)). Each limb generally carries part of the primary

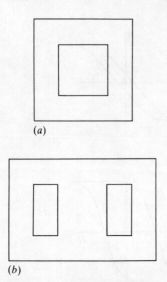

(*a*)

(*b*)

Fig. 17.6   (a) Core type; (b) Shell type

and part of the secondary windings. The shell type has three limbs, the centre limb having twice the cross-sectional area of the other two limbs (Fig. 17.6(b)). The centre limb carries both the primary and the secondary windings.

17.9 Transformer efficiency

The *efficiency* η (eta) of a transformer is

$$\text{efficiency} = \frac{\text{output power}}{\text{input power}} \qquad [17.10]$$

This is often expressed as a percentage,

$$\text{efficiency} = \frac{\text{output power}}{\text{input power}} \times 100\%$$

The output power is the input power minus the losses, with the power loss in the transformer being the iron losses and copper loss. The *copper loss* is the term used to describe the I^2R losses in the copper conductors of the primary and secondary windings due to currents flowing through them. Thus

$$\text{efficiency} = \frac{(\text{input power} - \text{losses})}{\text{input power}} \times 100\%$$

$$= \left(1 - \frac{\text{losses}}{\text{input power}}\right) \times 100\%$$

$$= \left(1 - \frac{\text{iron loss} + \text{copper losses}}{\text{input power}}\right) \times 100\% \qquad [17.11]$$

Example 17.6   *A 10-kVA transformer has an iron loss of 50 W and is operating at full load with a copper loss of 120 W and a power factor of 0.8. What is the efficiency of the transformer under these conditions?*

output power = output volt-amperes × power factor

$$= 10\,000 \times 0.8$$

$$= 8000 \text{ W}$$

total losses = 50 + 120

$$= 170 \text{ W}.$$

Hence

$$\text{efficiency} = \frac{\text{output power}}{\text{input power}} \times 100\%$$

$$= \frac{8000}{8000 + 170} \times 100\%$$

$$= 97.9\%.$$

Note that a transformer operating at full load is one that is delivering into a load the specified volt-amperes; the actual power will depend on the power factor.

Example 17.7   *A 40-kVA single-phase transformer has an iron loss of 400 W and a full-load copper loss of 600 W. What is (a) the total losses and (b) the efficiency at half load for a power factor of 0.8?*

(a) The copper loss varies as the square of the current, therefore the copper loss at half load is

copper loss on half load = $600 \times (0.5)^2$

$$= 150 \text{ W}.$$

Thus

total loss on half load $= 150 + 400 = 550$ W.

The iron loss is generally assumed to be constant and independent of the load.

(b) At 0.8 power factor the full load output is $40 \times 0.8 = 32$ kW. At half load the output is half this, i.e. 16 kW. Thus

input power $= 16\,000 + 550$

$= 16\,550$ W.

Hence the efficiency is given by

$$\text{efficiency} = \frac{\text{output power}}{\text{input power}} \times 100\%$$

$$= \frac{16\,000}{16\,550} \times 100\%$$

$$= 96.7\%.$$

Note that a transformer operating at half load is one that is operating at half the specified volt-amperes, this means that for a given load the secondary current is half that which would occur if the transformer was operating at full load.

17.10 Transformer regulation

The secondary voltage of a transformer decreases as the load is increased because of the increase in the losses that occur. The difference between the secondary voltage at no load and the secondary voltage at full load, when divided by the secondary no load voltage, is called the *voltage regulation per unit*.

Voltage regulation

$$= \frac{\text{No load voltage} - \text{full load voltage}}{\text{no load voltage}} \qquad [17.12]$$

This is also sometimes expressed as a percentage.

Example 17.8 *Calculate the voltage regulation per unit for a single-phase transformer which has an open circuit voltage of 250 V and a terminal voltage of 240 V when on full load.*

Using equation [17.12],

$$\text{voltage regulation} = \frac{250 - 240}{250}$$

$$= 0.04.$$

17.11 Impedance matching

One of the uses of a transformer is to alter the apparent value of an impedance. For an ideal transformer equation [17.1] gives

$$\frac{V_1}{V_2} = \frac{N_1}{N_2}$$

Also, if the transformer supplies a load taking a secondary current I_2, then according to equation [17.2]

$$\frac{I_1}{I_2} = \frac{N_2}{N_1}.$$

Thus

$$V_1 = \frac{N_1 V_2}{N_2}$$

$$I_1 = \frac{N_2 I_2}{I_1}$$

and so

$$\frac{V_1}{I_1} = \left(\frac{N_1}{N_2}\right)^2 \frac{V_2}{I_2}.$$

But V_1/I_1 is the impedance Z_1 looking in to the primary terminals of the transformer. It is the impedance into which the source generator is working. It is the effective load as seen by the source. V_2/I_2 is the impedance Z_2 of the secondary. Thus

$$Z_1 = \left(\frac{N_1}{N_2}\right)^2 Z_2. \qquad [17.13]$$

Thus the interposition of the transformer between the source and the load alters the apparent impedance of the load.

The maximum power transfer theorem (see chapter 20) states that the maximum power is transferred from one circuit to another when the impedance of the load circuit equals the internal impedance of the supply circuit. By using a transformer to alter the impedance it is possible to have maximum power transfer between circuits which otherwise would not have had equal impedances. The transformer is said to be *impedance matching*.

Example 17.9 *An amplifier has an output impedance of 50 Ω and is to be connected to a circuit which has an input impedance of 800 Ω. What should be the turns ratio of the transformer used to link the two and give maximum power transfer?*

Using equation [17.13],

$$50 = \left(\frac{N_1}{N_2}\right)^2 800.$$

Thus

$$\text{turns ratio} = \frac{N_1}{N_2} = \sqrt{\left(\frac{50}{800}\right)} = 0.25.$$

17.12 Summary of important formulae

$$\frac{V_1}{V_2} = \frac{I_2}{I_1} = \frac{N_1}{N_2} \qquad\qquad\qquad [17.2]$$

Induced e.m.f. per turn $= -2\pi ft\Phi_m \cos 2\pi ft$ [17.3]

R.m.s. value of induced e.m.f. per turn

$$= 4.44f\Phi_m \qquad\qquad\qquad [17.5]$$

$$\text{Iron loss} = I_c V_1 \qquad\qquad\qquad [17.6]$$

$$I_0 = \sqrt{(I_c^2 + I_{mag}^2)} \qquad\qquad\qquad [17.7]$$

$$\cos\Phi_0 = \frac{I_c}{I_0} \qquad\qquad\qquad [17.8]$$

Iron loss = hysteresis loss

$$+ \text{eddy current loss} \qquad [17.9]$$

$$\text{Efficiency} = \frac{\text{output power}}{\text{input power}} \qquad [17.10]$$

$$\text{Voltage regulation} = \frac{\text{No load voltage} - \text{full load voltage}}{\text{no load voltage}}$$

$$[17.12]$$

$$Z_1 = \left(\frac{N_1}{N_2}\right)^2 Z_2 \qquad\qquad\qquad [17.13]$$

17.13 Examples

1. A single-phase transformer has 1000 primary turns and 200 secondary turns. What is the secondary voltage and current, assuming the transformer to be ideal, for a primary input of 240 V at 3 A?
2. An ideal transformer has a step-up voltage ratio of 5:1. If the secondary voltage is 2000 V and its current 1.0 A, what is the primary voltage and current?

3. An ideal transformer has a step-down voltage ratio of 500/120 V. If the primary has 300 turns and the primary current is 1.2 A, what is (a) the number of turns on the secondary and (b) the secondary current?

4. A 10-kV single-phase transformer has 800 turns in the primary winding and 200 turns in the secondary. What is (a) the secondary voltage and current, (b) the primary current, when 250 V is connected to the primary?

5. The primary winding of an ideal transformer has 500 turns and takes a current of 6.0 A when supplied with an alternating input of 50 V. What is (a) the volts per turn, (b) the secondary voltage if the secondary winding has 200 turns, (c) the ampere-turns?

6. A 3000/240-V, 50-Hz, single-phase transformer has a core having an effective cross-sectional area of 120 cm^2 and a primary winding of 800 turns. What is (a) the number of secondary turns, (b) the maximum value of the flux density?

7. Derive an expression for the r.m.s. value of the e.m.f. induced in a coil of N turns by a sinusoidally varying flux with a maximum value of Φ_m and a frequency f.

   An ideal transformer, rated at 50 kVA, 1000/250 V, 50 Hz, has 100 secondary turns. Calculate (a) the values of the primary and secondary currents, (b) the maximum value of the flux.

8. A 3300/240-V, 50-Hz, single-phase core-type transformer is to have a maximum flux density of 1.2 T and a volts-per-turn of 12 V. What should be the number of primary and secondary turns and the cross-sectional area of the core?

9. A 240/20-V, 50-Hz, single-phase transformer takes a current of 1.2 A and dissipates 60 W. If the secondary is on open circuit, what is (a) the iron loss, (b) the power factor and (c) the magnetising current?

10. Distinguish between hysteresis and eddy current losses in the core of a transformer and explain how these losses depend on (a) the frequency and (b) the secondary load.

11. If the hysteresis loss in a transformer is 200 W when the frequency is 50 Hz, what will be the loss at 60 Hz if the maximum flux density in the core is unchanged?

12. Draw the phasor diagrams for a single-phase transformer operating (a) on no-load and (b) with a load when the voltage drop in the windings is negligible.

13. A single-phase transformer supplies an output power of 12 kW. If under these conditions the iron loss is 100 W and the copper loss 160 W, what is the efficiency of the transformer?

14. A 20 kVA transformer has an iron loss of 250 W and a copper loss on full load of 320 W. Determine the efficiency at full load if the power factor of the load is 0.8.

15. A 5 kVA transformer has an iron loss of 50 W and a full load copper loss of 120 W. What is the efficiency when the transformer is operating at half load, power factor unity?

16. Calculate the voltage regulation per unit for a single-phase transformer which has an open circuit voltage of 500 V and a terminal

voltage of 420 V when on full load.

17. A voltage source has an internal impedance of $30\,\Omega$ and is to be connected to a circuit which has an input impedance of $1000\,\Omega$ via a transformer. What should be the turns ratio of the transformer for maximum power transfer?

18. A circuit with an output impedance of $1.2\,k\Omega$ is to be matched for maximum power transfer with a circuit having an input impedance of $10\,\Omega$ by a transformer. What should be the turns ratio of that transformer?

CHAPTER 18

Electrical Measurements

18.1 Electrical indicating instruments

An analogue instrument is generally fitted with a pointer which indicates on a scale the value of the quantity being measured. The moving system of such an instrument is usually carried by a spindle of hardened steel, having its ends tapered and highly polished* to form pivots which rest in hollow-ground bearings, usually of sapphire, set in steel screws. In some instruments, the moving system is attached to two thin ribbons of spring material such as beryllium–copper alloy, held taut by tension springs mounted on the frame of the movement. This arrangement eliminates pivot friction and the instrument is less susceptible to damage by shock or vibration.

Analogue instruments possess three essential features:

(a) a *deflecting device* whereby a mechanical force is produced by, and is an analogue of, the electric current, voltage or power,
(b) a *controlling device* whereby the value of the deflection is dependent upon the magnitude of the quantity being measured, and
(c) a *damping device* to prevent oscillation of the moving system and enable the latter to reach its final position quickly.

The action of the deflecting device depends upon the type of instrument, and the principle of operation of each of the instruments most commonly used in practice will be described in later sections.

18.2 Controlling devices

There are two types of controlling devices, namely:

(a) spring control and
(b) gravity control (not used in modern instruments).

---

* The necessity of handling measuring instruments with care may be realized from the fact that if a pivot point has a circle of contact 0.05 mm in diameter and supports a mass of 3 grams (a normal mass for the moving system of an ammeter or voltmeter), the pressure is approximately 15 MN/m^2.

The most common arrangement of spring control utilises two spiral hairsprings, A and B (fig. 18.1), the inner ends of which are attached to the spindle S. The outer end of B is fixed, whereas that of A is attached to one end of a lever L, pivoted at P, thereby enabling zero adjustment to be easily

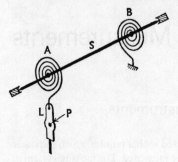

Fig. 18.1   Spring control

effected. The hairsprings are of non-magnetic alloy such as phosphor–bronze or beryllium–copper.

The two springs, A and B, are wound in opposite directions so that when the moving system is deflected, one spring winds up while the other unwinds, and the controlling torque is due to the combined torsions of the springs. Since the torsional torque of a spiral spring is proportional to the angle of twist, the controlling torque is directly proportional to the angular deflection of the pointer.

With gravity control, masses A and B are attached to the spindle S (fig. 18.2), the function of A being to balance the weight of pointer P. Mass

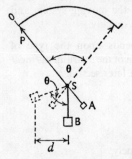

Fig. 18.2   Gravity control

B therefore provides the controlling torque. When the pointer is at zero, B hangs vertically downwards. When P is deflected through an angle θ, the controlling torque is equal to (weight of B × distance d) and is therefore proportional to the sine of the angular deflection. This has the disadvantage that with deflections of the order of 70° and 80°, the controlling torque increases very slowly with increase of deflection; thus:

$$\frac{\text{controlling torque for } 80° \text{ deflection}}{\text{controlling torque for } 70° \text{ deflection}} = \frac{\sin 80°}{\sin 70°} = 1\cdot048,$$

whereas with spring control:

$$\frac{\text{controlling torque for } 80° \text{ deflection}}{\text{controlling torque for } 70° \text{ deflection}} = \frac{80}{70} = 1.143.$$

Hence, with gravity control, the scale at the top end is more open than with spring control and the total deflection is limited to about 80°. A further disadvantage of gravity control is that the instrument must be correctly levelled before being used. The only advantage of the gravity-controlled instrument is that it is cheaper than the corresponding spring-controlled instrument.

18.3 Damping devices

The combination of the inertia of the moving system and the controlling torque of the spiral springs or of gravity gives the moving system a natural frequency of oscillation (section 13.8). Consequently, if the current through an under-damped ammeter were increased suddenly from zero to OA (fig. 18.3), the pointer would oscillate about its mean position, as

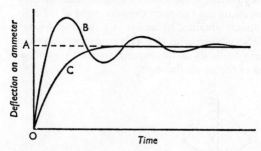

Fig. 18.3 Damping curves

shown by curve B, before coming to rest. Similarly every fluctuation of current would cause the pointer to oscillate and it might be difficult to read the instrument accurately. It is therefore desirable to provide sufficient damping to enable the pointer to reach its steady position without oscillation, as indicated by curve C. Such an instrument is said to be *dead-beat*.

The two methods of damping commonly employed are

(a) eddy-current damping and
(b) air damping.

One form of eddy-current damping is shown in fig. 18.4, where a copper or aluminium disc D, carried by a spindle, can move between the poles of a permanent magnet M. If the disc moves clockwise, the e.m.f.s

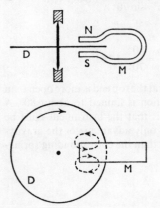

Fig. 18.4   Eddy-current damping

induced in the disc circulate eddy currents as shown dotted. It follows from Lenz's Law that these currents exert a force opposing the motion producing them, namely the clockwise movement of the disc.

Another arrangement, used in moving-coil instruments (section 18.5), is to wind the coil on an aluminium frame. When the latter moves across the magnetic field, eddy current is induced in the frame; and, by Lenz's Law, this current exerts a torque opposing the movement of the coil.

Air damping is usually obtained by means of a thin metal vane V attached to the spindle S, as shown in fig. 18.5. This vane moves in a sector-shaped box C, and any tendency of the moving system to oscillate is damped by the action of the air on the vane.

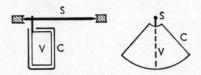

Fig. 18.5   Air damping

18.4 Types of ammeters, voltmeters and wattmeters

The principal types of electrical indicating instruments, together with the methods of control and damping, are summarized in Table 18.1.

Table 18.1

Type of instrument	Suitable for measuring	Method of control	Method of damping
Permanent-magnet moving-coil	Current and voltage d.c. only	Hairsprings	Eddy current
Moving-iron	Current and voltage d.c. and a.c.	Hairsprings	Air
Thermo-couple*	Current and voltage, d.c. and a.c.	As for moving coil	As for moving coil
Electro-dynamic† or dynamometer	Current, voltage and power, d.c. and a.c.	Hairsprings	Air
Electronic	Current and voltage, d.c. and a.c.	As for moving coil	As for moving coil
Rectifier	Current and voltage, a.c. only	As for moving coil	As for moving coil

* The hot-wire instrument is obsolete and only of historic interest.
† 'Electrodynamic' is the term recommended by the British Standards Institution, but 'dynamometer' is the term more frequently used in practice.

Apart from the electronic type of voltmeter, all voltmeters are in effect milliammeters connected in series with a non-reactive resistor having a high resistance. For instance, if a milliammeter has a full-scale deflection with 10 mA and has a resistance of 10 Ω and if this milliammeter is connected in series with a resistor of 9990 Ω, then the p.d. required for full-scale deflection is 0.01 × 10 000, namely 100 V, and the scale of the milliammeter can be calibrated to give the p.d. directly in volts.

18.5 Permanent-magnet moving-coil ammeters and voltmeters

The high coercive force (section 7.10) of modern steel alloys, such as Alcomax (iron, aluminium, cobalt, nickel and copper), allows the use of relatively short magnets and has led to a variety of arrangements of the magnetic circuit for moving-coil instruments. The front elevation and sectional plan of one arrangement are shown in fig. 18.6, where M represents a permanent magnet and PP are soft-iron pole-pieces. The hardness of permanet-magnet materials makes machining difficult, whereas soft iron can be easily machined to give exact airgap dimensions.

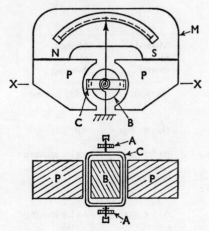

Fig. 18.6 Permanent-magnet moving-coil instrument

In one form of construction, the anisotropic* magnet and the pole-pieces are of the sintered type, i.e. powdered magnet alloy and powdered soft iron are compressed in a die to the required shape and heat-treated so that the magnet and the pole-pieces become alloyed, thereby eliminating airgaps at the junctions of the materials.

An alternative arrangement is to cast the magnet and attach the soft-iron plate, in one piece, to the two surfaces of M which have been rendered flat by grinding, the joints being made by a resin-bonding technique. This construction enables the drilling of the cylindrical hole and the machining of the gaps to be done with precision.

The rectangular moving-coil C in fig. 18.6 consists of insulated copper wire wound on a light aluminium frame fitted with polished steel pivots resting in jewel bearings. Current is led into and out of the coil by spiral hairsprings AA, which also provide the controlling torque. The coil is free† to move in airgaps between the soft-iron pole-pieces PP and a soft-iron cylinder B supported by a brass plate (not shown). The functions of the central core B are (a) to intensify the magnetic field by reducing the length of airgap across which the magnetic flux has to pass and (b) to gave a radial magnetic flux of uniform density, thereby enabling the scale to be uniformly divided.

An alternative arrangement of the magnet system is shown in fig. 18.7, where M represents the magnet and PP are soft-iron pole-pieces.

---

* A magnet is made *anisotropic* by being cooled at a particular rate in a powerful directional magnetic field. As it solidifies, the magnetic domain (section 7.11) remain aligned in the same direction, thereby giving the magnet a high coercive force in that direction. An *isotropic* magnet material, on the other hand, has equal magnetic properties in all directions.

† Students often form the impression that the moving coil is wound on the iron cylinder. It is important to realize that this cylinder is *fixed* and that the frame, on which the coil is wound, *does not touch* the cylindrical core.

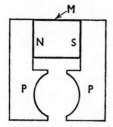

Fig. 18.7   An alternative arrangement of the magnet system

Figure 18.8 shows an arrangement where the central cylindrical core consists of an anisotropic magnet M, with soft-iron pole-pieces PP attached to M. The return path for the magnetic flux is provided by a soft-

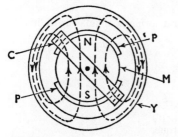

Fig. 18.8   Moving-coil instrument with centre-core magnet

iron ring or yoke, concentric with the core and separated from it by an airgap of uniform width, the distribution of the flux being as indicated by the dotted lines in fig. 18.8. The moving coil C, wound on an aluminium frame supported by jewel bearings and controlled by spiral hairsprings, is free to move in the airgap between the pole-pieces and the yoke. This arrangement gives a very compact movement and is particularly suitable for small instruments.

By using a more elaborate magnet system, it is possible to construct moving-coil instruments with scales extending over about 250°, but such instruments are beyond the scope of this volume.

The manner in which a torque is produced when the coil is carrying a current may be understood more easily by considering a single turn PQ, as in fig. 18.9. Suppose P to carry current outwards from the paper; then Q is carrying current towards the paper. Current in P tends to set up a magnetic field in a counterclockwise direction around P and thus strengthens the magnetic field on the lower side and weakens it on the upper side. The current in Q, on the other hand, strengthens the field on the upper side while weakening it on the lower side. Hence, the effect is to distort the magnetic flux as shown in fig. 18.9. Since the flux behaves like stretched elastic cords, it tries to take the shortest path between poles NS, and thus exerts forces *FF* on coil PQ, tending to move it out of the magnetic field.

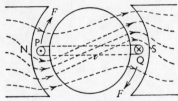

Fig. 18.9   Distribution of resultant magnetic field

The deflecting torque ∝ current through coil × flux density in gap

$$= kI \text{ for uniform flux density,}$$

where k = a constant for a given instrument

and I = current through coil.

The controlling torque of the spiral springs ∝ angular deflection

$$= c\theta$$

where c = a constant for given springs

and θ = angular deflection.

For a steady deflection,

controlling torque = deflecting torque,

and $$c\theta = kI$$

$$\therefore \quad \theta = \frac{k}{c}I,$$

i.e. the deflection is proportional to the current and the scale is therefore uniformly divided.

A numerical example on the calculation of the torque on a moving coil is given in example 5.3, p. 91.

As already mentioned in section 18.3, damping is effected by eddy currents induced in the metal frame on which the coil is wound.

Owing to the delicate nature of the moving system, this type of instrument is only suitable for measuring currents up to about 50 milliamperes directly. When a larger current has to be measured, a *shunt* S (fig. 18.10), having a low resistance, is connected in parallel with the moving coil MC, and the instrument scale may be calibrated to read directly the total current I. Shunts are made of a material, such as manganin (copper, manganese and nickel), having negligible temperature coefficient of resistance. A 'swamping' resistor r, of material having negligible temperature coefficient of resistance, is connected in series with the moving coil. The latter is wound with copper wire and the function of r is to reduce the error due to the variation of resistance of the moving coil with variation of temperature.

The shunt shown in fig. 18.10 is provided with four terminals, the

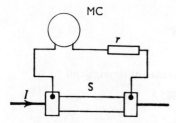

Fig. 18.10 Moving-coil instrument as an ammeter

milliammeter being connected across the potential terminals. If the instrument were connected across the current terminals, there might be considerable error due to the contact resistance at these terminals being appreciable compared with the resistance of the shunt.

The moving-coil instrument may be made into a voltmeter by connecting a resistor R of manganin or other similar material in series, as in fig. 18.11. Again, the scale may be calibrated to read directly the voltage applied to the terminals TT.

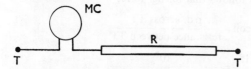

Fig. 18.11 Moving-coil instrument as a voltmeter

The main advantages of the moving-coil instrument are:

(i) high sensitivity,
(ii) uniform scale,
(iii) well-shielded from any stray magnetic field.

Its main disadvantages are:

(i) more expensive than the moving-iron instrument,
(ii) only suitable for direct currents and voltages.

Example 18.1 *A moving coil gives full-scale deflection with 15 mA and has a resistance of 5 Ω. Calculate the resistance required (a) in parallel to enable the instrument to read up to 1 A and (b) in series to enable it to read up to 10 V.*

(a) From Ohm's Law,

$$\frac{\text{current through coil}}{\text{(fig. 18.10)}} = \frac{\text{p.d. across coil}}{\text{resistance of coil}}$$

$$\therefore \quad \frac{15}{1000} = \frac{\text{p.d. (in volts) across coil}}{5}$$

so that

$$\text{p.d. across coil} = 0.075\,\text{V}.$$

From fig. 18.10 it follows that

current through S = total current − current through coil

$$= 1 - 0.015 = 0.985\,\text{A}.$$

Similarly,

$$\text{current through S} = \frac{\text{p.d. across S}}{\text{resistance of S}}$$

$$\therefore \quad 0.985 = \frac{0.075}{\text{resistance of S (in ohm)}}$$

and

$$\text{resistance of S} = \frac{0.075}{0.985} = 0.076\,14\,\Omega.$$

(b) From Ohm's Law it follows that for fig. 18.11,

$$\text{current through coil} = \frac{\text{p.d. across TT}}{\text{resistance between TT}}$$

$$\therefore \quad \frac{15}{1000} = \frac{10}{\text{resistance between TT}}$$

so that

$$\text{resistance between TT} = 666.7\,\Omega.$$

Hence,

resistance required in series with coil

= total resistance between TT − resistance of coil

$$= 666.7 - 5 = 661.7\,\Omega.$$

The moving-coil instrument can be arranged as a multi-range ammeter by making the shunt of different sections as shown in fig. 18.12, where A represents a milliammeter in series with a 'swamping' resistor r of material having negligible temperature coefficient of resistance.

With the selector switch S on, say, the 50-A stud, a shunt having a very low resistance is connected across the instrument, the value of its resistance being such that full-scale deflection is produced when $I = 50$ A. With S on the 10-A stud, the resistance of the two sections of the shunt is approximately five times that of the 50-A section, and full-scale deflection is obtained when $I = 10$ A. Similarly, with S on the 1-A stud, the total resistance of the three sections is such that full-scale deflection is obtained with $I = 1$ A. Such a multi-range instrument is provided with three scales so that the value of the current can be read directly.

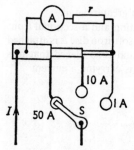

Fig. 18.12   Multi-range moving-coil ammeter

A multi-range voltmeter is easily arranged by using a tapped resistor in series with a milliammeter A, as shown in fig. 18.13. For instance, with

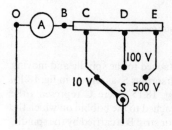

Fig. 18.13   Multi-range moving-coil voltmeter

the data given in example 18.1, the resistance of section BC would be 661.7 Ω for the 10-V range. If D be the tapping for, say, 100 V, the total resistance between O and D = 100/0.015 = 6666.7 Ω, so that the resistance of section CD = 6666.7 − 666.7 = 6000 Ω. Similarly, if E is to be 500-V tapping, section DE must absorb 400 V at full-scale deflection; hence the resistance of DE = 400/0.015 = 26 667 Ω. With the aid of selector switch S, the instrument can be used on three voltage ranges, and the scales can be calibrated to enable the value of the voltage to be read directly.

18.6 Moving-iron ammeters and voltmeters

Moving-iron instruments can be divided into two types:

(i) the *attraction* type, in which a sheet of soft iron is attracted towards a solenoid, and

(ii) the *repulsion* type, in which two parallel rods or strips of soft iron, magnetized inside a solenoid, are regarded as repelling each other.

These two types will now be described in greater detail.

Type (i). Figure 18.14 shows an end elevation and a sectional front view (taken on XX) of the attracted-iron type. A soft-iron disc A is attached to a spindle S carried by jewelled centres J, and is so placed that it is attracted towards solenoid C when the latter is carrying a current.

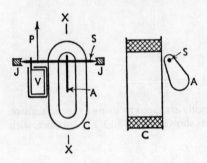

Fig. 18.14   Attraction-type moving-iron instrument

Damping is provided by vane V attached to the spindle and moving in an air chamber, and control is by hairsprings S, as shown in fig. 18.15.

Type (ii). This type is shown in fig. 18.15 where C represents the solenoid. A soft-iron rod or strip A is attached to the bobbin on which the coil is wound, and another soft-iron rod or strip B is carried by the spindle.

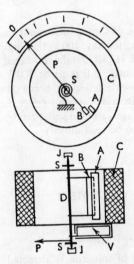

Fig. 18.15   Repulsion-type moving-iron instrument

When a current flows through coil A, A and B are magnetized in the same direction and it is usual to say that B tries to move away from A because poles of the same polarity repel each other. Such a statement, however, gives no indication as to how the deflecting force is actually produced.

In section 5.3, it was shown that when two *permanent* magnets are placed side by side, with the N poles pointing in the same direction as in fig. 5.6, the force of repulsion between the magnets is due to lateral pressure in the magnetic field occupying the space *between* the magnets. In the case of two parallel rods, A and B, situated inside a coil C carrying a current, as in fig. 18.16,* the distribution of the magnetic flux is roughly as shown dotted. There is practically no magnetic flux in the space between A and B, and therefore there cannot be any lateral pressure in that region.

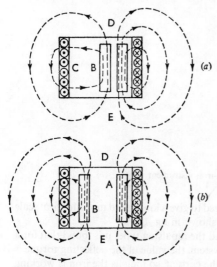

Fig. 18.16 Distribution of magnetic flux in the repulsion-type moving-iron instrument

All the magnetic flux must be linked with the whole or part of the coil, and fig. 18.16(a) shows the approximate distribution of the flux passing through the rods when they are close together. It was mentioned in section 5.3 that magnetic flux behaves as if it were in tension; hence the flux passing through rod B tends to shorten its paths by pulling B towards the left, as shown in fig. 18.16(b). (Rod A is assumed to be fixed.) In addition to this tension effect, there is also a lateral pressure in regions D and E tending to push apart the fluxes passing through A and B, thereby helping to urge rod B towards the left. These effects combine to produce a clockwise deflecting torque on the moving system in fig. 18.15.

As a result of the shortened air paths of the flux passing through B

---

* The rods are shown much wider than is the case in an actual instrument. This is done to enable the dotted lines representing the flux passing through the rods to be shown clearly.

when the latter has moved towards the left, the magnitude of the flux is increased, thereby increasing the inductance of the coil; but the implication of this effect is beyond the scope of this book.

In fig. 18.15 the controlling torque is supplied by two spiral hairsprings, S, each having its inner end attached to the spindle. Air damping is provided by vane V moving in an air chamber.

In commercial instruments, it is usual for the moving-iron B to be in the form of a thin curved plate and for the fixed iron A to be a tapered curved sheet, as shown (without control and damping devices) in fig. 18.17.

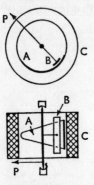

Fig. 18.17   Repulsion-type moving-iron instrument

This construction can be arranged to give a longer and more uniform scale than is possible with the rods shown in fig. 18.15.

For both the attraction and the repulsion types it is found that for a given position of the moving system, the value of the deflecting torque is proportional to the square of the current, so long as the iron is working below saturation. Hence, if the current wave-form is as shown in fig. 18.18,

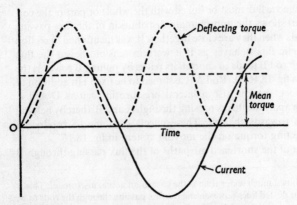

Fig. 18.18   Deflecting torque in a moving-iron instrument

the variation of the deflecting torque is represented by the dotted wave. If the supply frequency is, say, 50 Hz, the torque varies between zero and a maximum 100 times a second, so that the moving system—owing to its inertia—takes up a position corresponding to the mean torque, where

mean torque $\propto$ mean value of the square of the current

$$= kI^2$$

where k = a constant for a given instrument

and I = r.m.s. value of the current (section 10.5).

Hence the moving-iron instrument can be used to measure both direct current and alternating current, and in the latter case the instrument gives the r.m.s. value of the current. Owing to the deflecting torque being proportional to the square of the current, the scale divisions are not uniform, being cramped at the beginning and open at the upper end of the scale.

Since the strength of the magnetic field and therefore the magnitude of the deflecting torque depend upon the number of ampere-turns on the solenoid, it is possible to arrange different instruments to have different ranges by merely winding different number of turns on the solenoids. For example, suppose that full-scale deflection is obtained with 400 ampere-turns, then for

full-scale reading with 100 A, no. of turns = 400/100 = 4,

and

full-scale reading with 5 A, no. of turns = 400/5 = 80.

A moving-iron voltmeter is a moving-iron milliammeter connected in series with a suitable non-reactive resistor.

Example 18.2   *A moving-iron instrument requires 400 ampere-turns to give full-scale deflection. Calculate* (a) *the number of turns required if the instrument is to be used as an ammeter reading up to 50 A and* (b) *the number of turns and the total resistance if the instrument is to be arranged as a voltmeter reading up to 300 V with a current of 20 mA.*

(a) No. of turns = 400/50 = 8.
(b) No. of turns = 400/0.02 = 20 000.

   Total resistance = 300/0.02 = 15 000 Ω.

The advantages of moving-iron instruments are:

  (i) robust construction,
 (ii) relatively cheap,
(iii) can be used to measure direct and alternating currents and voltages.

The disadvantages of moving-iron instruments are:

(i) Affected by stray magnetic fields. Error due to this cause is minimized by the use of a magnetic screen such as an iron casing (section 5.4).

(ii) Liable to hysteresis error when used in a d.c. circuit; i.e. for a given current, the instrument reads higher with decreasing than with increasing values of current. This error is reduced by making the iron strips of nickel-iron alloy such as Mumetal (section 7.10).

(iii) Owing to the inductance of the solenoid, the reading on moving-iron voltmeters may be appreciably affected by variation of frequency. This error is reduced by arranging for the resistance of the voltmeter to be large compared with the reactance of the solenoid.

(iv) Moving-iron voltmeters are liable to a temperature error owing to the solenoid being wound with copper wire. This error is minimized by connecting in series with the solenoid a resistor of a material, such as manganin, having a negligible temperature coefficient of resistance.

18.7 Thermocouple instruments

This type of instrument utilizes the thermoelectric effect observed by Seebeck in 1821, namely that in a closed circuit consisting of two different metals, an electric current flows when the two junctions are at different temperatures. Thus, if A and B in fig. 18.19 are junctions of copper and iron

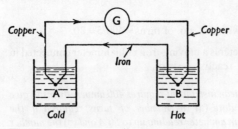

Fig. 18.19　A thermocouple

wires, each immersed in water, then if the vessel containing B is heated, it is found that an electric current flows from the iron to the copper at the cold junction and from the copper to the iron at the hot junction, as indicated by the arrowheads. A pair of metals arranged in this manner is termed a *thermocouple* and gives rise to a *thermo-e.m.f.* when the two junctions are at different temperatures.

This *thermoelectric effect* may be utilized to measure temperature. Thus, if the reading on galvanometer G be noted for different temperatures of the water in which junction B is immersed, the temperature of junction A being maintained constant, it is possible to calibrate the galvanometer in terms of the difference of temperature between A and B. The materials used in practice depend upon the temperature range to be measured; thus, copper–constantan couples are suitable for temperatures up to about 400°C and iron-constantan couples up to about 900°C, constantan being an alloy of copper and nickel. For temperatures up to about 1400°C, a couple made of platinum and platinum–iridium alloy is suitable.

A thermocouple can be used to measure the r.m.s. value of an alternating current by arranging for one of the junctions of wires of dissimilar material, B and C (fig. 18.20), to be placed near or welded to a resistor H carrying the current *I* to be measured. The current due to the thermo-e.m.f. is measured by a moving-coil microammeter A. The heater and the thermo-couple can be enclosed in an evacuated glass bulb D, shown dotted in fig. 18.20, to shield them from draughts. Ammeter A may

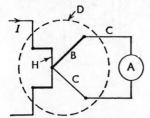

Fig. 18.20 A thermocouple ammeter

be calibrated by noting its reading for various values of direct current through H and it can then be used to measure the r.m.s. value of alternating currents of frequencies up to several megahertz.

18.8 Electrodynamic or dynamometer instruments

The action of this type of instrument depends upon the electromagnetic force exerted between fixed and moving coils carrying current. The upper diagram in fig. 18.21 shows a sectional elevation through fixed coils FF

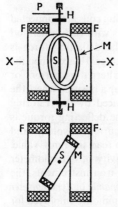

Fig. 18.21 Electrodynamic or dynamometer instrument

and the lower diagram represents a section plan on XX. The moving coil M is carried by a spindle S and the controlling torque is exerted by spiral hairsprings H, which may also serve to lead the current into and out of M.

The deflecting torque is due to the interaction of the magnetic fields produced by currents in the fixed and moving coils; thus fig. 18.22(a) shows the magnetic field due to current flowing through F in the direction indicated by the dots and crosses and fig. 18.22(b) shows that due to

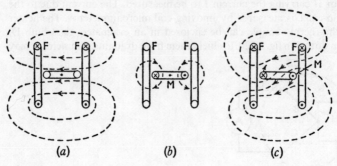

     (a)   (b)   (c)

Fig. 18.22

current in M. By combining these magnetic fields it will be seen that when currents flow simultaneously through F and M, the resultant magnetic field is distorted as shown in fig. 18.22(c) and the effect is to exert a clockwise torque on M.

Since M is carrying current at right angles to the magnetic field produced by F

deflecting force on each side of M

    $\propto$(current in M)

     $\times$(density of magnetic field due to current in F)

    $\propto$current in M $\times$ current in F.

In dynamometer ammeters, the fixed and moving coils are connected in parallel, whereas in voltmeters they are in series with each other and with the usual high resistance. In each case, the deflecting force is proportional to the square of the current or the voltage. Hence, when the dynamometer instrument is used to measure an alternating current or voltage, the moving coil—due to its inertia—takes up a position where the average deflecting torque over one cycle is balanced by the restoring torque of the spiral springs. For that position, the deflecting torque is proportional to the mean value of the square of the current or voltage, and the instrument scale can therefore be calibrated to read the r.m.s. value.

Owing to the highest cost and lower sensitivity of dynamometer ammeters and voltmeters compared with moving-iron instruments, the former are seldom used commercially, but *electrodynamic* or *dynamometer wattmeters* are very important because they are commonly employed for measuring the power in a.c. circuits. The fixed coils F are connected in

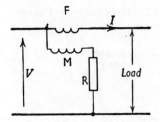

Fig. 18.23 Wattmeter connections

series with the load, as shown in fig. 18.23. The moving coil M is connected in series with a non-reactive resistor R across the supply, so that the current through M is proportional to and practically in phase with the supply voltage V; hence:

instantaneous force on each side of M

$\propto$(instantaneous current through F)

$\times$(instantaneous current through M)

$\propto$(instantaneous current through load)

$\times$(instantaneous p.d. across load)

$\propto$instantaneous power taken by load

$\therefore$ average deflecting force on M

$\propto$average value of the power over a
complete number of cycles.

When the instrument is used in an a.c. circuit, the moving coil—due to its inertia—takes up a position where the average deflecting torque over one cycle is balanced by the restoring torque of the spiral springs; hence the instrument can be calibrated to read the mean value of the power in an a.c. circuit.

18.9 Electronic voltmeters

Electronic voltmeters, unlike moving coil and moving iron instruments, can give very high instrument resistances. An instrument with a very high resistance takes little current from the circuit and can thus be used to measure the potential difference between points in a circuit where the current taken by other types of voltmeter might considerably modify the value of that potential difference.

Essentially an electronic voltmeter consists of an amplifier which is used to amplify the voltage being measured before applying it to a moving coil instrument, the amplifier taking negligible current from the circuit in which the measurement is made. Figure 18.24 shows the type of circuit that might be used for the electronic voltmeter.

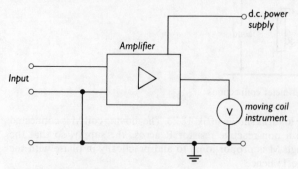

Fig. 18.24 The basis of an electronic voltmeter

Care has to be taken in using an instrument, such as the electronic voltmeter, which has part of its circuit connected to earth. If the circuit in which the measurement is being made has also an earth connection there is the possibility of short circuiting some of the components and so considerably affecting the readings given by the instrument and also possibly damaging the circuit. Electronic voltmeters are available with either earthed terminals or floating without earthing.

18.10 Digital voltmeters

These instruments involve an analogue-to-digital convertor to change the input signal, which is in continuous form, to a digital form, which is non-continuous. One method that is used is to count the number of pulses from an accurate electronic clock system that occur during the time taken from a ramp voltage to rise from zero to the value of the voltage being measured. A ramp voltage is one that rises at a constant rate and has a value at any instant which is proportional to the elapsed time.

18.11 Rectifier ammeters and voltmeters

In this type of instrument rectifiers (see Chapter 23) are used to convert the alternating current into a unidirectional current, the mean value of which is measured on a moving-coil instrument.

Rectifier ammeters usually consist of four rectifier elements arranged in the form of a bridge, as shown in fig. 18.25, where the apex of the black triangle indicates the direction in which the resistance is low, and A represents a moving-coil ammeter. During the half-cycles that the current is flowing from left to right in fig. 18.25, current flows through elements B

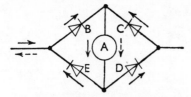

Fig. 18.25   Bridge circuit for full-wave rectification

and D, as shown by the full arrows. During the other half-cycles, the current flows through C and E, as shown by the dotted arrows. The waveform of the current through A is therefore as shown in fig. 18.26.

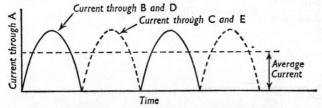

Fig. 18.26   Waveform of current through moving-coil ammeter

Consequently, the deflection of A depends upon the average value of the current, and the scale of A can be calibrated to read the r.m.s. value of the current on the assumption that the waveform of the latter is sinusoidal with a form factor of 1.11.

In a rectifier voltmeter, A is a milliammeter and the bridge circuit of fig. 18.25 is connected in series with a suitable resistor.

The main advantage of the rectifier voltmeter is that it is far more sensitive than other types of voltmeter suitable for measuring alternating voltages. Also, rectifiers can be incorporated in universal instruments, such as the Avometer, thereby enabling a moving-coil milliammeter to be used in combination with shunt and series resistors to measure various ranges of direct current and voltage, and in combination with a bridge rectifier and suitable resistors to measure various ranges of alternating current and voltage.

18.12 Measurement of resistance by the voltmeter–ammeter method

The most obvious way of measuring resistance is to measure the current through and the p.d. across the resistor and then apply Ohm's Law. This method, however, must be used with care; for instance if the instruments be connected as in fig. 18.27, and if the voltmeter V be other than the electronic type, the current taken by V passes through A and may be

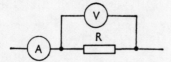

Fig. 18.27 Measurement of resistance

comparable with that through R if the resistance of the latter is fairly high. If the resistance of V is known, its current can be calculated and subtracted from the reading on A to give the current through R.

A better method is to connect the voltmeter across R and A as in fig. 18.28, then:

$$\frac{\text{reading on V}}{\text{reading on A}} = \text{resistance of } (R + A).$$

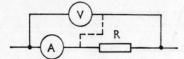

Fig. 18.28 Measurement of resistance

The resistance of A can be easily calculated from the p.d. across A—obtained by connecting V as shown dotted—and the corresponding current through A.

18.13 Measurement of resistance by substitution

(a) *Series method.* The unknown resistor X (fig. 18.29) is connected in series with a variable resistor R, such as a decade resistance box, which can be varied in steps of 1 or 0.1 Ω. The total resistance of R must exceed that of

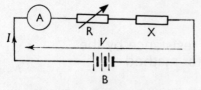

Fig. 18.29 Measurement of resistance

X. R is adjusted to give a convenient reading on ammeter A—preferably a reading such that the pointer is exactly over a scale mark near the top end of the scale. X is then removed and R readjusted to give the same reading on A. The increase of R gives the resistance of X.

If the resistance of R is fixed, it is necessary to note the current I_1 with R and X in series and the current I_2 with X removed. If V be the terminal voltage, which may be assumed constant if the current is very small compared with the rated current of battery B, then:

$$I_1 = \frac{V}{R+X}, \quad \text{and} \quad I_2 = \frac{V}{R}$$

$$\frac{I_2}{I_1} = \frac{R+X}{R} = 1 + \frac{X}{R}$$

$$\therefore \quad \frac{X}{R} = \frac{I_2}{I_1} - 1 = \frac{I_2 - I_1}{I_1}$$

and

$$X = R\left(\frac{I_2 - I_1}{I_1}\right).$$

If the resistance of the ammeter is appreciable, it must be included with that of R in the above calculation.

A modification of this arrangement is frequently used in universal instruments. A moving-coil milliammeter A is connected in series with a variable resistor r and a cell B to terminals TT, as in fig. 18.30. The

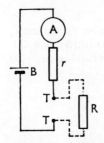

Fig. 18.30 An ohmmeter circuit

procedure is to short-circuit the terminals TT and adjust r to give full-scale deflection on A. The unknown resistor R is then connected across TT, and A can be calibrated to give the resistance of R directly, thereby making the instrument into an ohmmeter. Thus, suppose A to have full-scale deflection with 1 mA and B to have an e.m.f. of 1.5 V. For full-scale deflection with TT short-circuited, total resistance of r and B must be 1.5 ÷ 1/1000, namely 1500 Ω. If a 1000-Ω resistor is connected across TT, the current is $1.5 \times 1000/2500$, namely 0.6 mA; i.e. a scale deflection of 0.6 mA corresponds to a resistance of 1000 Ω, etc., as shown in fig. 18.31.

The resistance of r in fig. 18.30 is variable to allow for variation of the e.m.f. and internal resistance of cell B. It should be mentioned, however, that if the e.m.f. of B falls appreciably below its rated value, the resistance scale will only be approximately correct.

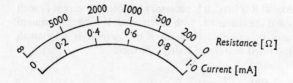

Fig. 18.31   Ammeter and ohmmeter scales

(b) *Alternative-circuit method.* In fig. 18.32, X is the unknown resistor, R a known variable resistor and S a two-way switch. The reading on A is noted when S is on *a*. The switch is then moved over to *b* and R adjusted to give the same reading on A. The resistance of X is obviously the same as that of R.

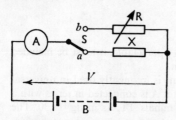

Fig. 18.32   Measurement of resistance

If the resistance of R is fixed, it is necessary to note the current I_1 through X when S is on *a* and the current I_2 through R when S is on *b*. If V is the terminal voltage and if the resistances of the battery and ammeter are negligible compared with those of X and R, then:

$$I_1 = V/X, \qquad \text{and} \qquad I_2 = V/R$$

$$\therefore \quad I_2/I_1 = X/R$$

so that

$$X = R \times I_2/I_1.$$

18.14 Measurement of resistance by the Wheatstone bridge

The four branches of the network, CDFEC, in fig. 18.33, have two known resistances P and Q, a known variable resistance R and the unknown resistance X. A battery B is connected through a switch S, to junctions C and F; and a galvanometer G, a variable resistor A and a switch S_2 are in series across D and E. The function of A is merely to protect G against an excessive current should the system be seriously out of balance when S_2 is closed.

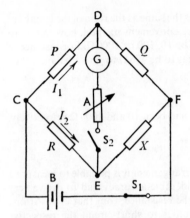

Fig. 18.33 Wheatstone bridge

With S_1 and S_2 closed, R is adjusted until there is no deflection on G even with the resistance of A reduced to zero. Junctions D and E are then at the same potential, so that the p.d. between C and D is the same as that between C and E, and the p.d. between D and F is the same as that between E and F.

Suppose I_1 and I_2 to be the currents through P and R respectively when the bridge is balanced. From Kirchhoff's First Law it follows that since there is no current through G, the currents through Q and X are also I_1 and I_2 respectively. But

p.d. across $P = PI_1$

and

p.d. across $R = RI_2$

$$\therefore \quad PI_2 = RI_2 \qquad [18.1]$$

Also

p.d. across $Q = QI_1$

and

p.d. across $X = XI_2$

$$\therefore \quad QI_1 = XI_2. \qquad [18.2]$$

Dividing [18.2] by [18.1], we have:

$$Q/P = X/R$$

and

$$X = R \times Q/P. \qquad [18.3]$$

The resistances P and Q may take the form of the resistance of a slide-wire, in which case R may be a fixed value and balances obtained by moving a sliding contact along the wire. If the wire is homogeneous and of uniform section, the ratio of P to Q is the same as the ratio of the lengths of wire in the respective arms. A more convenient method, however, is to arrange P and Q so that each may be 10, 100 or 1000 Ω. For instance, if $P = 1000\,\Omega$ and $Q = 10\,\Omega$, and if R has to be 476 Ω to give a balance, then from [18.3]:

$$X = 476 \times 10/1000 = 4.76\,\Omega.$$

On the other hand, if P and Q had been 10 and 1000 Ω respectively, then for the same value of R:

$$X = 476 \times 1000/10 = 47\,600\,\Omega.$$

Hence it is seen that with this arrangement it is possible to measure a wide range of resistance with considerable accuracy and to derive very easily and accurately the value of the resistance from that of R. At one time, resistance boxes fitted with plugs to short-circuit the respective resistance elements were commonly employed; and in one pattern, known as the Post Office box, the ratio arms P and Q and the variable resistance R were constructed on this principle in one compact unit, complete with battery and galvanometer switches. Due partly to the trouble experienced with badly-fitting plugs or plugs inadequately pressed into their sockets and partly to the labour involved in removing and replacing plugs when balancing the bridge and in the subsequent adding up of the resistances left in circuit, plugs have been superseded by rotary dial switches.

18.15 The potentiometer

One of the most useful instruments for the accurate measurement of p.d., current and resistance is the potentiometer, the principle of action being that an unknown e.m.f. or p.d. is measured by balancing it, wholly or in part, against a known difference of potential.

In its simplest form, the potentiometer consists of a wire MN (fig. 18.34) of uniform cross-section, stretched alongside a scale and connected

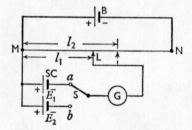

Fig. 18.34 A simple potentiometer

across an accumulator B of ample capacity. A standard cell SC of known e.m.f. E_1, for example a cadmium cell having an e.m.f. of 1.018 59 V at 20°C (section 3.8), is connected between M and terminal *a* of a two-way switch S, care being taken that the corresponding terminals of B and SC are connected to M.

Slider L is then pressed momentarily against wire MN and its position adjusted until the galvanometer deflection is zero when L is making contact with MN. Let l_1 be the corresponding distance between M and L. The fall of potential over length l_1 of the wire is then the same as the e.m.f. E_1 of the standard cell.

Switch S is then moved over to *b*, thereby replacing the standard cell by another cell, such as a Leclanché cell, the e.m.f. E_2 of which is to be measured. Slider L is again adjusted to give zero deflection on G. If l_2 be the new distance between M and L, then:

$$E_1/E_2 = l_1/l_2$$

$$\therefore \quad E_2 = E_1 \times l_2/l_1. \quad\quad\quad [18.4]$$

18.16 A commercial form of a potentiometer

The simple arrangement described in the preceding section has two disadvantages: (i) the arithmetical calculation involved in expression [18.4] may introduce an error, and in any case takes an appreciable time; (ii) the accuracy is limited by the length of slide-wire that is practicable and by the difficulty of ensuring exact uniformity over a considerable length.

In the commercial type of potentiometer shown in fig. 18.35 these

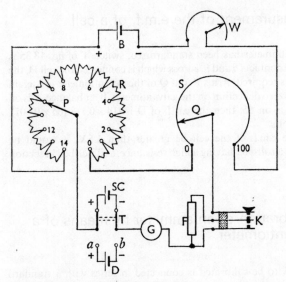

Fig. 18.35 A commercial potentiometer

disadvantages are practically eliminated. R consists of fourteen resistors in series, the resistance of each resistor being equal to that of the slide-wire S. The value of the current supplied by accumulator B is controlled by a slide-wire resistor W. A double-pole change-over switch T, closed on the upper side as in fig. 18.35, connects the standard cell SC between arm P and galvanometer G. A special key K, when slightly depressed, inserts a resistor F in series with G; but when K is further depressed, F is short-circuited. The galvanometer is thereby protected against an excessive current should the potentiometer be appreciably out of balance when K is first depressed.

18.17 Standardization of the potentiometer

Suppose the standard cell SC to be of the cadmium type having an e.m.f. of 1.018 59 V at 20°C. Arm P is placed on stud 10 and Q on 18.59—assuming the scale alongside S to have 100 divisions. The value of W is then adjusted for zero deflection on G when K is fully depressed. The p.d. between P and Q is then exactly 1.018 59 V, so that the p.d. between two adjacent studs of R is 0.1 V and that corresponding to each division of S's scale is 0.001 V. Consequently if P is moved to, say, stud 4 and Q to 78.4 on the slide-wire scale, the p.d. between P and $Q = (4 \times 0.1) + (78.4 \times 0.001) = 0.4784$ V. It is therefore a simple matter to read the p.d. directly off the potentiometer.

Since most potentiometers have fourteen steps on R, it is usually not possible to measure directly a p.d. exceeding 1.5 V.

18.18 Measurement of the e.m.f. of a cell

After the potentiometer has been standardized, switch T of fig. 18.35 is changed over to contacts *a* and *b*, across which is connected the cell D, the e.m.f. of which is required. Arms P and Q of the potentiometer are again adjusted to give zero deflection on the galvanometer. If, at balance, P is on stud 14 and Q is on 65, then the e.m.f. of D is $(14 \times 0.1) + (65 \times 0.001)$, namely 1.465 volts.

Should the e.m.f. of the cell be greater than 1.5 V, it would be necessary to use a *volt-box* having a high resistance, as described in section 18.20.

18.19 Calibration of an ammeter by means of a potentiometer

The ammeter A to be calibrated is connected in series with a standard resistor H and a variable resistor J across an accumulator L of ample current capacity, as in fig. 18.36. The standard resistor H is usually provided with four terminals, namely two heavy current terminals CC and two potential terminals PP. The resistance between the potential terminals is known with a high degree of accuracy and its value must be

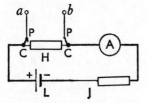

Fig. 18.36 Calibration of an ammeter

such that with the maximum current through the ammeter, the p.d. between terminals PP does not exceed 1.5 V. For instance, suppose A to be a 10-A ammeter; then the resistance of H must not exceed 1.5/10, namely 0.15 Ω. Further, the resistance of H should preferably be a round figure, such as 0.1 Ω in this case, in order that the current may be quickly and accurately deduced from the potentiometer readings.

Terminals PP of the standard resistor are connected to terminals *ab* (fig. 18.35) of the potentiometer (cell D having been removed). After the potentiometer has been standardized, switch T is changed over to *ab*; and with the current adjusted to give a desired reading on the scale of ammeter A, arms P and Q are adjusted to give zero deflection on the galvanometer. For instance, suppose the current to be adjusted to give a reading of, say, 6 A on the ammeter scale, and suppose the readings on P and Q, when the potentiometer is balanced, to be 5 and 86.7 respectively, then the p.d. across terminals PP is 0.5867 V; and since the resistance between the potential terminals PP is assumed to be 0.1 Ω, the true value of the current through H is 0.5867/0.1, namely 5.867 A. Hence, the ammeter is reading high by 0.133 A.

18.20 Calibration of a voltmeter by means of a potentiometer

Suppose the voltmeter to be calibrated to have a range of 0–100 V. It is therefore necessary to use a *volt-box* to enable an accurately known fraction—not exceeding 1.5 V—to be obtained. The high-resistance volt-box *ae*, fig. 18.37, has tappings at accurately determined points. This arrangement enables voltmeters of various ranges to be calibrated; thus the 100-V voltmeter is connected acorss the 150-V tappings, the resistance between *ad* being 100 times that between *ab*. The 1.5-V tappings are connected to terminals *ab* of the potentiometer (fig. 18.35). Various

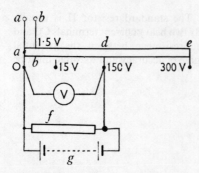

Fig. 18.37   Calibration of a voltmeter

voltages can be applied to the voltmeter by moving the slider along a
resistor f connected across a suitable battery g.

Let us suppose that the voltmeter reading has been adjusted to 70 V
and that the corresponding readings on P and Q to give a balance are 7
and 8.4 respectively. The p.d. across ab is 0.7084 V, and the true value of
the p.d. across ad is therefore 70.84 V. Hence the voltmeter is reading low
by 0.84 V.

18.21   Measurement of resistance by means of a potentiometer

The resistor X (fig. 18.38), whose resistance is to be determined, is
connected in series with a known standard resistor R, an ammeter A and a
variable resistor j across an accumulator h. The purpose of A is simply to
check that the value of the current is not excesssive.

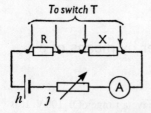

Fig. 18.38   Measurement of resistance

Connections are taken from R, to, say, the upper pair of terminals of
switch T in fig. 18.35 (the standard cell having been removed), and those
from X are taken to the lower pair of terminals of T. With a constant
current through R and X, potentiometer readings are noted, first with
switch T on the upper side to measure the p.d. across R, and then with T on

the lower side to measure the p.d. across X. If these readings were 0.648 V and 0.1242 V respectively and if the resistance of R were 0.1 Ω, then for a current I amperes through R and X:

$$I \times 0.1 = 0.648, \quad \text{and} \quad IX = 0.1242,$$

$$\therefore \quad X/0.1 = 0.1242/0.648$$

and

$$X = 0.019\,17\,\Omega.$$

This method is particularly suitable for the accurate measurement of low resistances, in which case it may be necessary to use knife-edge contacts, as indicated by the arrowheads in fig. 18.38, to give the precise points between which the resistance is being determined.

18.22 Alternating current bridges

Bridge circuits similar in form to that of the Wheatstone bridge, section 18.14, can be used for the measurement of capacitance and inductance. The general principle is the same, the essential differences being that an a.c. source is used and the detector has to be one that can detect alternating current, e.g. a pair of headphones or an electronic voltmeter (fig. 18.39). At

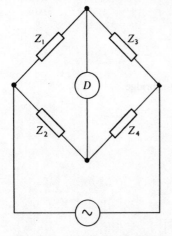

Fig. 18.39 The basic a.c. bridge

the balanced condition when no current is detected by the detector, then

$$\frac{Z_1}{Z_2} = \frac{Z_3}{Z_4} \qquad\qquad [18.5]$$

where the above are the impedances in the various arms of the bridge.

There are many variations of the a.c. bridge. An example of such a bridge for the measurement of capacitance is the *De Sauty bridge*. In this bridge resistors are used for Z_1 and Z_2 and capacitors for Z_3 and Z_4. Thus $Z_1 = R_1$; $Z_2 = R_2$; $Z_3 = 1/\omega C_3$ and $Z_4 = 1/C_4$. Hence

$$R_1 C_3 = R_2 C_4.$$

Thus if C_3 is the capacitor being measured, the other components are variable ones for which the values at balance are determined.

An example of an a.c. bridge used for the measurement of inductance

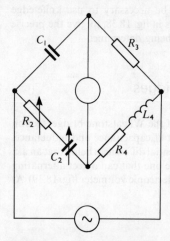

Fig. 18.40   The Owen bridge

is the *Owen bridge* (fig. 18.40). For such a bridge

$$Z_1 = 1/\omega C_1$$
$$Z_2 = \sqrt{[R_2^2 + (1/\omega C_2)^2]}$$
$$Z_3 = R_3$$
$$Z_3 = \sqrt{(R_4^2 + (\omega L_4)^2)}.$$

Thus, using equation (18.5)

$$\frac{1/\omega C_1}{\sqrt{[R_2^2 + (1/\omega C_2)^2]}} = \frac{R_3}{\sqrt{[R_4^2 + (\omega L_4)^2]}}$$

Squaring both sides of the equation and rearranging the terms gives

$$R_4^2 + (\omega L_4)^2 = (\omega C_1 R_3)^2 [R_2^2 + (1/\omega C_2)^2]$$
$$= (\omega C_1 R_3 R_2)^2 + (C_1 R_3 / C_2).$$

At balance the resistive terms and the reactance terms must balance. Thus

$$R_4 = C_1 R_3 / C_2$$

and $(\omega L_4)^2 = (\omega C_1 R_3 R_2)^2$

$$L_4 = C_1 R_3 R_2.$$

The components R_2 and C_2 are adjusted to obtain these balance conditions.

18.23 The cathode ray oscilloscope

The cathode ray oscilloscope (often abbreviated to c.r.o.) is an instrument that enables instantaneous values of voltages to be displayed by a spot of light on a phosphor-coated screen. Figure 18.41 shows the basic

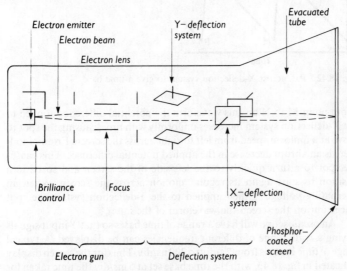

Fig. 18.41 The cathode ray oscilloscope

construction of the c.r.o. Essentially there is an electron gun which produces a beam of electrons focused onto the screen. The impact of the electrons on the screen causes light to be emitted. The brilliance of this light is controlled by the brilliance control which controls the number of electrons passing down the tube per second. The sharpness of the spot of light on the screen is controlled by varying the voltage to one of the electron lens elements. The beam of electrons, and hence the spot of light on the screen, can be deflected by applying potential differences to the Y-deflection system, for deflections in the vertical direction, or the X-deflection system, for deflections in the horizontal direction.

The screen of the oscilloscope is marked with a grid of squares, each square being referred to as one division. Potential differences are applied to the deflection plates through an amplifier system which allows different sensitivities to be obtained. The type of range available is typically from about 5 mV/div to 20 V/div. Thus if, for instance, the Y-amplifier system was set to 5 mV/div and a signal applied to the Y-input caused the spot of light on the screen to be deflected through 2 divisions then the applied potential difference would have been 10 mV.

While external potential differences can be applied to the X-deflection system, in most applications of the c.r.o. an internal signal is applied which gives a deflection of the spot in the horizontal direction which is proportional to time. This internal signal is referred to as the time base. The waveform used to give a time base is a sawtooth form, fig. 18.42.

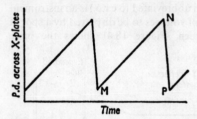

Fig. 18.42   P.d. across X-deflection system to give a time base

From point M to N on this waveform the potential difference applied to the X-deflection system increases uniformly with time, causing the spot to move at a uniform speed from left to right across the screen. From N to P there is an abrupt decrease in the applied potential difference. This causes the spot to return quickly to the left side of the screen and so be in a position to start again its regular motion across the screen. When an alternating voltage is then applied to the Y-deflection system the spot traces out on the screen the waveform of the input.

An oscilloscope will have a range of time bases so that Y-input signals having a wide range of different frequencies can be displayed. A typical range of time bases is from 2 s/div to 200 ns/div. Thus for the screen display indicated in fig. 18.43, with the time base set at 5 ms/div the time taken for

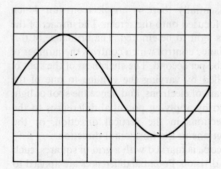

Fig. 18.43

one complete cycle is 8×5 ms $= 40$ ms. Hence the frequency is $1/0.004 =$ 250 Hz.

To obtain a stationary trace on the screen it is necessary to introduce some synchronizing arrangement in order to ensure that the successive traces exactly superimpose on each other. This is done by a trigger circuit. This is sensitive to the level of voltage applied to it and only initiates the sweep of the time base at a particular voltage level. This means that the trace on the screen always starts from the same voltage point in its waveform.

18.24 Applications of the c.r.o.

(a) Displaying waveforms

This involves the waveform concerned being applied to the Y-input and the timebase being used. If the waveform is repetitive a stationary trace can be obtained on the screen by the use of the trigger control.

(b) Measurement of d.c. voltages

The unknown voltage is connected to the Y-input, the time base generally being switched off as it plays no role with a direct voltage. The spot on the screen is displaced by the input and the amount of displacement is then measured. If a calibrated Y-amplifier setting is being used the deflection can be directly converted into a voltage. If not, a known voltage should be supplied to the Y-input in order to calibrate the deflection.

(c) Measurement of direct current

The deflection of the spot on a c.r.o. screen is directly related to the input potential difference. If current is to be determined then the current should be passed through a resistor, of known resistance, and the potential difference across the resistor measured, as in (b) above. Then using $I = V/R$ the current can be calculated.

(d) Measurement of a.c. voltages

The alternating voltage is applied to the Y-input with the time base operating on an appropriate range. To hold the trace stationary the trigger control should be used. The peak-to-peak distance can be measured and converted to the peak-to-peak voltage using a calibrated Y-amplifier setting or by calibrating the Y-deflection using a known voltage to the Y-input. The time taken for one cycle can be determined, using the calibrated time base, and hence the frequency of the voltage determined.

(e) Measurement of alternating currents

The alternating current is passed through a resistor, of known resistance, and the potential difference across the resistor measured using the oscilloscope, as in (d). Then using $I = V/R$ the current can be calculated.

If the alternating potential difference displayed on the screen has a peak-to-peak distance of 3 divisions and the Y-amplifier sensitivity is 20 mV/div then the peak-to-peak potential difference is 60 mV and hence the maximum voltage is 30 mV. If this is the potential difference across a resistor of resistance 50 Ω then the maximum current is $0.030/50 = 6.0 \times 10^{-14}$ A = 0.60 mA.

(f) Measurement of frequency

As has already been indicated, the frequency of a waveform can be determined by applying the waveform to the Y-input and using the calibrated time base to determine the time difference between two identical points on the waveform. A more accurate method is to compare the waveform with another waveform of known frequency using *Lissajous' figures*. If the unknown waveform is applied to the Y-input then the known frequency waveform is applied to the X-input, the time base not being used. The known frequency is adjusted until one of the traces shown in fig. 18.44 appears on the screen. When such a trace appears there is an exact relationship between the two frequencies.

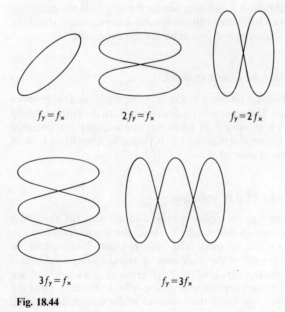

$f_Y = f_X$ $2f_Y = f_X$ $f_Y = 2f_X$

$3f_Y = f_X$ $f_Y = 3f_X$

Fig. 18.44

If a vertical and a horizontal line are imagined as being drawn at the side and top of the trace then

$$\frac{f_x}{f_y} = \frac{\text{number of loops touching vertical line}}{\text{number of loops touching horizontal line}}$$

(g) Measurement of phase differences

If a double beam c.r.o. is available then the potential differences being compared can be applied to the two Y-inputs and two traces obtained on the screen. The separation of two adjacent peaks can be determined using the grid of the oscilloscope screen and compared with the length of a complete cycle. Thus if a complete cycle has a length of 2.2 divisions and the separation between the two peaks is 0.3 divisions then the phase difference is $0.3/2.2$ fraction of a complete cycle, i.e. $(0.3/2.2) \times 2\pi = 0.27\pi$ or $(0.3/2.2) \times 360° = 49.1°$.

A more accurate method is to apply one of the signals to the Y-input and the other to the X-input, the time base being switched off. Since the frequencies are the same the Lissajous figure obtained is an ellipse. The phase angle ϕ is given by, for equal amplitude inputs,

$$\sin \phi = \frac{a}{b}$$

where b and a are as indicated in fig. 18.45. Figure 18.46 shows some of the forms of traces for particular phase angles.

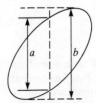

Fig. 18.45

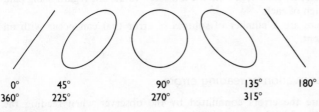

0°	45°	90°	135°	180°
360°	225°	270°	315°	

Fig. 18.46

18.25 Decibel measurements

The *bel* is a logarithmic unit used when two amounts of power are compared, the *decibel* being one tenth of a bel. Thus on a non-logarithmic scale we might have

$$\text{power ratio} = \frac{\text{power 1}}{\text{power 2}}.$$

On a logarithmic scale we have

$$\text{power ratio in bels} = \log_{10}\left(\frac{\text{power 1}}{\text{power 2}}\right)$$

$$\text{power ratio in decibels} = 10\log_{10}\left(\frac{\text{power 1}}{\text{power 2}}\right).$$

Thus if the power input to an amplifier is 2 mW and the power output is 150 mW then

$$\text{power ratio} = \frac{150}{2}$$

$$= 75.$$

In decibels

$$\text{power ratio in dB} = 10\log_{10}75$$

$$= 18.8\,\text{dB}.$$

18.26 Accuracy and errors

The British Standard definition of *accuracy* is: a general term describing the degree of closeness with which the indications of an instrument approach the true values of the quantities measured. The more accurate an instrument the closer will be the indicated values to the true values. The term *error* is used for the differences obtained by subtracting the true value of the quantity measured from the indicated value, due regard being paid to the sign of each.

There are a number of sources of error that can occur with an instrument.

(a) Observation or reading errors

These are the errors committed by the observer when reading the indication of the instrument. There might thus be an error due to the observer reading the indication of a pointer on a scale at an angle other than from directly above the pointer, this being referred to as an error due to parallax. Another possibility is the error due to an observer having to estimate a fractional part of a scale interval.

(b) Construction errors

These are errors inherent within an instrument as a result of its manufacture. They could be due to the tolerances on the dimensions of components.

(c) Environmental effects

Errors can arise due to environmental effects on the components of an instrument. Thus, for example, a change in temperature will affect the resistances of components.

(d) Ageing effects

Errors can occur as a result of components getting older and their values changing slightly, perhaps the build up of deposits on surfaces affecting contact resistances or surface leakage resistance.

(e) Insertion errors

The connection of, say, a voltmeter or an ammeter into a circuit can, because of the resistance of the meter, markedly affect the current in the circuit and hence the readings obtained by the instrument may not be the readings that occurred prior to the insertion of the instrument.

(f) Approximation errors

In, for example, bridge circuits the expression used may be only an approximation and thus there will be some error due to its use.

The above errors can be termed *systematic errors* in that they are predictable errors. Their causes can be determined and either eliminated or the errors allowed for. However, in any measurement process there are always *random errors*. These errors are unpredictable. Thus there might be an error in a reading due to a transient or a surge on the power supply. The random errors will differ from reading to reading and thus the taking of the average of a number of readings rather than just a single reading can reduce the effects of such errors, it being assumed that random errors are as likely to lead to high values as low values and the average will thus be nearer to the true value than any single reading.

18.27 Examples

1. A milliammeter gives full-scale deflection with 5 mA and has a resistance of 12 Ω. Calculate the resistance necessary (a) in parallel, to enable the instrument to read up to 10 A and (b) in series, to enable it to read up to 100 V.

2. If the shunt for Question 1(a) is to be made of manganin strip having a resistivity of 0.5 $\mu\Omega$ m, a thickness of 0.5 mm and a length of 60 mm, calculate the width of the strip.

3. Why is spring control to be preferred to gravity control in an electrical measuring instrument?

    The coil of a moving-coil meter has a resistance of 5 Ω and gives full-scale deflection when a current of 15 mA passes through it. What modification must be made to the instrument to convert it into (a) an ammeter reading to 15 A and (b) a voltmeter reading to 15 V?

    (U.L.C.I.)

4. The rectangular moving-coil of an ammeter is wound with $30\frac{1}{2}$ turns. The effective axial length of the magnetic field is 20 mm and the effective radius of the coil is 8 mm. The flux density in the gap is 0.12 T and the controlling torque of the hairsprings is 0.5×10^{-6} N m per degree of deflection. Calculate the current to give a deflection of 60°.

5. Describe, with a sketch, the main constructional features of a moving-coil instrument.

    The coil of such an instrument is wound with $40\frac{1}{2}$ turns and has an effective axial length of 25 mm and an effective breadth of 18 mm. The flux density in the airgap is 0.125 T. Calculate the torque, in newton metres, on the coil when it is carrying a current of 20 mA.

    (E.M.E.U.)

6. A moving-iron ammeter is wound with 40 turns and gives full-scale deflection with 5 A. How many turns would be required on the same bobbin to give full-scale deflection with 20 A?

7. Describe with the aid of a diagram the construction of a repulsion-type moving-iron instrument with particular reference to the means used for (a) deflection, (b) control and (c) damping.

    A moving-iron voltmeter, in which full-scale deflection is given by 100 V, has a coil of 10 000 turns and a resistance of 2000 Ω. Calculate the number of turns required on the coil if the instrument is converted for use as an ammeter reading 20 A full-scale deflection.

    (U.E.I.)

8. The coil of a moving-iron voltmeter has an inductance of 0.35 H and the total resistance of the instrument when made to read up to 20 V is 400 Ω. If the instrument is calibrated with direct voltage, calculate the percentage error when it is used on (a) a 50-Hz supply, (b) a 500-Hz supply.

9. Give a summary of four different types of voltmeters commonly used in practice. State whether they can be used on a.c. or d.c. circuits.

    In *one* case give a sketch showing the construction with the method of control and damping employed. A d.c. voltmeter has a resistance of 28 600 Ω. When connected in series with an external resistor across a 480-V d.c. supply, the instrument reads 220 V. What is the resistance of the external resistor? (U.E.I.)

10. An unknown resistor R is measured by means of an ammeter (resistance 0.5 Ω), a voltmeter (resistance 65 Ω) and a battery, the voltmeter being connected directly across R. Draw a diagram of connections. If the ammeter reads 5.2 A and the voltmeter 13 V, calculate the value of R (a) approximately, (b) accurately. What change in connections would you make in order to measure a resistance of about 100 Ω, and why? (E.M.E.U.)

11. In a Wheatstone bridge arranged as in fig. 18.33, $P = 100 \Omega$, $Q = 40 \Omega$ and $R = 156 \Omega$ when the bridge is balanced. Calculate the value of the unknown resistance. Also find the value of the current in each circuit if the battery B has an e.m.f. of 4 V and an internal resistance of 20 Ω.

12. Draw a neat diagram showing how you would connect a slide-wire Wheatstone bridge in order to measure the value of an unknown resistance.

   A resistor of unknown value is placed in the left-hand side and a standard resistor of 12 Ω in the right-hand side of a metre slide-wire Wheatstone bridge. Balance is obtained at a point on the slide-wire 41.7 cm from the left-hand-side.
   (a) Calculate the value of the unknown resistance.
   (b) If the standard resistance were to be reduced to 8 Ω, at what distance from the left-hand end of the slide-wire would balance then be obtained? (E.M.E.U.)

13. Four resistors are connected in series to form a closed circuit ABCD. The resistances AB, BC, CD and DA respectively are 16, 14, 7 and 3 Ω. If a p.d. of 3 V is applied between A and C, being positive with respect to C, calculate the value and the direction of the p.d. between points B and D. What value of resistance should be connected in parallel with resistor AB to give zero p.d. between B and D? (U.E.I.)

14. The arms of a Wheatstone bridge have the following resistances: AB, 10 Ω; BC, 20 Ω; CD, 30 Ω; and DA, 10 Ω. A galvanometer having a resistance of 40 Ω is connected between B and D and a cell having an e.m.f. of 2 V and negligible internal resistance is connected across A and C, its positive end being connected to A. Calculate the current through the galvanometer and state its direction.

15. The current through an ammeter connected in series with a standard shunt of 0.01 Ω, as in fig. 18.36, is adjusted to 40 A; and the readings on the studs and the slide-wire of the potentiometer, when balanced, are 4 and 12.3 respectively. Calculate the percentage error of the ammeter and state whether the instrument is reading high or low.

16. The reading on a voltmeter connected across the 0/300-V range of a volt-box (fig. 18.37) is adjusted to 250 V, and the readings on the studs and the slide-wire of the potentiometer, when balanced, are 12 and 34.7 respectively. Calculate the percentage error of the voltmeter and state whether the instrument is reading high or low.

17. Explain the principle of operation of a simple potentiometer when used to measure the e.m.f. of a cell.

    In an experiment with a simple potentiometer, a balance was obtained on a length of 60 cm of the wire when using a standard cell of electromotive force 1.0183 V. When a dry cell was substituted for the standard cell, a balance was obtained with a length of 85 cm. Calculate the e.m.f. of the dry cell.   (U.L.C.I.)

18. Explain the theory of the simple potentiometer. What are the advantages of this method of measuring a p.d.?

    A standard cell (e.m.f. 1.018 V) gives a balance of 50.9 cm, and a cell X of unknown e.m.f. gives a balance at 93 cm, when each is tested on a certain potentiometer. When the cell X is connected to a voltmeter whose resistance is 200 Ω, the voltmeter reads 1.84 V. Determine the internal resistance of X.   (E.M.E.U.)

19. Explain the theory of an a.c. bridge used for the measurement of inductance. What type of instrument can be used as a detector for such a bridge?

    An Owen bridge has a 200-pF capacitor in arm AB, a 120 Ω resistor in arm BC, an inductor with resistance in arm CD and in arm DA a 350-pF capacitor in series with a 1100-Ω resistor. The detector is connected between B and D, the oscillator between A and C. What is the value of the inductance and resistance of the inductor?

20. Explain the theory of an a.c. bridge used for the measurement of capacitance.

    A De Sauty bridge has a 400-Ω resistor in arm AB, a 220-Ω resistor in arm BC, a 2.2-μF capacitor in arm CD and an unknown capacitor in arm DA. The detector is connected between B and D, the oscillator between A and C. What is the value of the unknown capacitor?

21. Explain the basic principle of the cathode ray oscilloscope and describe the function of the various controls found on such an instrument.

    Explain how a cathode ray oscilloscope can be used to determine (a) the frequency of a waveform and (b) its peak voltage.

22. Explain the various sources of errors that can occur with an instrument and how the effects of random errors can be minimized.

CHAPTER 19

Electrolysis:
Primary and Secondary Cells

19.1 A simple voltaic cell

It was in 1789 that a chance observation by Luigi Galvani* (1737–98) led to the idea of generating an electric current from a source of chemical energy. Galvani, a professor of anatomy at Bologna, noticed that recently-skinned frogs' legs, hung by copper wire to an iron balcony, were convulsed whenever they touched the iron. Another Italian, Alessandro Volta* (1745–1827), a professor of physics at Pavia, subsequently showed that if a rod of two dissimilar metals, such as copper and iron, was placed so that one end was in contact with a nerve on a frog's leg and the other end in contact with a muscle of the foot, muscular contraction took place. Following his investigations of this phenomenon, Volta, in 1799, constructed a simple battery—known as Volta's pile—by assembling discs of zinc, cloth soaked with brine, and copper, piled one upon another in that order. By this means, a large number of cells were obtained in series, giving a high electromotive force, but having the disadvantage of high internal resistance.

It was a comparatively short step to replace the wet cloth of Volta's pile by an electrolyte to give the simple voltaic cell shown in fig. 19.1. In this cell, plates of copper and zinc are immersed in dilute sulphuric acid contained in a glass vessel.

When a resistor R and an ammeter A are connected in series across the terminals, it is found that current flows through R from the copper electrode to the zinc electrode and that the difference of potential between the plates is about 0.8 volt, the copper plate being the positive electrode.

This type of cell is referred to as a *primary cell* since it can only transform chemical energy into electrical energy, and the cell can be replenished only by renewal of the active materials. In a *secondary cell*, or *accumulator*, the electrode materials can be re-activated, i.e. the chemical

---

* Galvani's name is perpetuated in such terms as 'galvanize' and 'galvanometer', and the unit of e.m.f. and of potential difference, the 'volt', has been named after Volta.

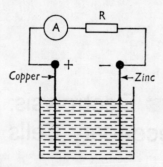

Fig. 19.1   A simple voltaic cell

action is reversible. This means that chemical energy is converted into electrical energy when the cell is discharging, and electrical energy is converted into chemical energy when the cell is being charged, as already referred to in section 2.1.

19.2 Electrolytic dissociation

Before we consider the various types of cells, it will be helpful to discuss what happens when an electric current flows through an electrolyte. In section 2.6 we dealt with Faraday's Laws of electrolysis. These are the facts of electrolysis; now we will consider the mechanism of electrolysis.

A molecule of, say, common salt consists of one atom of sodium and one atom of chlorine. When these atoms combine to form a molecule, an electron is attracted from the sodium atom to join the chlorine atom. Consequently the chlorine atom has a surplus *negative* charge and is therefore termed a *negative ion*. The sodium atom, on the other hand, is left with an excess positive charge and is therefore termed a *positive ion*. It is thought that the stability of the molecule is largely due to the electrostatic attraction between the positive sodium ion and the negative chlorine ion.

When common salt is dissolved in water, the sodium and chlorine atoms of many of the molecules become separated—a phenomenon known as *electrolytic dissociation*. The separated ions are free to wander at random in the electrolyte and to recombine with other oppositely-charged free ions. In other words, dissociation goes on continuously, some molecules breaking up and oppositely-charged free ions recombining to form new molecules, thus maintaining the number of free ions constant.

The effect of dissociation in a solution of common salt in water can be represented thus:

$$NaCl \rightleftarrows Na^+ + Cl^-.$$

The arrow pointing towards the right indicates dissociation and that pointing towards the left indicates that sodium and chlorine ions can recombine to form neutral molecules of sodium chloride.

The electrical charge on an ion is the same as the chemical valency* of the atom. For instance, it will be seen from the table on p. 22 that the valency of both sodium and chlorine is one; hence an ion of sodium is an atom which has lost *one* electron and is represented by Na^+, whereas an ion of chlorine is an atom which has gained *one* electron and is represented by Cl^-. On the other hand, if we consider copper sulphate ($CuSO_4$), the valency of copper is two; hence an ion of copper is an atom which has lost *two* electrons and is represented by Cu^{++}.

Let us next consider a dilute solution of sulphuric acid (H_2SO_4) in water. In this case, the dissociation of one molecule of the sulphuric acid produces two positive ions of hydrogen, each having a surplus positive charge equal in *magnitude* to that of an electron, and one negative sulphate (SO_4) ion, carrying a surplus of two electrons. This dissociation can therefore be represented thus:

$$H_2SO_4 \rightleftharpoons H^+ + H^+ + SO_4^{--}.$$

Lastly, let us consider a solution of copper sulphate ($CuSO_4$) in water. As mentioned above, the copper ion has lost two electrons so that its surplus positive charge is equal in magnitude to that on two electrons. The sulphate (SO_4) ion has, as stated in the preceding paragraph, a surplus of two electrons; hence:

$$CuSO_4 \rightleftharpoons Cu^{++} + SO_4^{--}.$$

19.3 Electrolytic cell with carbon electrodes

Figure 19.2 shows two carbon plates, C and D, immersed in dilute sulphuric acid. Carbon is used partly because it is not affected by the products of electrolysis and partly because carbon electrodes can easily be obtained from disused Leclanché cells. The electrodes are connected to a battery B through an ammeter A, a resistor R and a switch S, and a moving-coil voltmeter is connected across the electrodes.

With switch S closed, current flows through the circuit, including the electrolyte. In the latter, the positive hydrogen ions are attracted towards the negative electrode, or *cathode*, D. When they reach this electrode, each ion absorbs an electron from the electrode to form a neutral atom of hydrogen. But neutral atoms of hydrogen cannot continue to exist independently, and pairs of such atoms combine to form molecules (H_2) of hydrogen. These molecules appear as a gas, some of which adheres to electrode D as a gaseous layer and the remainder rises to the surface in the form of tiny bubbles.

---

* The valency of an element or a group of elements is the number of hydrogen atoms which will combine with or replace one atom of that element or group of elements.

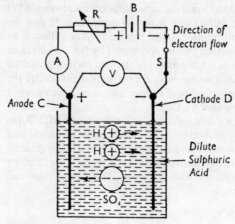

Fig. 19.2 Electrolytic cell with carbon electrodes

The negative sulphate (SO_4) ions are attracted towards the positive electrode, or *anode*, C. The ions which reach C give up their surplus electrons to that electrode; but SO_4 cannot exist uncharged and therefore attacks a water molecule to re-form a molecule of sulphuric acid and release an atom of oxygen. The latter combines with another oxygen atom to form an oxygen molecule (O_2). Thus we have:

$$2SO_4 + 2H_2O = 2H_2SO_4 + O_2.$$

The net result of electrolysis in this type of cell is to decompose water into its constituents, hydrogen and oxygen.

It will be seen that in the electrolyte, the positive and negative ions, by migrating towards the respective electrodes, act as charge carriers; but in the external circuit, the current is due to the movement of free electrons from the positive electrode, via battery B, to the negative electrode.

When switch S is closed, it is found that the current instantly rises to its maximum value and then falls off, at first fairly rapidly. But when the switch is opened after the current has been flowing for several minutes, it is found that the voltmeter reads about 1.7 V, electrode C being positive relative to D. This p.d. falls to zero as the gases rise to the surface or are absorbed by the electrolyte.

The presence of this p.d. is due to the gaseous layers on the electrodes harbouring hydrogen and sulphate ions which have not made contact with their respective electrodes and therefore still retain their charges. These accumulated charges repel other hydrogen and sulphate ions that are approaching the cathode and anode respectively, thus giving rise to an e.m.f. that is in opposition to that of battery B. This phenomenon is known as *polarization*; and in order to send current through the cell, the p.d. applied across the electrodes has to neutralize this back e.m.f. due to polarization in addition to providing the IR drop due to the resistance of the electrolyte. The product of the current and the back e.m.f. is the power required to maintain the corresponding rate of electrolysis.

19.4 Electrolytic cell with copper electrodes

Let us next consider the case of two copper plates immersed in a solution of copper sulphate ($CuSO_4$) in water, the electrical circuit being as shown in fig. 19.3. As already explained in section 19.2, the effect of dissociation in

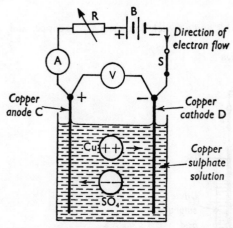

Fig. 19.3 Electrolytic cell with copper electrodes

a copper sulphate solution can be represented thus:

$$CuSO_4 \rightleftharpoons Cu^{++} + SO_4^{--}.$$

With switch S closed, the positive copper ions are attracted towards cathode D. On reaching this electrode, each ion absorbs two electrons from the electrode to form neutral atoms of copper and these are deposited on the surface of the copper plate.

The negative sulphate ions are simultaneously attracted towards anode C and each ion, on reaching the latter, gives up two electrons to that electrode, thereby compensating for the two electrons taken by each copper ion from the cathode. The neutral sulphate SO_4 cannot exist uncharged and therefore combines with an atom of copper from anode C to form a molecule of copper sulphate which goes into solution. Thus the density of the electrolyte remains constant, and the net effect is to transfer copper from the anode to the cathode.

No gas is released at either electrode. Consequently there is no polarization and the reading on voltmeter V is zero when switch S is opened. This experiment shows that one method of eliminating polarization is to substitute a harmless ion in place of the harmful hydrogen ion, i.e. to arrange for the neutralized ion to be of the same basic metal as the positive plate. Thus in the cadmium standard cell (section 3.8), the current in the region adjacent to the positive mercury electrode is carried by positive mercury ions and negative sulphate (SO_4) ions, thereby eliminating polarization. The same principle is applied in the mercury cell described in section 19.7.

19.5 Action of the simple voltaic cell

Figure 19.4 shows a resistor R and a switch S in series across copper and

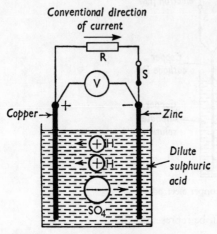

Fig. 19.4   A simple voltaic cell

zinc electrodes immersed in dilute sulphuric acid, with a voltmeter connected across the electrodes. Since the copper electrode is positive relative to the zinc electrode, current flows through R in the direction shown by the arrow. This is the conventional direction. Actually, the current in this external circuit is a flow of electrons in the opposite direction.

With switch S open, the p.d. between the two electrodes is about 0.8 V. Immediately S is closed, there is a displacement of electrons in the external circuit with the result that some electrons are withdrawn from the zinc electrode and the same number of electrons move on to the copper electrode. This leaves the zinc plate with a surplus of positive charge and the copper plate with a surplus of negative charge. Consequently the negative sulphate ions in the electrolyte are attracted towards the zinc and the positive hydrogen ions are attracted towards the copper, as indicated in fig. 19.4.

The sulphate ions, after giving up their surplus electrons to the zinc electrode, combine with the zinc to form zinc sulphate ($ZnSO_4$) which goes into solution. The hydrogen ions, on the other hand, after absorbing electrons from the copper plate, form a gaseous layer on the surface of the copper. This gaseous layer has two disadvantages:

(a) it gives rise to polarization, i.e. it sets up a back e.m.f.,
(b) it acts as a shield reducing the active area of the electrode and thus increases the internal resistance of the cell.

The result is that the terminal voltage of the cell falls very rapidly and this type of cell is therefore only suitable for intermittent use. Polarization can be reduced by introducing a depolarizing agent to combine with the hydrogen, e.g. manganese dioxide in the Leclanché cell (section 19.6), or it can be eliminated by arranging for the neutralized ions to be of the same basic metal as the positive electrode, e.g. the mercury cell (section 19.7).

Local action. If the negative electrode of the simple voltaic cell is made of commercial zinc, it is found to dissolve rapidly. This is due to impurities, such as lead, giving rise to local action, i.e. the zinc and the impurities form the electrodes of tiny cells, and localized currents flow through the electrolyte between these electrodes and complete their paths through the body of the zinc electrode.

Local action can be practically eliminated by bringing the zinc plate or rod into contact with mercury in dilute acid, thereby forming a protective film on the surface of the electrode. The zinc is then said to be *amalgamated.*

19.6 Leclanché cell

The 'wet' type of Leclanché cell (now practically obsolete) consists of a carbon plate, surrounded by a mixture of manganese dioxide (MnO_2) and powdered carbon, in an unglazed earthenware pot. This pot and an amalgamated zinc rod are immersed in a saturated solution of sal-ammoniac (ammonium chloride, NH_4Cl) in water. The carbon plate and the zinc rod form the positive and negative electrodes respectively.

The function of the manganese dioxide is to reduce polarization by combining with the hydrogen released at the carbon plate to form water and a brown oxide of manganese (Mn_2O_3), thus:

$$H_2 + 2MnO_2 = Mn_2O_3 + H_2O.$$

In the 'dry' type of Leclanché cell, the same ingredients are present, and fig. 19.5 is a sectional view of the construction most commonly used. A carbon rod A is surrounded by a black depolarizing paste B, consisting of manganese dioxide, powdered carbon, salammoniac, zinc chloride and water, the paste being usually contained in a bag of coarse linen. Around this depolarizer is a mixture P of flour, plaster of Paris, salammoniac and zinc chloride, with water added to form a white paste. The latter need only be thick enough to prevent the black paste B touching the zinc container Z. Above the depolarizer and the white paste is a layer S of sawdust or similar porous material, the cell being sealed with a layer of pitch T in

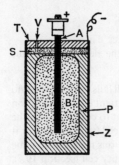

Fig. 19.5 'Dry' Leclanché cell

which there is a vent tube V. The zinc container Z is usually covered with a cardboard case.

A Leclanché cell has an e.m.f. of about 1.5 V when new, but this e.m.f. falls fairly rapidly if the cell is in continuous use. This fall is due to polarization—the hydrogen film at the carbon electrode forms faster than can be dissipated by the depolarizer. However, if the cell is disconnected from the external circuit, depolarization continues and the e.m.f. recovers its normal value. Hence the Leclanché cell is only suitable for intermittent use, e.g. for electric torches, radio receivers, etc.

The life of a dry Leclanché cell is reduced by local action and even the shelf-life is limited to about two years owing to the local action that goes on continuously in the cell. Also, this type of cell does not lend itself to miniaturization since the number of ampere hours obtainable from the cell falls off rapidly as the size is reduced.

19.7 Mercury cell*

This type of cell was developed to meet the requirements of miniaturization, e.g. for guided missiles, medical electronics, hearing aids, etc., where it is necessary to reduce the size of the cell but at the same time obtain (a) a high ratio of output energy/mass, (b) a constant e.m.f. over a relatively long period and (c) a long shelf life, i.e. absence of local action.

A cross-section of the basic type of mercury cell is shown in fig. 19.6. The negative electrode is zinc, either as a foil or as powder compressed into a hollow cylinder. This electrode is surrounded by a layer of electrolyte consisting of a concentrated aqueous solution of potassium hydroxide (KOH) and zinc oxide (ZnO). Surrounding the electrolyte is a layer of mercuric oxide (HgO). This oxide contains a small percentage of finely

---

* The authors are indebted to Mallory Batteries Ltd., for information concerning this cell.

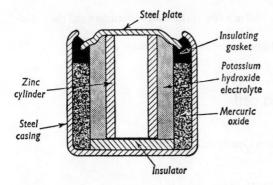

Fig. 19.6   A mercury cell

powdered graphite to reduce the internal resistance of the cell.

The above constituents are assembled in a nickel-plated or stainless steel container which forms the positive electrode. The zinc cylinder and the electrolyte are supported on a disc of insulating material; and the cell is sealed by an insulating gasket between the container and a nickel-plated steel plate resting on top of the zinc cylinder.

When the cell is connected to an external circuit, the positive potassium ions (K^+) in the electrolyte move towards the mercuric oxide whilst the negative hydroxide ions, $(OH)^-$, move towards the negative electrode and give up their surplus electrons to the zinc, after which they combine with the zinc to form zinc oxide and water.

Simultaneously, in the mercuric oxide, the positive mercury ions (Hg^{++}) move towards the positive electrode and their charges are neutralized by the electrons absorbed from that electrode, after which the mercury is deposited on the inside of the casing. At the same time, the negative oxygen ions (O^{--}) of the mercuric oxide move towards the electrolyte and combine with the potassium ions and some water to re-form potassium hydroxide.

There is no polarization since no gases are evolved at either electrode, except under abnormal operating conditions. Owing to the absence of polarization, this type of cell maintains its terminal voltage practically constant at about 1.2 to 1.3 volts (depending upon the value of the load current) for a relatively long time. Also, owing to local action being practically negligible, the cell can be stored for a long period at normal atmospheric temperature without appreciable loss of capacity.

19.8 Secondary cells

Secondary cells may be divided into two types: (a) the lead-acid cell in which lead plates covered with compounds of lead are immersed in a dilute solution of sulphuric acid in water, (b) the nickel-cadmium and the nickel-iron alkaline cells.

These cells will now be considered in greater detail.

19.9 Lead-acid cell

The plates used in this type of cell may be grouped thus:

(a) Formed or Planté plates

These are formed from lead plates by charging, discharging, charging in the reverse direction, etc., a number of times, the forming process being accelerated by the use of suitable chemicals. The main difficulty with this type of construction is to secure as large a working surface as possible for a given mass of plate. One method of increasing the surface area is to make the plates with deep corrugations, as in fig. 19.7, with reinforcing ribs at intervals.

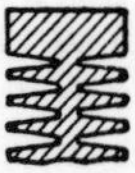

Fig. 19.7

(b) Pasted or Faure plates

Plates in which a paste of the active material is either pressed into recesses in a lead-antimony grid or held between two finely perforated lead sheets cast with ribs and flanges so that the two sheets, when riveted together, form in effect a number of boxes which hold the paste securely in position. The paste is usually sulphuric acid mixed with red lead (Pb_3O_4) for the positives and with litharge (PbO) for the negatives. A small percentage of a material such as powdered pumice is added to increase the porosity of the paste. For a given ampere hour capacity (section 19.11), the mass of a pasted plate is only about a third of that of a formed plate.

Students are advised to examine specimen plates or plates taken from disused accumulators.

When weight is of no importance it is common practice to make the positive plates of the 'formed' type and the negative plates of the 'pasted' type. The active material on the positive plates expands when it is subjected to chemical changes; consequently it is found that the greater mechanical stiffness of the 'formed' construction is an important advantage in reducing the tendency of the plates to buckle. This tendency to buckle is reduced still further by constructing the cell with an odd

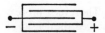

Fig. 19.8 Arrangement of plates

number of plates, as shown in fig. 19.8, the outer plates being always negative. This arrangement enables both sides of each positive plate to be actively employed, and the tendency of one side of a plate to expand and cause buckling is neutralized by a similar tendency on the other side.

The plates are assembled in glass, polystyrene, vulcanized-rubber or resin-rubber containers and separated by a special grade of paper or microporous sheets of a plastic material.

The most suitable relative density of the acid depends upon the type of cell and the state of charge of the cell. An average value, however, is about 1.21.

19.10 Chemical reactions in a lead-acid cell

The chemical reactions taking place during charge and discharge are complicated, and all we can do here is to indicate the most important reactions and to account for the variation in the density of the electrolyte.

When the cell is fully charged, the active material on the positive plate is lead dioxide (PbO_2) and that on the negative plate is spongy or porous lead (Pb). During discharge, the positive hydrogen ions (H^+) travel towards the positive electrode (PbO_2), and after their charges have been neutralized by the absorption of electrons from this electrode, the hydrogen atoms combine with lead dioxide and sulphuric acid to form lead sulphate and water, thus:

$$H_2 + PbO_2 + H_2SO_4 = PbSO_4 + 2H_2O.$$

The negative sulphate ions (SO_4^{--}) travel towards the negative electrode (Pb) and after giving up their surplus electrons, they combine with lead to form lead sulphate, thus:

$$Pb \times SO_4 = PbSO_4.$$

During charge the chemical reactions are reversed and the active material is converted back to lead dioxide on the positive electrode and to spongy lead on the negative electrode.

The above reactions may be summarized thus:

	Positive plate	Electrolyte	Negative plate	
	Lead dioxide	Sulphuric acid	Lead	
Dis-	(PbO_2)	($2H_2SO_4$)	(Pb)	↑ Charge
charge ↓	Lead sulphate	Water	Lead sulphate	
	($PbSO_4$)	($2H_2O$)	($PbSO_4$)	

It will be seen that for every two molecules of sulphuric acid decomposed during discharge, two molecules of water are formed; hence the density of the electrolyte falls as the cell discharges. The reverse process occurs during charging so that when the cell is fully charged, the density of the electrolyte is restored to its initial value.

19.11 Characteristics of a lead-acid cell

When a cell is discharged, its terminal voltage falls at a rate that depends upon the discharge current; and the normal capacity of the cell is taken as the number of ampere hours it can give on a 10-hour discharge at a constant current before its p.d. falls to about 1.85 V. Such a condition is represented by curve A in fig. 19.9. Curve B represents the variation of p.d.

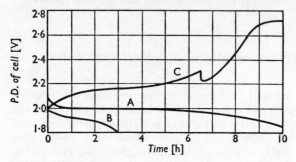

Fig. 19.9 Characteristics of a lead-acid cell

when the cell is discharged at about 2.5 times the normal rate. The higher the rate of discharge, the smaller is the number of ampere hours obtainable from the cell before its p.d. falls below the permissible value—a value that depends upon the discharge rate. The lower curve in fig. 19.10 represents the capacity in ampere hours obtainable at different discharge rates, and the upper curve represents the corresponding rates, the values for both curves being expressed as percentages of the 10-hour values.

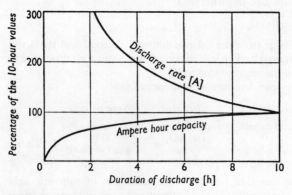

Fig. 19.10 Capacity of a lead-acid cell

The reduction in the capacity at high rates of discharge is due to the chemical reactions being at first confined to the outer layers of the active material. Consequently the acid in the pores of these outer layers is used up before fresh acid can take its place. Owing to the relatively high resistance of very weak acid, the voltage falls rapidly, though there may be plenty of unchanged active material still left in the inner layers. If a cell is discharged at a very high rate until the terminal voltage falls to the minimum permissible value and then left on open circuit, the weakened acid becomes strengthened by diffusion of the electrolyte and a further discharge can be obtained from the cell.

The first portion of curve C in fig. 19.9 represents the variation of p.d. during charge at normal rate, but the actual value of the p.d. varies appreciably for different types of cells. It is seen that after about 5 hours the p.d. increases more rapidly, and in order to prevent the temperature exceeding about 40°C, it is necessary to reduce the charging rate; thus the second portion of curve C represents the variation of p.d. with the charging current reduced to half the normal rate.

The ampere hour efficiency of an accumulator is the ratio of the number of ampere hours obtainable during discharge to that required to restore it to its original condition. The value for a lead-acid cell is about 90 per cent. The watt hour efficiency takes the voltage variation into account, and its value is about 75 per cent.

19.12 Care of a lead-acid battery

The principal difficulty in the maintenance of this type of cell in a healthy condition is the tendency for the lead sulphate to become hard and form a white incrustation on the plates. This white sulphate has a relatively high resistance and reduces the porosity of the plates so that the acid has greater difficulty in penetrating into the active material. The capacity of the cell is thereby reduced.

The following are the chief precautions to be taken in the maintenance of a lead-acid battery:

(a) The cells should be recharged as soon as possible after discharge. White sulphate forms more readily the lower the state of charge of a cell.
(b) The rates of charge and discharge specified by the makers should not be exceeded.
(c) Records should be kept of the p.d. of each cell and of the relative density of the electrolyte; and if either becomes abnormal, the trouble should be immediately rectified.
(d) If it is possible, the plates should be examined periodically for colour and for clearance between the positive and negative plates and between the plates and any sediment at the bottom of the container. When healthy, the positive plates are chocolate-brown and the negative plates are slate-grey.

(e) The level of the electrolyte must be kept above the tops of the plates, and the loss of water by evaporation and decomposition during gassing should be made up by the addition of distilled water. Tap water should only be used if it has been found by analysis to be free of injurious chemicals. The 'make-up' water should preferably be added when the cells are gassing, so that it may mix thoroughly with the electrolyte.

(f) Any sulphuric acid added to the electrolyte should be of the pure colourless variety. No acid should be added until it has been checked that the low density is not due to white sulphate on the plates or to the cell being undercharged.

(g) The cells should be given a periodic overcharge at about half the normal rate, the latter being maintained until the cells are gassing freely and until half-hourly readings of the terminal voltage and of the relative density of each cell show no further increase. This ensures the removal of any white sulphate and restores the whole of the active material to its normal condition.

(h) If a battery has a relatively light duty it is advisable to discharge it occasionally through an artificial load at its normal rate and then immediately recharge it.

(i) If the battery is being used as a standby source to give emergency lighting, etc., it should be connected across a d.c. supply so as to receive a continuous trickle-charge.

(j) The temperature of the cells should not exceed about 40°C, otherwise the plates deteriorate rapidly.

19.13 Alkaline cells

In both the nickel-iron and the nickel-cadmium types, the positive plates are made of nickel hydroxide enclosed in finely perforated steel tubes or pockets, the electrical resistance being reduced by the addition of flakes of pure nickel or graphite. These tubes or pockets are assembled in nickelled-steel plates. In the nickel-iron cell the negative plate is made of iron oxide with a little mercuric oxide to reduce the resistance, the mixture being enclosed in perforated steel pockets, also assembled in nickelled-steel plates. In the nickel-cadmium cell the active material is cadmium mixed with a little iron, the purpose of the latter being to prevent the active material caking and losing its porosity.

In both types of cell, the electrolyte is a solution of potassium hydroxide (KOH) having a relative density of about 1.15 to 1.2, depending upon the type of cell and the conditions of service. The electrolyte does not undergo any chemical change; consequently the quantity of electrolyte can be reduced to the minimum necessitated by adequate clearance between the plates.

The plates are separated by insulating rods and assembled in sheet-steel containers, the latter being mounted in non-metallic crates to insulate the cells from one another.

19.14 Chemical reactions in an alkaline cell

When the nickel-cadmium cell is in a charged condition the active material on the positive plates appears to be a hydroxide of nickel having the chemical formula $Ni(OH)_3$ and that on the negative is pure cadmium. During discharge, the $Ni(OH)_3$ is converted into the lower hydroxide $Ni(OH)_2$ and the cadmium is converted into cadmium hydroxide $Cd(OH)_2$. During charge, the chemical reactions are reversed. These reactions may be summarized thus:

Positive	*Negative*		*Positive*	*Negative*
plate	*plate*		*plate*	*plate*
		Discharge		
$2Ni(OH)_3$	$+Cd$	$\rightleftarrows$	$2Ni(OH)_2 + Cd(OH)_2$	
		Charge		

In the nickel-iron cell the reactions are exactly similar, except that iron replaces cadmium.

19.15 Characteristics of the alkaline cell

Curve A in fig. 19.11 represents the terminal voltage of a nickel-cadmium

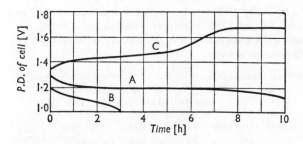

Fig. 19.11 Characteristics of an alkaline cell

cell during discharge at the 10-hour rate, while curve B shows the variation for a 3-hour rate. Owing to the fact that no change occurs in the composition of the electrolyte during discharge, the number of ampere hours obtainable from an alkaline cell is much less affected by the discharge rate than is the case with the lead-acid cell. Figure 19.12 shows the capacity of a nickel-cadmium cell—expressed as a percentage of the 10-hour capacity—for different rates of discharge. It is seen that for a 1-hour rate it is possible to obtain about 84 per cent of the 10-hour capacity,

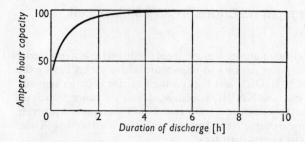

Fig. 19.12   Capacity of an alkaline cell

compared with about 50 per cent for the lead-acid cell (fig. 19.10).

Curve C of fig. 19.11 represents the variation of the terminal voltage when the cell is charged at 1.5 times the 10-hour discharge rate.

The ampere hour efficiency of the nickel-cadmium cell is about 75–80 per cent, while its watt hour efficiency is about 60–65 per cent.

The advantages of the alkaline accumulator are: (a) its mechanical construction enables it to withstand considerable vibration and (b) it is free from 'sulphating' or any similar trouble and can therefore be left in any state of charge without damage. Its disadvantages are: (a) its cost is greater than that of the corresponding lead cell, (b) its average discharge p.d. is about 1.2 V compared with 2 V for the lead cell, so that for a given voltage the number of alkaline cells is about 67 per cent greater than that of lead cells.

Example 19.1 *A terminal voltage of approximately 120 V is to be maintained by a battery of alkaline cells. The initial and final values of the e.m.f. per cell are 1.3 V and 1.15 V respectively. The internal resistance per cell is 0.01 Ω and the discharge current is 10 A. Calculate the initial and final number of cells required in series.*

Initial terminal voltage/cell = initial e.m.f./cell − voltage drop/cell
due to internal resistance
$$= 1.3 - (10 \times 0.01) = 1.2 \text{ V},$$

∴   initial number of cells = 120/1.2 = 100.

Final terminal voltage/cell = 1.15 − (10 × 0.01) = 1.05 V,

∴   final number of cells = 120/1.05 = 114.

Example 19.2   *A battery of 30 lead-acid cells is to be charged at a constant current of 8 A from a 110-V d.c. supply. The terminal voltage per cell is 1.9 V at the commencement of charging and 2.6 V at the end. Calculate the maximum and minimum values of the resistor required in series with the battery.*

Figure 19.13 represents a variable resistor R connected in series with the battery of 30 cells across a 110-V supply.

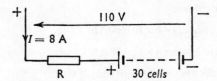

Fig. 19.13 Example 19.2

At commencement of charging,

$$\text{total p.d. across battery} = 1.9 \times 30 = 57\,\text{V},$$

$$\therefore \quad \text{corresponding p.d. across } R = 110 - 57 = 53\,\text{V}$$

and

$$\text{corresponding resistance of } R = 53/8 = 6.625\,\Omega.$$

At end of charging,

$$\text{total p.d. across battery} = 2.6 \times 30 = 78\,\text{V},$$

$$\text{corresponding p.d. across } R = 110 - 78 = 32\,\text{V}$$

and

$$\text{corresponding resistance of } R = 32/8 = 4\,\Omega.$$

Example 19.3 *A battery of 80 cells is charged through a resistor from a 240-V supply. At the beginning of the charge, the e.m.f. per cell is 1.9 V and the charging current is 5 A. The internal resistance of each cell is 0.06 Ω. Calculate the value of the resistor.*

$$\text{Terminal voltage/cell} = \text{e.m.f./cell} + \text{voltage drop/cell due to}$$
$$\text{internal resistance}$$

$$= 1.9 + (5 \times 0.06) = 2.2\,\text{V}.$$

$$\therefore \quad \text{total voltage across battery} = 2.2 \times 80 = 176\,\text{V}.$$

$$\text{Corresponding voltage across resistor} = 240 - 176 = 64\,\text{V},$$

so that

$$\text{value of resistor} = 64/5 = 12.8\,\Omega.$$

Example 19.4 *A fully-charged lead-acid cell was completely discharged in 10 h, the discharge current being constant at 6 A. The average terminal voltage during discharge was 1.95 V. A charging current of 4 A, maintained constant for 17 h, was required to restore the cell to its initial state of charge, the average terminal voltage being 2.3 V. Calculate (a) the ampere hour* *efficiency and* (b) *the watt hour efficiency.*

(a)   Output of cell, in ampere hours $= 6 \times 10 = 60 \, \text{A h}$.

Input to cell, in ampere hours $= 4 \times 17 = 68 \, \text{A h}$.

$\therefore$ ampere hour efficiency $= 60/68 = 0.882$ per unit

$= 88.2$ per cent.

(b)   Output of cell, in watt hours $= 6 \times 1.95 \times 10 = 117 \, \text{W h}$

Input to cell, in watt hours $= 4 \times 2.3 \times 17 = 156.4 \, \text{W h}$.

$\therefore$ watt hour efficiency $= 117/156.4 = 0.748$ per unit

$= 74.8$ per cent.

19.16 Examples

1. What is the difference between a primary and a secondary cell? Six cells, each of 2 V and internal resistance 2 Ω, are connected in two groups of three in series, and the two groups connected in parallel to an external resistor of 30 Ω. Sketch the arrangement and calculate the current which will flow through the external resistor.   (U.E.I.)

2. An accumulator is overcharged by 4 A for 25 h. If the electrochemical equivalents of hydrogen and oxygen are 0.010 45 and 0.082 95 mg/C respectively, find the volume of water required to be added to compensate for gassing. Assume 1 mm^3 of water to have a mass of 1 mg.

3. An accumulator has a terminal voltage of 1.9 V when supplying a current of 8 A. The terminal voltage rises to 2.03 V immediately the load is switched off. Calculate the internal resistance of the accumulator.

4. A certain alkaline cell had a terminal voltage of 1.24 V when supplying a current of 3 A. Immediately after the load was disconnected, the terminal voltage was 1.28 V. Calculate the internal resistance of the cell.

5. If the terminal voltage of a lead-acid cell varies between 2.1 V and 1.85 V during discharge, calculate the number of cells required to give 230 V (a) at the beginning of discharge, (b) at the end of discharge.

6. Calculate the number of lead-acid cells to be connected in series to give a terminal voltage of 240 V when the battery is supplying a current of 12 A. Assume each cell to have an e.m.f. of 2 V and an internal resistance of 0.025 Ω.

7. A battery of 50 cells in series is charged through a resistor of 4 Ω from a 230-V supply. If the terminal voltage per cell is 2 V and 2.7 V respectively at the beginning and end of the charge, calculate the charging current (a) at the beginning and (b) at the end of the charge.

8. A battery of 40 cells in series is to be charged from a 220-V supply. If the average p.d. per cell during charge is 2.2 V, calculate the resistance in series with the battery to give an average charging current of 5 A.

9. Tabulate the relative advantages and disadvantages of lead-acid and alkaline cells.

    Explain the difference between the *constant-current* and the *constant-voltage* methods of charging and give any disadvantages of each system.

    A battery of 10 cells is to be charged at a constant current of 8 A from a generator of e.m.f. 35 V and internal resistance 0.2 Ω. The internal resistance of each cell is 0.08 Ω and its e.m.f. ranges from 1.85 V discharged to 2.15 V charged. Determine the maximum and minimum values of the series resistor which must be included in the circuit. (U.E.I.)

10. Explain the terms *ampere hour efficiency* and *watt hour efficiency* with reference to secondary cells. Which is the greater and why?

    A battery of 12 cells is charged through a resistor from a constant 30-V supply. When charging commences, the e.m.f. per cell is 1.85 V and the charging current is 4 A. At the end of the charge, the e.m.f. per cell has risen to 2.25 V. If each cell has a constant internal resistance of 0.05 Ω, calculate the value of the external resistor and the current at the end of the charge. (E.M.E.U.)

11. A lead-acid cell is discharged at a constant current of 5 A for 10 h, the average value of the terminal voltage being 1.92 V. The cell is then recharged at a constant rate of 3 A for 20 h, with an average terminal voltage of 2.2 V. Calculate (a) the ampere hour efficiency and (b) the watt hour efficiency.

12. An alkaline cell is discharged at a constant current of 5 A for 10 h, the average terminal voltage being 1.2 V. A charging current of 3 A, maintained for 21 h, is required to bring the cell back to the initial state of charge, the average terminal voltage being 1.48 V. Calculate the ampere hour and watt hour efficiencies.

13. The 10-h capacity of a certain accumulator is 60 A h. From figs. 19.10 and 19.12, calculate the number of ampere hours obtainable on a half-hour discharge rate if the accumulator is of (a) the lead-acid type and (b) the alkaline type.

CHAPTER 20

Circuit Theorems

20.1 Kirchhoff's laws

Section 4.9 of this book gives Kirchhoff's laws and illustrates their application to relatively simple circuits. The laws can be stated as:

First law: If several conductors meet at a point, the total current flowing towards that point is equal to the total current flowing away from it, i.e. the algebraic sum of the currents is zero.

Second law: In any closed circuit, the algebraic sum of the products of the current and the resistance of each part of the circuit is equal to the resultant e.m.f. in the circuit.

Example 20.1 *Determine the current through the 10-Ω resistor for the circuit shown in fig. 20.1.*

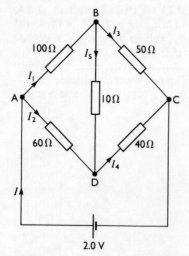

Fig. 20.1

The current directions in the figure have been arbitrarily chosen. If the direction chosen is wrong the answer will work out to give a negative value for that current, indicating it is in the opposite direction to that chosen.

Applying Kirchhoff's first law to junctions A, B and D gives the following equations:

$$I = I_1 + I_2 \tag{20.1}$$

$$I_3 = I_1 - I_5 \tag{20.2}$$

$$I_4 = I_2 + I_5. \tag{20.3}$$

The above equations indicate that we only need to find three currents, .e.g. I_1, I_2 and I_5, to be able to calculate all the currents.

Applying Kirchhoff's second law to circuit loops, we have for loop ABD, when we go clockwise round the loop:

$$100I_1 + 10I_5 - 60I_5 = 0. \tag{20.4}$$

The zero is because there is no source of e.m.f. in that loop. The negative sign with the $60I_2$ is because the postulated current direction gives an IR product in the opposite direction to the way we are proceeding round the loop.

For loop BCD, we have:

$$50I_3 - 30I_4 - 10I_5 = 0. \tag{20.5}$$

For the loop containing the source and ADC,

$$60I_2 + 40I_4 = 2.0. \tag{20.6}$$

We can, using equations [20.1], [20.1] and [20.3], rewrite [20.4], [20.5] and [20.6] so that they are all in terms of just the three currents I_1, I_2 and I_5. Equation [20.4] is already in terms of just these three currents.

Equation [20.5] becomes:

$$50(I_1 - I_5) - 40(I_2 + I_5) - 10I_5 = 0$$
$$50I_1 - 40I_2 - 100I_5 = 0. \tag{20.7}$$

Equation [20.6] becomes

$$60I_2 + 40(I_2 + I_5) = 2.0$$
$$100I_2 + 40I_5 = 2.0. \tag{20.8}$$

We now have three equations, [20.4], [20.7] and [20.8], which involve just three currents. Equation [20.8] gives us a relationship between just two of the currents and so we can use it to eliminate I_2 from [20.4] and [20.7],

$$I_2 = \frac{2.0 - 40I_5}{100}.$$

Hence [20.4] becomes:

$$100I_1 + 10I_5 - 60\left(\frac{2.0 - 40I_5}{100}\right) = 0$$

$$10\,000I_1 + 1000I_5 - 120 + 2400I_5 = 0. \qquad [20.9]$$

Equation [20.7] becomes:

$$50I_1 - 40\left(\frac{2.0 - 40I_5}{100}\right) - 100I_5 = 0$$

$$5000I_1 - 80 + 1600I_5 - 10\,000I_5 = 0. \qquad [20.10]$$

We now have two simultaneous equations with just two unknowns. To eliminate I_1 from these equations we can double [20.10] and subtract it from [20.9]. Our two equations, with some slight simplification, are thus:

$$10\,000I_1 + 3400I_5 - 120 = 0 \qquad\qquad [20.9]$$

$$10\,000I_1 - 16\,800I_5 - 80 = 0 \qquad\qquad [20.10] \times 2$$

$$\overline{\;\; 20\,200I_5 - 40 = 0} \qquad \text{result of subtraction.}$$

Hence $I_5 = 1.98 \times 10^{-3}$ A.

We could find any of the other currents by substituting this value of current in the appropriate equation.

20.2 Thévenin's theorem

In section 4.3 of this book the concept of a cell as having an internal resistance was introduced. All signal sources and generators have internal resistance. Because of this internal resistance the potential difference between the terminals of the cell, signal source or generator, will depend on the load current. This potential difference equals the e.m.f. when there is no load current, i.e. there is open circuit, when there is a load current the potential difference is lower than the e.m.f. due to the potential drop occurring across the internal resistance. The greater the load current the greater the potential drop across the internal resistance and so the lower the potential difference between the terminals.

Example 20.2   *A battery has an e.m.f. of 12 V and an internal resistance of 1.0 Ω. Calculate the battery terminal potential difference when a current of 2.0 A is taken from it.*

Figure 20.2 shows the situation when we consider the battery to be replaced by a constant voltage source in series with the internal resistance. The term constant voltage source is taken to mean one for which the terminal voltage does not change when the load current changes.

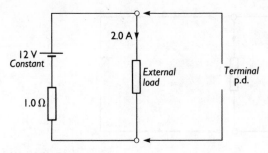

Fig. 20.2

The potential difference across the external load, i.e. the terminal potential difference, is 12 V minus the potential drop across the internal resistance of $IR = 2.0 \times 1.0 = 2.0$ V. Thus the terminal potential difference is $12 - 2.0 = 10$ V.

Thévenin's theorem states that: the current in any impedance connected to two terminals of a network is the same as if the impedance had been connected to a source consisting of a constant voltage supply whose e.m.f. is the open-circuit voltage measured at the terminals and whose internal impedance is the impedance of the network with all the sources replaced by impedances equal to their internal impedances.

To illustrate the use of this theorem, consider the circuit shown in fig. 20.3(*a*). To find the current through the load, R_L, we first consider there to be terminals on either side of the load. Then the open circuit voltage between these terminals is determined, i.e. the voltage when the load resistor is not present, fig. 20.3(*b*). For the example shown: current through $R_1 = E/(R_1 + r)$, hence the potential difference across $R_1 = ER_1(R_1 + r)$. Because there is open circuit there will be no current through R_2 and so no potential drop across it. This means that the open circuit voltage between the two terminals is the same as the potential difference across R_1, i.e. $V = ER_1/(R_1 + r)$. Having found the open-circuit voltage we now need to find the internal resistance. Using the theorem, this is the resistance of the circuit shown in fig. 20.3(*c*), all sources having been replaced by their internal resistances. Hence we have R_2 in series with a parallel arrangement of R_1 and r. The resistance is thus $R_2 + R_1r/(R_1 + r)$. Figure 20.3(*d*) now shows the resulting simplified circuit which we can then use to determine the current through the load.

Example 20.3 *Determine the current flowing through the 12-Ω resistor in the circuit shown in fig. 20.4.*

To use Thévenin's theorem we will consider there to be terminals on either side of the 12-Ω resistor. Now we can consider the open-circuit voltage between these terminals.

The current through the 10-Ω resistor is given by

$E = 6 = I(2 + 5 + 10)$
$I = 0.35$ A.

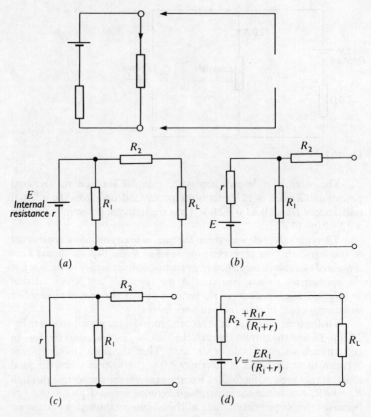

Fig. 20.3

(a)   (b)   (c)   (d)

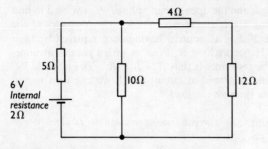

Fig. 20.4

The potential difference across the 10-Ω resistor is thus

$$V = IR = 0.35 \times 10 = 3.5\,\text{V}.$$

Since there is open circuit and thus no current through the 4-Ω resistor, the open circuit voltage between the terminals must be 3.5 V.

To find the internal resistance of the network between the terminals we have to find the resistance of the circuit shown in fig. 20.5. The

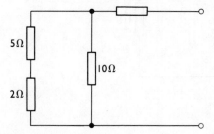

Fig. 20.5

resistance of the parallel arrangement is given by

$$\frac{1}{R} = \frac{1}{10} + \frac{1}{(2+5)}$$

and so

$$R = 4.1 \ \Omega.$$

When we take into account that there is a 4-Ω resistor in series with this, we have a total resistance of 8.1 Ω.

The circuit we are now left to consider is that shown in fig. 20.6. Hence the current through the 12-Ω resistor is given by

$$I = \frac{3.5}{(8.1 + 12)}$$

$$= 0.17 \ \text{A}.$$

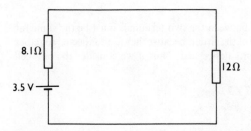

Fig. 20.6

Example 20.4 *Calculate the current flowing through the 4-Ω resistor in the circuit shown in fig. 20.7.*

To calculate the open circuit voltage between terminals located on either side of the 4-Ω resistor the circuit is as shown in fig. 20.8. The two

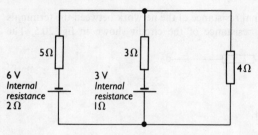

Fig. 20.7

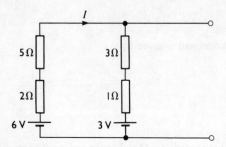

Fig. 20.8

batteries are acting in opposition in driving a current through the circuit, therefore the net e.m.f. is $(6-3)=3$ V. Hence the current I is given by

$$I = \frac{3}{(5+2+3+1)}$$

$$I = 0.27 \text{ A}.$$

The potential difference between the two terminals must be the same for each of the two branches of the circuit because they are connected together to the two terminal points concerned. Thus if we consider the 6-V cell branch:

$$V = 6 - IR = 6 - 0.27 \times 7$$

$$= 4.1 \text{ V}.$$

If however we had considered the other branch we would need to have taken into account the fact that the current direction is in the opposite direction to that which the 3-V cell would have produced and so

$$V = 3 + IR = 3 + 0.27 \times 4$$

$$= 4.1 \text{ V}.$$

To calculate the internal resistance between the terminals we have to consider the circuit in fig. 20.9. This is just $7\,\Omega$ in parallel with $4\,\Omega$ and so

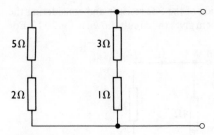

Fig. 20.9

the resistance is given by

$$\frac{1}{R} = \frac{1}{7} + \frac{1}{4}$$

$$R = 2.5 \, \Omega.$$

Hence the simplified circuit is as in fig. 20.10. The current is therefore given by

$$I = \frac{4.1}{2.5 + 4}$$

$$= 0.63 \, \text{A}.$$

Fig. 20.10

Example 20.5 *Calculate the current flowing through the 5-Ω resistor in the circuit shown in fig. 20.11.*

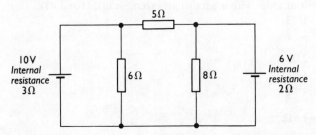

Fig. 20.11

Putting terminals either side of the 5-Ω resistor and considering the open-circuit voltage between them gives the circuit shown in fig. 20.12. For

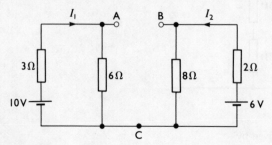

Fig. 20.12

the left-hand end of the circuit we can calculate the current I_1,

$$V = 10 = I_1(3+6)$$

$$I = 1.1\,\text{A}.$$

Hence the potential difference between A and C is

$$V_{AC} = IR = 1.1 \times 6 = 6.6\,\text{V}.$$

Similarly we can calculate the current I_2 for the right-hand circuit.

$$V = 6 = I_2(2+8)$$

$$I_2 = 0.6\,\text{A}.$$

Hence the potential difference between B and C is

$$V_{BC} = IR = 0.6 \times 8 = 4.8\,\text{V}.$$

We now have the potentials at A and at B relative to a common point C, hence

$$V_{AV} = V_{AC} - V_{BC} = 6.6 - 4.8$$

$$= 1.8\,\text{V}.$$

This is the open circuit voltage between the two terminals.

To calculate the internal resistance, we have a $3\,\Omega$ and $6\,\Omega$ parallel arrangement in series with a parallel arrangement of $2\,\Omega$ and $8\,\Omega$.

$$\frac{1}{R_{AC}} = \frac{1}{3} + \frac{1}{6}$$

$$R_{AC} = 2\,\Omega$$

$$\frac{1}{R_{BC}} = \frac{1}{2} + \frac{1}{8}$$

$$R_{BC} = 1.6\,\Omega.$$

Hence total resistance $= R_{AC} + R_{BC} = 3.6\,\Omega.$

Thus to calculate the current through the 5-Ω resistor we have the simplified circuit shown in fig. 20.13. The current I is thus

$$I = \frac{1.8}{3.6 + 5}$$

$$= 0.21 \text{ A.}$$

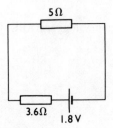

Fig. 20.13

Example 20.6 *Use Thévenin's theorem to determine the current through the 10-Ω resistor in fig. 20.1, a repeat of the example 20.1 that was solved using Kirchhoff's laws.*

Putting terminals either side of the 10-Ω resistor and considering the open-circuit voltage between them gives the circuit shown in fig. 20.14.

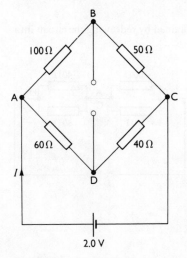

Fig. 20.14

The circuit now consists of two parallel arms, ABC and ADC. Arm ABC has a total resistance of 150 Ω and ADC 100 Ω. Hence the total resistance is given by

$$\frac{1}{R} = \frac{1}{150} + \frac{1}{100}$$

$$R = 60\,\Omega.$$

The current I is thus

$$I = V/R = 2.0/60 = 0.0333\,\text{A}.$$

This current at the junction A divides itself with part going along arm ABC and part along ADC.

The current along ABC, I_{ABC}, is given by

$$I_{ABC} = \frac{2.0}{150} = 0.0133\,\text{A}.$$

Hence the current along ADC, I_{ADC}, is given by

$$I_{ADC} = 0.0333 - 0.133 = 0.0200\,\text{A}.$$

The potential difference A and B is thus

$$V_{AB} = 0.0133 \times 100 = 1.33\,\text{V}.$$

The potential difference between A and D is

$$V_{AD} = 0.020 \times 60 = 1.20\,\text{V}.$$

The potential difference between B and C is therefore

$$V_{BD} = V_{AB} - V_{AD} = 1.33 - 1.20 = 0.13\,\text{V}.$$

This is the open circuit voltage.

The internal resistance can be obtained by redrawing the circuit into

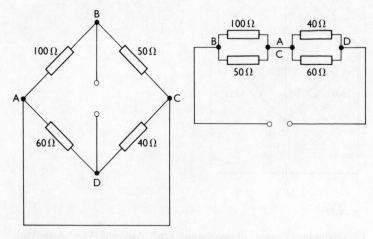

Fig. 20.15 Alternative forms of the same circuit

the form shown in fig. 20.15. The resistance between B and A/C is given by

$$\frac{1}{R_B} = \frac{1}{100} + \frac{1}{50}$$

$R_B = 33.3\ \Omega.$

The resistance between D and A/C is given by

$$\frac{1}{R_D} = \frac{1}{60} + \frac{1}{40}$$

$R_D = 24\ \Omega.$

The total resistance is therefore $33.3 + 24 = 57.3\ \Omega.$

The simplified circuit is therefore as in fig. 20.16. The current is therefore

$$I = \frac{0.133}{57.3 + 10}$$

$$= 1.93 \times 10^{-3}\ A.$$

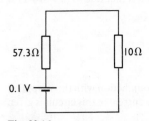

Fig. 20.16

Because of slightly different rounding errors on the data in this version of the solution and that in example 20.1 the results are not perfectly equal.

Example 20.7 *Determine the current through the 10-Ω resistor in the circuit shown in fig. 20.17.*

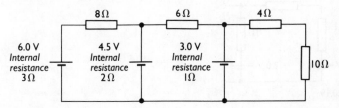

Fig. 20.17

With such a complex circuit we use Thévenin's theorem on only pieces of the circuit at a time. Thus fig. 20.18 shows the first piece of the

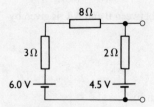

Fig. 20.18

circuit to which Thévenin's theorem can be applied. The current in the circuit is given by

$$I = \frac{(6.0 - 4.5)}{(3 + 8 + 2)}$$

$$= 0.115 \, \text{A}.$$

Hence the open circuit voltage between the terminals is

$$V = 4.5 + 0.115 \times 2$$

$$= 4.73 \, \text{V}$$

The internal resistance between the terminals is that due to $(8 + 3) = 11 \, \Omega$ in parallel with $2 \, \Omega$.

$$\frac{1}{R} = \frac{1}{11} + \frac{1}{2}$$

$$R = 1.69 \, \Omega.$$

If we now consider the second piece in conjunction with the simplified first piece of the circuit we have fig. 20.19. The current in this circuit is given by

$$I = \frac{(4.73 - 3.0)}{(1.69 + 6 + 1)}$$

$$= 0.199 \, \text{A}.$$

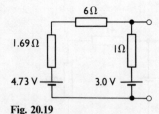

Fig. 20.19

Hence the open circuit voltage between the terminals is

$$V = 3.0 + 0.199 \times 1$$

$$= 3.199 \, \text{V}.$$

The internal resistance between the terminals is that due to $(6 + 1.69) =$ 7.69 Ω in parallel with 1 Ω.

$$\frac{1}{R} = \frac{1}{7.69} + \frac{1}{1}$$

$$R = 0.855 \, \Omega.$$

The final circuit is thus as in fig. 20.20. Hence the current in the circuit, and hence through the 10-Ω resistor, is

$$I = \frac{3.199}{(0.885 + 4 + 10)}$$

$$= 0.215 \, \text{A}.$$

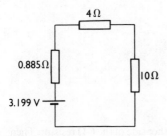

Fig. 20.20

20.3 Norton's theorem

Thévenin's theorem replaced a complicated circuit with a constant voltage supply and a resistance in series with it. Norton's theorem replaces a complicated circuit with a constant current supply and a resistor in parallel with it.

Norton's theorem states that: the current in any impedance connected to two terminals of a network is the same as if the impedance had been connected to a source consisting of a constant current supply whose current is equal to the short circuit current measured at the terminals and whose internal impedance is infinite, but which has an impedance, equal to the impedance of the network with all the sources replaced by impedances equal to their internal impedances, placed in parallel with it.

Thus with Norton's theorem a d.c. network is replaced by a constant current source with a resistor in parallel with it, as in fig. 20.21. Note the symbol used for a constant current source. As with Thévenin's theorem, Norton's theorem can be used with both a.c. or d.c. With d.c. we can replace the word impedance in the theorem by resistance.

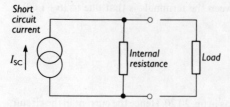

Fig. 20.21

Example 20.8   *Determine the current flowing through the 6-Ω resistor in the circuit shown in fig. 20.22.*

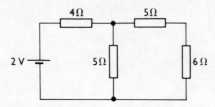

Fig. 20.22

Consider there to be terminals on either side of the 6-Ω resistor. Then short circuit the terminals, as in fig. 20.23. The circuit now consists of a 4-Ω

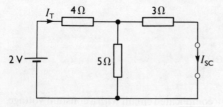

Fig. 20.23

resistor in series with a parallel arrangement of 5 Ω and 3 Ω. The parallel arrangement has a resistance of

$$\frac{1}{R} = \frac{1}{5} + \frac{1}{3}$$

$$R = 1.88 \, \Omega.$$

Hence the circuit resistance is $4 + 1.88 = 5.88 \, \Omega$. The current I_T is given by

$$I_T = \frac{2}{4 + 5.88} = 0.202 \, \text{A}.$$

Thus the potential drop across the 4-Ω resistor is $4 \times 0.202 = 0.808 \, \text{V}$ and so the potential drop across the parallel arrangement is $2 - 0.808 =$

1.192 V. Hence the short circuit current, i.e. the current through the 3-Ω resistor, is

$$I_{SC} = \frac{1.192}{3} = 0.397\,\text{A}.$$

The internal resistance of the network is the resistance of the circuit shown in fig. 20.24. This is the resistance of a 3-Ω resistor in series with a

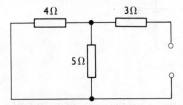

Fig. 20.24

parallel arrangement of 4 Ω and 3 Ω. The parallel arrangement has a resistance given by

$$\frac{1}{R} = \frac{1}{4} + \frac{1}{5}$$

$$R = 2.22\,\Omega.$$

Hence the internal resistance is $3 + 2.22 = 5.22\,\Omega$.

The circuit simplified as a result of using Norton's theorem is as shown in fig. 20.25. To find the current through the 6-Ω resistor we need to

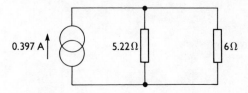

Fig. 20.25

determine the potential difference across the parallel arrangement. The resistance of this parallel arrangement is

$$\frac{1}{R} = \frac{1}{5.22} + \frac{1}{6}$$

$$R = 2.79\,\Omega.$$

Hence the potential difference is $V = IR = 0.397 \times 2.79 = 1.11$ V. Hence the current through the 6-Ω resistor is given by

$$I = \frac{1.11}{6} = 0.185\,\text{A}.$$

Example 20.9   *Calculate the current through the 4-Ω resistor in the circuit shown in fig. 20.26.*

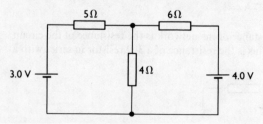

Fig. 20.26

We can consider this circuit in two pieces and the short circuit current arising from each part when terminals placed either side of the 4-Ω resistor are short circuited. Figure 20.27 shows the two pieces. For fig. 20.27(a) the

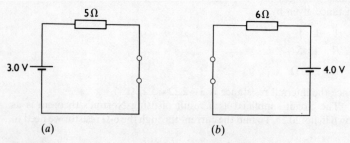

Fig. 20.27

short circuit current is given by

$$I_{SC} = \frac{3.0}{5} = 0.6\,\text{A}.$$

For fig. 20.27(b) the short circuit current is given by

$$I_{SC} = \frac{4.0}{6} = 0.667\,\text{A}.$$

The internal resistances of the two pieces of circuit are 5 Ω and 6 Ω and thus the circuit resulting from the application of Norton's theorem is as shown in fig. 20.28(a). What we have is a total current of $0.6 + 0.667 = 1.267\,\text{A}$ being supplied to a parallel arrangement of resistors, fig. 20.28(b). The parallel arrangement has a total resistance given by

$$\frac{1}{R} = \frac{1}{5} + \frac{1}{4} + \frac{1}{6}$$

$$R = 1.62\,\Omega.$$

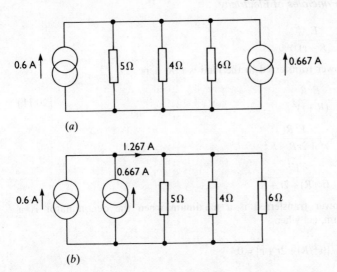

Fig. 20.28

Hence the potential difference across the arrangement is

$$V = 1.267 \times 1.62 = 2.053 \, \text{V}$$

and so the current through the $4 \, \Omega$ resistor is

$$I = \frac{2.053}{4} = 0.513 \, \text{A}.$$

20.4 Maximum power transfer

Consider a source having an e.m.f. E and an internal resistance r connected to a load of resistance R, as in fig. 20.29. Then

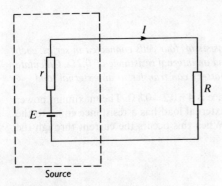

Fig. 20.29

$$I = \frac{E}{(R+r)}.$$

The power transferred to the load is I^2R, hence

$$P = \frac{E^2 R}{(R+r)^2} \qquad\qquad [20.11]$$

$$= \frac{E^2 R}{r^2 + 2rR + R^2}$$

$$= \frac{E^2}{(r^2/R) + 2r + R}.$$

The power transferred is a maximum when the denominator is a minimum, i.e. when

$$\frac{\mathrm{d}}{\mathrm{d}R}\{(r^2/R) + 2r + r\} = 0$$

$$-(r^2/R^2) + 1 = 0$$

$$R = r.$$

Hence the power transferred from the source to the load is a maximum when the resistance of load is equal to the internal resistance of the source. This is referred to as resistance matching.

The maximum power transfer theorem states that when one circuit is supplying another, the maximum power is transferred when the resistance of the load circuit equals the internal resistance of the supply circuit.

The maximum power transferred is given by substituting $r = R$ in equation [20.11],

$$\text{maximum power} = \frac{E^2 R}{(R+R)^2}$$

$$= \frac{E^2}{4R}.$$

Example 20.10 *A battery consists of four cells connected in series, each cell having an e.m.f. of 1.3 V and an internal resistance of 0.2 Ω. Calculate the maximum power that the battery can transfer to an external load.*

The total internal resistance is $4 \times 0.2 = 0.8\,\Omega$. The maximum power will be transferred when the external load has a resistance equal to the internal resistance, i.e. $0.8\,\Omega$. When this occurs the current through the load is given by

$$I = \frac{4 \times 1.3}{0.8 + 0.8}$$

$$= 3.25\,\text{A}.$$

The power dissipated in the load is then

$$P = I^2 R = 3.25^2 \times 0.8$$

$$= 8.45 \, \text{W}.$$

Example 20.11   *For the circuit shown in fig. 20.30 calculate the load resistance for maximum power transfer and the value of this power.*

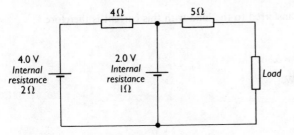

Fig. 20.30

Using Thévenin's theorem to simplify the circuit, fig. 20.31 shows the circuits for calculating the open-circuit voltage and the internal resistance. The net e.m.f. in fig. 20.31(a) is $4.0 - 2.0 = 2.0 \, \text{V}$. This drives a current

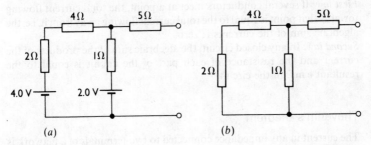

Fig. 20.31

through a resistance of $2 + 4 + 1 = 7 \, \Omega$. Hence the current is $2.0/7 = 0.29 \, \text{A}$. The potential difference across the load plus the 5-Ω resistor will be

$$V = 4.0 - 0.29(2 + 4) = 2.26 \, \text{V}.$$

This will be the open circuit voltage since there is no current through the 5 Ω resistor and hence no potential drop across it. The internal resistance is given by $(R + 5)$, where R is

$$\frac{1}{R} = \frac{1}{(2+4)} + \frac{1}{1}$$

$$R = 0.86 \, \Omega.$$

Hence internal resistance $= 5.86 \, \Omega$.

Maximum power will occur when the load resistance has the same value as the internal resistance, i.e. $5.86\,\Omega$.

The current in the circuit when the load is $5.86\,\Omega$ is

$$I = \frac{2.26}{(5.86 + 5.86)}$$

$$= 0.19\,\text{A}.$$

The power transferred to the load with this current is therefore

$$P = I^2 R$$

$$= 0.19^2 \times 5.86$$

$$= 0.22\,\text{W}.$$

20.5 Summary of important laws

Kirchhoff's laws

First law: If several conductors meet at a point, the total current flowing towards that point is equal to the total current flowing away from it, i.e. the algebraic sum of the currents is zero.

Second law: In any closed circuit, the algebraic sum of the products of the current and the resistance of each part of the circuit is equal to the resultant e.m.f. in the circuit.

Thévenin's theorem

The current in any impedance connected to two terminals of a network is the same as if the impedance had been connected to a source consisting of a constant voltage supply whose e.m.f. is the open-circuit voltage measured at the terminals and whose internal impedance is the impedance of the network with all the sources replaced by impedances equal to their internal impedances.

Norton's theorem

The current in any impedance connected to two terminals of a network is the same as if the impedance had been connected to a source consisting of a constant current supply whose current is equal to the short circuit current measured at the terminals and whose internal impedance is infinite, but which has an impedance, equal to the impedance of the network with all the sources replaced by impedances equal to their internal impedances, placed in parallel with it.

The maximum power transfer theorem

When one circuit is supplying another, the maximum power is transferred when the resistance of the load circuit equals the internal resistance of the supply circuit.

20.6 Examples

1. Use Kirchhoff's laws to determine the current through the 5-Ω resistor in the bridge circuit shown in fig. 20.32.

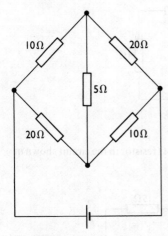

Fig. 20.32

2. Use Kirchhoff's laws to determine the current through the 50-Ω resistor in the circuit shown in fig. 20.33

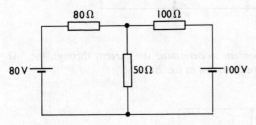

Fig. 20.33

3. A 12-V battery of internal resistance 3 Ω is connected in parallel with a 6 V battery of internal resistance 2 Ω to a load of 40 Ω. By the use of Thévenin's theorem calculate the current through the load.

4. Use Thévenin's theorem to determine the current through the 5 Ω resistor in fig. 20.32.

5. A d.c. generator giving an e.m.f. of 12 V with an internal resistance of 2.0 Ω is connected in parallel with another generator giving 20 V with an internal resistance of 5.0 Ω. What current will this arrangement give through a common load of 10.0 Ω?

6. A battery having an e.m.f. of 12 V and an internal resistance of 1.0 Ω is connected in parallel with another battery having an e.m.f. of 6 V and an internal resistance of 0.5 Ω. If they supply a common load of 8.0 Ω what is (a) the current through the load and (b) the currents in each battery?

7. Calculate the current through the 4-Ω resistor in the circuit shown in fig. 20.34.

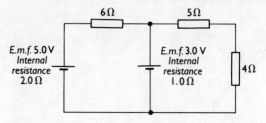

Fig. 20.34

8. Calculate the current through the 20-Ω resistor in the circuit shown in fig. 20.35.

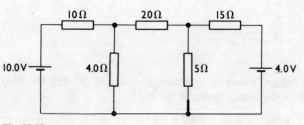

Fig. 20.35

9. Use Norton's theorem to determine the current through the 5-Ω resistor in the circuit shown in fig. 20.36.

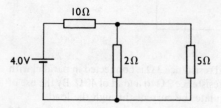

Fig. 20.36

10. Use Norton's theorem to determine the current through the 5-Ω resistor in the circuit shown in fig. 20.37.

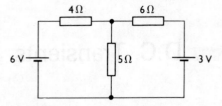

Fig. 20.37

11. A battery consists of six cells in series, each cell having an e.m.f. of 1.2 V and an internal resistance of 0.3 Ω. What is the maximum power the battery can transfer to an external load?

12. For the circuit shown in fig. 20.37, what resistor should be used to replace the 5-Ω resistor if maximum power is to be dissipated in it?

13. A battery having an e.m.f. of 3.0 V and an internal resistance of 1.0 Ω is connected in parallel with another battery of e.m.f. 1.5 V and internal resistance 0.8 Ω. What is the value of the resistance to which the combination will deliver maximum power and what is the value of that power?

14. A battery having an e.m.f. of 12 V and an internal resistance of 2.0 Ω is connected across a 20-Ω resistor. A variable resistor is then connected in parallel with the 20-Ω resistor. What must be the value of this variable resistor if the battery is to deliver maximum power?

CHAPTER 21

▬▬▬▬▬▬ **D.C. Transients**

21.1 Transients in a CR circuit

When a voltage V is applied to a capacitor in series with a resistor, a transient current flows as the capacitor charges up. The current drops to zero when the capacitor is fully charged (see section 9.17). If at some instant the potential difference across the resistor R is v_R and the potential difference across the capacitor C is v_C, then

$$V = v_R + v_C.$$ [21.1]

If at this instant the transient current is i, then

$$i = \frac{dq}{dt}$$

where q is the charge on the capacitor at that instant, current being the rate of movement of charge. But $q = Cv_C$, hence

$$i = C\frac{dv_C}{dt}.$$ [21.2]

But $v_R = iR$, hence $v_R = CR(dv_C/dt)$.
Thus equation [21.1] becomes

$$V = CR\frac{dv_C}{dt} + v_C.$$

Rearranging this equation gives

$$V - v_C = CR\frac{dv_C}{dt}$$

$$\frac{dt}{CR} = \frac{dv_C}{V - v_C}.$$

If we take the value of v_C to be zero when $t = 0$, and the potential difference across the capacitor to be v_C at time t, then

$$\int_0^t \frac{dt}{CR} = \int_0^{v_C} \frac{dv_C}{V - v_C}$$

$$\frac{t}{CR} = \ln\left(\frac{V}{V - v_C}\right)$$

$$\frac{V}{V - v_C} = e^{t/CR}$$

$$v_C = V(1 - e^{-t/CR}). \qquad [21.3]$$

This equation is described by the graph shown in fig. 21.1. The graph also

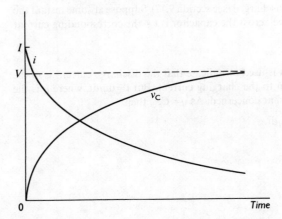

Fig. 21.1

shows the way the current varies with time. This can be obtained from equation [21.3] by using equation [21.2].

$$i = C \frac{dv_C}{dt} = C \frac{d}{dt}\{V(1 - e^{-t/CR})\}$$

$$= \frac{CV}{CR} e^{-t/CR}$$

$$= (V/R)e^{-t/CR}.$$

When the current is switched on, i.e. $t = 0$, the current will have the value (V/R) since $e^{-0} = 1$. This is the maximum current, I. Hence

$$i = I e^{-t/CR}. \qquad [21.4]$$

The product CR is known as the *time constant* of the circuit, the unit of the time constant being seconds when C is in farads and R in ohms. The following table shows how the potential difference across the capacitor, v_C, and the circuit current, i, vary in time intervals equal to the time constant.

Time	v_C	i
1CR	$v_C = V(1 - e^{-1}) = 0.632\,\text{V}$	$i = Ie^{-1} = 0.368I$
2CR	$v_C = V(1 - e^{-2}) = 0.865\,\text{V}$	$i = Ie^{-2} = 0.135I$
3CR	$v_C = V(1 - ^{-3}) = 0.950\,\text{V}$	$i = Ie^{-3} = 0.049I$
4CR	$v_C = V(1 - e^{-4}) = 0.982\,\text{V}$	$i = Ie^{-4} = 0.018I$
5CR	$v_C = V(1 - e^{-5}) = 0.993\,\text{V}$	$i = Ie^{-5} = 0.007I.$

When a charged capacitor C is allowed to discharge through a resistance R a transient current flows. The current drops to zero when the capacitor is fully discharged (see section 9.17). Suppose at some instant the potential difference across the capacitor is v_C, the corresponding current will be

$$i = -v_C/R.$$

The negative sign is used to indicate that the discharge current is in the opposite direction to the charging current. But $i = \mathrm{d}q/\mathrm{d}t$, where q is the charge at the instant concerned. As $q = Cv_c$, then

$$i = C\,\mathrm{d}v_C/\mathrm{d}t.$$

Hence

$$-\frac{v_C}{R} = C\frac{\mathrm{d}v_C}{\mathrm{d}t}$$

$$\frac{\mathrm{d}t}{CR} = -\frac{\mathrm{d}v_C}{v_C}.$$

At time $t = 0$ then $v_C = V$, the voltage applied to the capacitor to charge it up. Hence

$$\int_0^t \frac{\mathrm{d}t}{CR} = -\int_V^{v_c} \frac{\mathrm{d}v_C}{v_C}$$

$$\frac{t}{CR} = \ln\left(\frac{V}{v_C}\right)$$

$$\frac{V}{v_C} = e^{t/CR}$$

$$v_C = V\,e^{-t/CR}. \qquad\qquad [21.5]$$

This equation describes how the potential difference across the capacitor changes with time as the capacitor discharges (fig. 21.2).

As $i = -v_C/R$ then we can derive a relationship for the current variation with time.

$$i = -\frac{v_C}{R} = -\frac{V}{R}e^{-t/CR}.$$

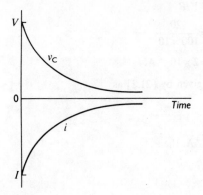

Fig. 21.2

The initial current I is V/R, hence

$$i = -Ie^{-t/CR}. \qquad [21.6]$$

Figure 21.2 shows how this current i varies with time.

The following table shows how the potential difference across the capacitor, v_C, and the circuit current, i, vary with time intervals equal to the time constant CR.

Time	v_C	i
$1CR$	$v_C = Ve^{-1} = 0.368\,\text{V}$	$i = -Ie^{-1} = -0.368I$
$2CR$	$v_C = Ve^{-2} = 0.135\,\text{V}$	$i = -Ie^{-2} = -0.135I$
$3CR$	$v_C = Ve^{-3} = 0.049\,\text{V}$	$i = -Ie^{-3} = -0.049I$
$4CR$	$v_C = Ve^{-4} = 0.018\,\text{V}$	$i = -Ie^{-4} = -0.018I$
$5CR$	$v_C = Ve^{-5} = 0.007\,\text{V}$	$i = -Ie^{-5} = -0.007I.$

Example 21.1 *A 4-μF capacitor in series with a 100-kΩ resistor is to be charged by being connected to a 20-V d.c. supply. Calculate* (a) *the time constant of the circuit,* (b) *the initial charging current,* (c) *the initial rate of growth of the potential difference across the capacitor,* (d) *the potential difference across the capacitor and the circuit current after a time equal to twice the time constant,* (e) *the potential difference across the capacitor and the circuit current 0.5 s after the supply has been switched on.*

(a) Time constant $= CR$

$$= 4 \times 10^{-6} \times 100 \times 10^3$$

$$= 0.4\,\text{s}.$$

(b)   Initial charging current $I = V/R$

$$= \frac{20}{100 \times 10^3}$$

$$= 2 \times 10^{-4}\,\text{A}.$$

(c)   The rate of growth of v_C is given by [21.2] as

$$i = C\frac{dv_C}{dt}.$$

Hence as initially $i = I = 2 \times 10^{-4}\,\text{A}$, then

$$\frac{dv_C}{dt} = \frac{2 \times 10^{-4}}{4 \times 10^{-6}}$$

$$= 50\,\text{V s}^{-1}.$$

(d)   Equation [21.3] gives

$$v_C = V(1 - e^{-t/CR}).$$

When $t = 2CR$, then

$$v_C = V(1 - e^{-2}) = 0.865\,\text{V}$$

$$= 0.865 \times 10$$

$$= 17.3\,\text{V}.$$

The current is given by equation [21.4] as

$$i = I\,e^{-t/CR}.$$

When $t = 2CR$, then

$$i = I\,e^{-2} = 0.135I$$

$$= 0.135 \times 2 \times 10^{-4}$$

$$= 2.7 \times 10^{-5}\,\text{A}.$$

(e)   Equation [21.3] gives

$$v_C = V(1 - e^{-t/CR})$$

$$= 20(1 - e^{-0.5/0.4})$$

$$= 14.3\,\text{V}.$$

Equation [21.4] gives

$$i = I\,e^{-t/CR}$$

$$= 2 \times 10^{-4}\,e^{-0.5/0.4}$$

$$= 5.7 \times 10^{-5}\,\text{A}.$$

Example 21.2   *A 8-μF capacitor is fully charged to a potential difference of 20 V. The capacitor is then discharged through a 1-MΩ resistor. Calculate*

(a) *the initial discharge current,* (b) *the initial rate at which the potential difference across the capacitor is changing,* (c) *the current and potential difference at a time equal to the time constant after the start of the discharge,* (d) *the current and potential difference 12 s after the start of the discharge.*

(a) As $i = -v_C R$ and initial $v_C = V$, then the initial current is $20/(1 \times 10^6) = 20 \times 10^{-6}$ A.

(b) As $i = C(dv_C/dt)$, then when the current is 20×10^{-6} A, then

$$\frac{dv_C}{dt} = \frac{20 \times 10^{-6}}{8 \times 10^{-6}}$$

$$= 2.5 \, V \, s^{-1}.$$

(c) Equation [21.6] gives

$$i = -I e^{-t/CR}.$$

When $t = CR$, then

$$i = -20 \times 10^{-6} e^{-1}$$

$$= 7.4 \times 10^{-6} \, A.$$

Equation [21.5] gives

$$v_C = V e^{-t/CR}.$$

When $t = CR$, then

$$v_C = 20 e^{-1}$$

$$= 7.4 \, V.$$

(d) Equation [21.6] gives

$$i = -I e^{-t/CR}$$

$$= -20 \times 10^{-6} e^{-12/8}$$

$$= -4.5 \times 10^{-6} \, A.$$

Equation [21.5] gives

$$v_C = V e^{-t/CR}$$

$$= 20 e^{-12/8}$$

$$= 4.5 \, V.$$

21.2 Transients in an LR circuit

When a voltage V is applied to an inductor in series with a resistor the current does not immediately rise to the steady value but a transient condition occurs while the current grows to this value (see section 8.6).

When the current through an inductor L changes there is a back e.m.f. E, where (see equation [8.1]):

$$E = -L\frac{di}{dt} \qquad [21.7]$$

di/dt is the rate of change of current with time. Hence the net e.m.f. is $(V - E)$ and so

$$V - L\frac{di}{dt} = iR \qquad [21.7]$$

$$\frac{V}{R} - \frac{L}{R}\frac{di}{dt} = i.$$

But the final steady current I is V/R, hence

$$I - i = \frac{L}{R}\frac{di}{dt}.$$

At $t = 0$ then $i = 0$, hence

$$\int_0^t \frac{R}{L}\,dt = \int_0^i \frac{di}{I - i}$$

$$\frac{R}{L}t = -\ln\left(\frac{I}{I - i}\right)$$

$$\frac{I - i}{I} = e^{-Rt/L}$$

$$i = I(1 - e^{-Rt/L}). \qquad [21.9]$$

The potential difference across the inductor v_L is given by

$$v_L = V - iR$$

$$= -IR(1 - e^{-Rt/L}).$$

But $V = IR$, hence

$$v_L = V - V + V e^{-Rt/L}$$

$$v_L = V e^{-Rt/L}. \qquad [21.10]$$

Figure 21.3 shows how the circuit current and the potential difference across the inductor vary with time when a voltage is applied to an inductor in series with a resistor. These graphs are described by equations [21.9] and [21.10].

L/R is known as the *time constant* for a circuit involving an inductor in series with a resistor. The following table shows how the current in the circuit and the potential difference across the inductor vary with time intervals equal to the time constant.

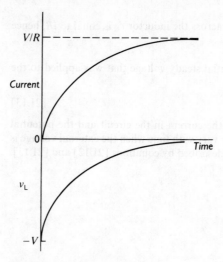

Fig. 21.3

Time	i	v_L
$1L/R$	$i = I(1 - e^{-1}) = 0.632I$	$v_L = V e^{-1} = 0.368\,V$
$2L/R$	$i = I(1 - e^{-2}) = 0.865I$	$v_L = V e^{-2} = 0.135\,V$
$3L/R$	$i = I(1 - e^{-3}) = 0.950I$	$v_L = V e^{-3} = 0.049\,V$
$4L/R$	$i = I(1 - e^{-4}) = 0.982I$	$v_L = V e^{-4} = 0.018\,V$
$5L/R$	$i = I(1 - e^{-5}) = 0.993I$	$v_L = V e^{-5} = 0.007\,V$

When the externally applied voltage to a circuit involving an inductor in series with a resistor is switched off the current does not immediately drop to zero but decays with time. A back e.m.f. is produced in the inductor due to the current changing and thus since there is now no source of external voltage equation [21.8] becomes

$$0 - L\frac{di}{dt} = iR. \qquad [21.11]$$

Hence since at $t = 0$ we have $i = I$, the original steady current, then

$$\int_0^t \frac{R}{L}\,dt = -\int_I^i \frac{di}{i}$$

$$\frac{Rt}{L} = \ln I - \ln i = \ln(I/i)$$

$$i = I e^{-Rt/L}. \qquad [21.12]$$

The potential difference across the inductor v_L is equal to iR, hence

$$v_L = IR\,e^{-Rt/L}.$$

But IR is the value of the initial steady voltage that was applied to the circuit. Hence

$$v_L = V\,e^{-Rt/L}. \qquad [21.13]$$

Figure 21.4 shows how the current in the circuit and the potential difference across the inductor vary with time when the external voltage is switched off. The graphs are described by equations [21.12] and [21.13].

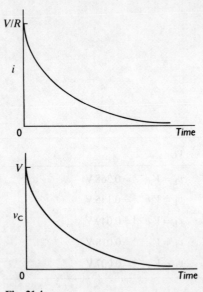

Fig. 21.4

The following table shows how the current in the circuit and the potential difference across the inductor vary with time intervals equal to the time constant.

Time	i	v_L
$1L/R$	$i = I\,e^{-1} = 0.368I$	$v_L = V\,e^{-1} = 0.368\,\text{V}$
$2L/R$	$i = I\,e^{-2} = 0.135I$	$v_L = V\,e^{-2} = 0.135\,\text{V}$
$3L/R$	$i = I\,e^{-3} = 0.049I$	$v_L = V\,e^{-3} = 0.049\,\text{V}$
$4L/R$	$i = I\,e^{-4} = 0.018I$	$v_L = V\,e^{-4} = 0.018\,\text{V}$
$5L/R$	$i = I\,e^{-5} = 0.007I$	$v_L = V\,e^{-5} = 0.007\,\text{V}$

In the derivations of both the growth and decay equations for the current in an LR circuit it has been assumed that we have a pure inductor in series with a pure resistor, i.e. the inductor has no resistance and the resistor has no inductance. In practice an inductor has both inductance and resistance, likewise a resistor is likely to have both resistance and inductance. An inductor can be considered to be composed of a pure inductance in series with a pure resistance. Thus the equations derived above can be applied to a real inductor. However, it is not possible to separate out as measurable quantities the potential difference across the inductor and that across the resistor when both apply to just a single component. The term v_L in such a situation refers therefore to the induced e.m.f. component of the voltage across the inductor, there is however another voltage component due to the resistance of the inductor.

Example 21.3 *A 120-V supply is connected across a coil having an inductance of 0.5 H and a resistance of 20 Ω. Calculate (a) the rate of growth of the current when the supply voltage is connected, (b) the final steady current, (c) the value of the current after a time equal to the time constant, (d) the value of the current after 0.04 s, (e) the induced e.m.f. in the inductor after 0.04 s.*

(a) Using equation [21.8]

$$\frac{di}{dt} = \frac{V - iR}{L}.$$

But initially $i = 0$, hence

$$\frac{di}{dt} = \frac{V}{L} = \frac{120}{0.5} = 240 \, \text{A s}^{-1}.$$

(b) The final steady current $I = V/R = 120/20 = 6.0 \, \text{A}$.

(c) Using equation [21.9]

$$i = I(1 - e^{-Rt/L}) = I(1 - e^{-1})$$

$$= 0.632I = 0.632 \times 6.0$$

$$= 3.8 \, \text{A}.$$

(d) Using equation [21.9]

$$i = 6.0(1 - e^{-20 \times 0.04/0.5})$$

$$= 4.8 \, \text{A}.$$

(e) Using equation [21.10]

$$v_L = V e^{-20 \times 0.04/0.5}$$

$$= 24.2 \, \text{V}.$$

Example 21.4 *A coil with an inductance of 0.6 H and a resistance of 100 Ω is supplied with a voltage from a d.c. source of 30 V and internal resistance 10 Ω. Calculate (a) the time constant for the circuit, (b) the steady state current, (c) the current 3 ms after the voltage is applied to the coil.*

(a) Time constant $= L/R = 0.6/(100 + 10)$

$$= 5.5 \times 10^{-3} \, \text{s}.$$

(b) The steady state current $I = V/R = 30/(100 + 10)$

$$= 0.27 \, \text{A}.$$

(c) Using equation [21.9],

$$i = I(1 - e^{-Rt/L})$$

$$= 0.27(1 - e^{-111 \times 0.003/0.6})$$

$$= 0.11 \, \text{A}.$$

Example 21.5 *A 0.5-H inductor in series with a 2-kΩ resistor is connected to a 12-V d.c. supply. (a) What is the steady current? (b) After reaching the steady current the d.c. supply is switched off and the inductor plus resistor short-circuited, what is the current after 0.1 ms?*

(a) The steady current $= V/R = 12/(2 \times 10^3) = 6 \times 10^{-3} \, \text{A}.$

(b) Using equation [21.12],

$$i = I e^{-Rt/L}$$

$$= 6 \times 10^{-3} e^{-2 \times 10^3 \times 0.1 \times 10^{-3}/0.5}$$

$$= 4.0 \times 10^{-3} \, \text{A}.$$

Example 21.6 *A relay coil has a resistance of 100 Ω and an inductance of 1.0 H. If the relay is connected in series with a 120-Ω resistor and a 50-V d.c. supply, how long will the relay take to operate if it needs 100 mA before the contacts close?*

The steady current in the circuit is $V/R = 50/(100 + 120) = 0.227 \, \text{A}.$ Using equation [21.9] for the growth of current in an LR circuit,

$$i = I(1 - e^{-Rt/L})$$

$$0.100 = 0.227(1 - e^{-(220/1.0)t})$$

$$0.441 = 1 - e^{-220t}$$

$$-220t = \ln 0.559$$

$$t = 2.64 \times 10^{-3} \, \text{s}.$$

21.3 Rectangular waveforms in circuits

When a rectangular waveform input is applied to a CR circuit the capacitor is charged while the waveform is increasing to its maximum value and while it is at its maximum value, discharging when the waveform decreases to its minimum value and while it is at this minimum value. Figure 21.5 shows the situation.

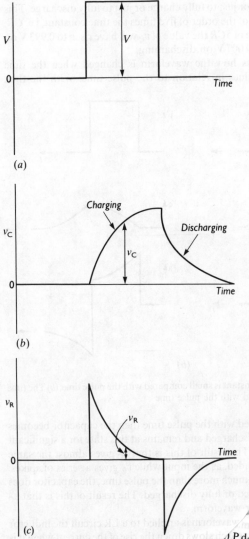

Fig. 21.5 (a) Input form; (b) P.d. across capaci[...] P.d. equation [21.1],

At any instant we must have

$$V = v_C + v_R.$$

Thus during the charging of the capacitor, as v_C increases so v_R decreases but the sum of v_C and v_R remains constant, being V. During the discharging of the capacitor when $V = 0$, then $(v_C + v_R)$ must equal zero at every instant of time.

For the graphs shown in fig. 21.5 the duration of the pulse is long enough for the capacitor just to fully charge or just to fully discharge. This means that the time is of the order of five times the time constant, i.e. CR, for the circuit. In a time of $5CR$ the value of v_C will have risen to 0.993 V on charging or fallen to 0.007 V on discharging.

Figure 21.6 shows how the waveform is changed when the time constant has other values in relation to the pulse time. With the time

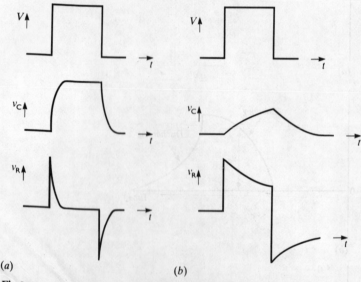

(a) (b)

Fig. 21.6 (a) The time constant is small compared with the pulse time; (b) The time constant is long compared with the pulse time

constant small compared with the pulse time then the capacitor becomes fully charged or fully discharged and remains at this state for a significant part of the pulse time. The result of this is that v_C gives almost the same waveform, slightly rounded, as the input while v_R gives a series of spikes. With the time constant much more than the pulse time, the capacitor does not become fully charged or fully discharged. The result of this is that v_C gives rise to a triangular waveform.

rising and a ... waveform.
falling. Figure ... waveform is applied to a LR circuit the inductor
At any instant ... down the rise of the current when it is
... fall of the current when it is

$$V = v_L + v_R.$$

quation [21.7],

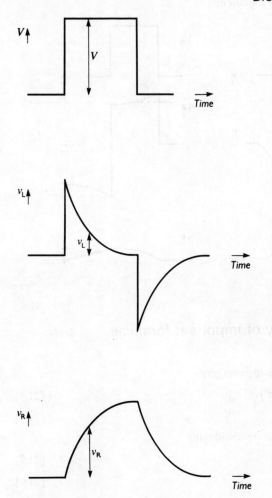

Fig. 21.7

Thus during the current change, as v_R increases so v_L decreases but the sum of v_R and v_L remains constant, being V. During the collapse of the current when $V = 0$, then $(v_L + v_R)$ must equal zero at every instant of time.

For the graphs shown in fig. 21.7 the duration of the pulse is long enough for the current just to reach its steady value or just to collapse to zero. This means that the time is of the order of five times the time constant, i.e. L/R, for the circuit. In a time of $5L/R$ the value of v_L will have dropped to $0.007\,V$.

Figure 21.8 shows how the waveform is changed when the time constant has other values in relation to the pulse time.

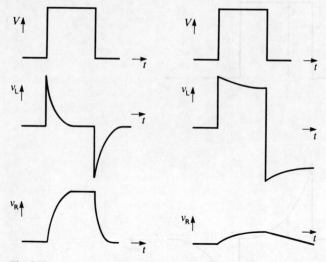

Fig. 21.8

21.4 Summary of important formulae

For a CR circuit during charging

$$v_C = V(1 - e^{-t/CR})$$ [21.3]

$$i = I e^{-t/CR}.$$ [21.4]

For a CR circuit during discharging

$$v_C = V e^{-t/CR}$$ [21.5]

$$i = -I e^{-t/CR}.$$ [21.6]

For a CR circuit,

time constant $= CR$.

For a LR circuit during current growth

$$i = I(1 - e^{-Rt/L})$$ [21.9]

$$v_L = V e^{-Rt/L}.$$ [21.10]

For a LR circuit during current collapse

$$i = I e^{-Rt/L}$$ [21.11]

$$v_L = V e^{-Rt/L}.$$ [21.13]

For a LR circuit,

time constant $= L/R$.

21.5 Examples

1. A 16-μF capacitor in series with a 1-MΩ resistor is to be charged by being connected to a 12-V d.c. supply. Calculate (a) the time constant of the circuit, (b) the initial charging current, (c) the initial rate of growth of the potential difference across the capacitor, (d) the potential difference across the capacitor and circuit current after a time equal to three times the time constant, (e) the potential difference across the capacitor and the circuit current 2.0 s after the supply has been switched on.

2. A 16-μF capacitor is fully charged to a potential difference of 24 V. The capacitor is then discharged through a 200-kΩ resistor. Calculate (a) the time constant of the discharge circuit, (b) the initial discharge current, (c) the initial rate at which the potential difference across the capacitor is changing, (d) the current and the potential difference at a time equal to twice the time constant after the start of the discharge, (e) the current and the potential difference 4.0 s after the start of the discharge.

3. A capacitor is fully charged to a potential difference of 6.0 V. When the capacitor is discharged through a 12-kΩ resistor the potential difference across the capacitor falls to 2.0 V in 60 ms. Calculate the capacitance of the capacitor.

4. A 2-μF capacitor is in series with a 1 kΩ resistor and a d.c. supply of 120 V. Calculate the current flowing in the circuit (a) 1 ms, (b) 2 ms, (c) 3 ms, after the supply is connected.

5. A 100-μF capacitor is connected in series with a 2-kΩ resistor. Determine the potential differences across the capacitor and the resistor 0.1 s after a supply of 12 V is connected to the arrangement.

6. A 12-V supply is connected across a coil having an inductance of 2 H and a resistance of 120 Ω. Calculate (a) the time constant of the circuit, (b) the rate of growth of the current when the voltage is first connected, (c) the final steady current, (d) the value of the current after a time equal to twice the time constant, (e) the value of the current after 50 ms, (f) the induced e.m.f. in the inductor at that time.

7. A coil with an inductance of 0.4 H is in series with a 200-Ω resistor. If a 20-V d.c. supply is connected across this arrangement, what is the current (a) 1 ms, (b) 2 ms, (c) 3 ms, after the supply is connected?

8. A relay coil has an inductance of 300 mH and a resistance of 10 Ω. If the relay is connected in series with a 100-Ω resistor and a 6-V d.c. supply, how long will it take the relay to operate if it needs 40 mA before the contacts close?

9. A coil of inductance 2 H and resistance 50 Ω is connected across a 20-V supply. Calculate the initial rate of change of current, the final steady current and the time taken for the current to reach 0.2 A.

10. The field windings of a machine have an inductance of 12 H and take a current of 2.5 A when steady current conditions occur after connection to a 120-V d.c. supply. Calculate (a) the resistance of the

windings, (b) the initial rate of growth of current, (c) the time taken for the current to reach 90% of its steady value.

11. The coil of a relay has an inductance of 1.0 H and a resistance of 30 Ω. The relay contacts close when a rising current reaches 150 mA and open when a decreasing current reaches 100 mA. A voltage pulse which rises from 0 to 6 V instantaneously, remains constant for 100 ms, and then falls instantaneously to zero is applied to the relay. What is the total time during which the relay contacts are closed?

CHAPTER 22

███████ **Three-Phase Supply**

22.1 Introductory

The mains supply in your home, in the United Kingdom, is a single-phase supply in that the waveform is a single sinusoidal wave. However, the generating, transmission and distribution in the United Kingdom is by a *three-phase supply*. This consists of three separate voltages of the same amplitude and frequency but separated in phase by $120°$ ($\frac{2}{3}\pi$), each of these voltages being referred to as a phase of the supply. Each phase is referred to by a number 1, 2 or 3, or a colour red, yellow or blue (fig. 22.1).

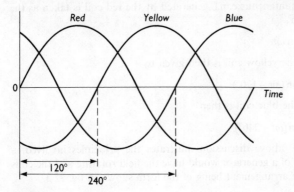

Fig. 22.1

Three-phase systems have advantages over single-phase systems in that:

1. A higher efficiency is possible in the generation.
2. Less conductors and hence aluminium or copper is required for transmission and distribution.
3. Three-phase motors have better characteristics than single-phase ones.
4. The same source can be used for both industrial users, requiring three-phase supplies, and domestic users requiring single-phase supplies.

22.2 Generation of three phases

A single-phase supply can be generated by a single coil rotating in a magnetic field, as in fig. 10.2 and described in section 10.2. A three-phase supply can be generated by having three such rotating coils on the same drive shaft, the coils being at angles of 120° to each other (fig. 22.2).

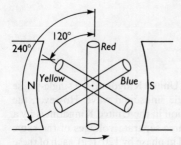

Fig. 22.2   The coils in the magnetic field

If the instantaneous e.m.f. generated by the red coil is taken as the reference, then

$e_R = E_m \sin \omega t.$

The e.m.f. of the yellow coil is then given by

$e_Y = E_m \sin (\omega t - 120°).$

The e.m.f. of the blue coil is then

$e_B = E_m \sin (\omega t - 240°).$

While the above discussion illustrates the principles involved, a practical form of a generator would have the field rotating and the coils stationary, the arrangement being of the form shown in fig. 22.3.

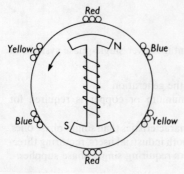

Fig. 22.3

The above equations however still describe the way the e.m.f.s vary with time, these equations being the ones describing the graphs shown in fig. 22.1. Such equations and graphs can be generated by the phasor diagram shown in fig. 22.4.

Figure 22.1 and the phasor diagram fig. 22.4 both show what is called *normal phase rotation*, the sequence being red, yellow, blue. If the phase

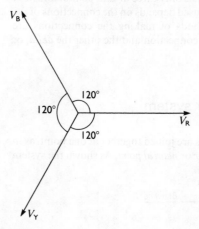

Fig. 22.4

sequence were red, blue, yellow then the phasor diagram would be as in fig. 22.5 and the sequence would be termed *reverse phase direction*. The

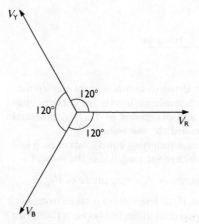

Fig. 22.5

phasors can be considered to be rotating anticlockwise.

22.3 Star and delta connections

With a single-phase supply two conductors are normally required for the transmission, an out and a return conductor for the current. With three-phase supply, while six conductors could be used with two conductors for each phase this is unnecessary in that only three or four conductors need be used. Whether three or four are used depends on the connections of the coils at the generator. Two methods of making the connections are possible, one being called the *star* connection and the other the *delta*, or *mesh*, connection.

22.4 The three-wire star system

In the star connection the windings are joined together at one point, as in fig. 22.6. This point is called the *star* or *neutral point*. As shown the system

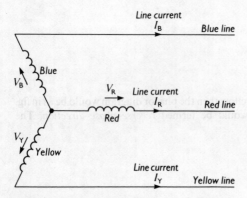

Fig. 22.6   The star connection

involves three conductors and is referred to as a *three-wire star system*.

The potential difference for each phase, i.e. from a line to the neutral point, is called the *phase voltage* and designated as V_P. The potential difference across any two lines is called the *line voltage*, V_L.

When the star-connected system is supplying equal loads in each line then the magnitude of the phase voltage for each line is the same, i.e.

V_P = magnitude of V_R = magnitude of V_B = magnitude of V_Y.

Such a system is said to be *balanced*. If the line voltage is taken from, say, the red and yellow lines, then the potential difference between those two lines is

$V_L = V_{RY}$ = phasor V_R − phasor V_Y

= phasor V_R + (− phasor V_Y).

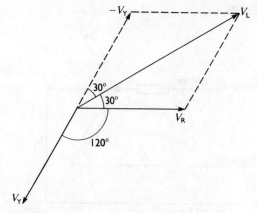

Fig. 22.7

Figure 22.7 shows this phasor addition. The magnitude of V_L is thus given by

$$V_L = V_R \cos 30° + V_Y \cos 30°.$$

But $\cos 30° = \frac{1}{2}\sqrt{3}$. Thus as $V_R = V_Y = V_P$, then

$$V_L = \sqrt{3}V_P. \tag{22.1}$$

This same relationship applied to V_{RY}, V_{YB} and V_{BR}.

From fig. 22.6 it is clear that the line current is equal to the phase current in the phase winding to which the line is connected. For a balanced load the phase currents will be all equal in magnitude to each other.

$$I_L = I_P. \tag{22.2}$$

Example 22.1   *What is the line voltage for a three-phase, balanced star-connected system having a phase voltage of 240 V?*

$$V_L = \sqrt{3}V_P$$
$$= \sqrt{3} \times 240$$
$$= 415.7 \text{ V}.$$

22.5 The four-wire star system

A four-wire system (fig. 22.8) is used when three separate loads have to be supplied, e.g. some houses in a road will obtain their supply from one phase of the supply, other houses from another phase. The supply for a

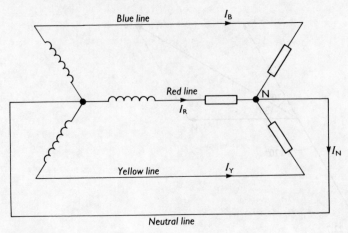

Fig. 22.8

house might therefore be the red-to-neutral phase, another the blue-to-neutral phase. Generally the loads are not equal and thus the loads are not balanced. Hence the current in any one line can differ from that in another line.

If Kirchhoff's law is applied to the neutral or star point N of the load then

$$I_N = \text{phasor sum of } I_R, I_B \text{ amd } I_Y. \qquad [22.3]$$

Figure 22.9 shows this phasor addition for the situation where the loads are balanced and the currents equal. The result is that the current in the

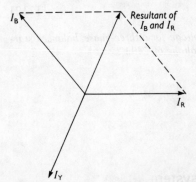

Fig. 22.9   For a balanced load the resultant of I_B and I_R is equal in magnitude but opposite to I_Y

neutral line I_N is zero. However, this is not the case when the load is not balanced.

Example 22.2 *In a three-phase, four-wire system the lines currents are:* $I_R = 10\,A$, $I_B = 12\,A$ *leading* I_R *by* $120°$, $I_Y = 15\,A$ *lagging* I_R *by* $120°$. *Determine the neutral current and its phase angle relative to* I_R.

Figure 22.10 shows the phasor diagram. The result can be obtained

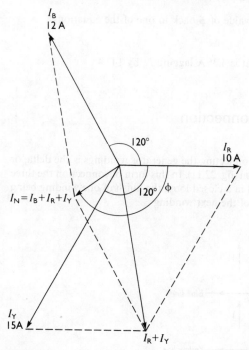

Fig. 22.10

by drawing the phasors to scale. The sequence adopted in the figure was to find the resultant phasor when I_R and I_Y are added. This resultant is then considered with I_B and the final resultant, I_N, obtained. An alternative to drawing is to calculate I_N by considering the sums of the vertical and the horizontal components of the phasors. Vertical components:

$$15 \sin 60° - 12 \sin 60° = I_N \sin (180° - \phi).$$

Horizontal components:

$$12 \cos 60° + 15 \cos 60° - 10 = I_N \cos (180° - \phi).$$

Thus

$$I_N \sin (180° - \phi) = 2.598$$

$$I_N \cos (180° - \phi) = 3.5.$$

Dividing these equations gives

$$\tan (180° - \phi) = 0.742$$

and so

$$180° - \phi = 36.6°.$$

Hence

$$\phi = 143.4°.$$

Thus, substituting this value of ϕ back in one of the equations,

$$I_N = 4.36 \text{ A.}$$

Thus the neutral current is 4.36 A lagging I_R by 143.4°.

22.6 The delta connection

An alternative way of connecting the generator windings is the delta, or mesh, form of connection (fig. 22.11). In this form of connection the three windings are connected in a closed loop, the finish of one winding being connected to the start of the next winding.

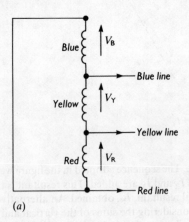

(a)

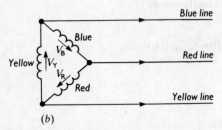

(b)

Fig. 22.11 Two ways of showing delta connection, (b) being the conventional way

The potential difference across a phase is the same as the potential difference between two lines, e.g. the potential difference between the red and blue lines is the phase voltage across the blue winding. Hence

$$\text{line voltage } V_L = \text{phase voltage } V_P. \qquad [22.4]$$

The line current is the phasor difference between the two phase currents connected to that line. Thus the line current for the red line is the phasor difference between the blue phase current and the red phase current (fig. 22.12). For a balanced load the phase currents are the same for

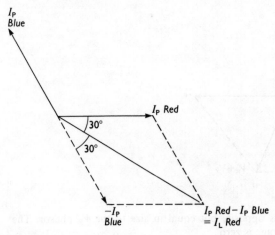

Fig. 22.12

each phase. Hence

$$I_L = I_P \cos 30° + I_P \cos 30°.$$

But $\cos 30° = \frac{1}{2}\sqrt{3}$. Thus

$$I_L = \sqrt{3}I_P. \qquad [22.5]$$

Example 22.3 *What is the current in each phase when the line current drawn by a balanced load from a delta connected generator is 50 A?*

$$I_L = \sqrt{3}I_P.$$

Thus

$$I_P = \frac{50}{\sqrt{3}}$$

$$= 28.9 \text{ A}.$$

22.7 An incorrectly closed delta system

With a correctly closed delta connection the direction of each of the phase voltages round the loop is in the same direction when they are positive. It must be pointed out that they are not all positive at the same instant of time, as indicated by the arrows for the phase voltages in fig. 22.11(b).

   When each of the phases has the same maximum voltage, then the three phasors are equal in length and the phasor diagram for the voltages is as shown in fig. 22.13. If the phasors for V_R and V_Y are added then the

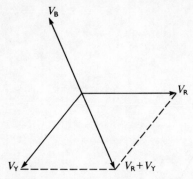

Fig. 22.13

resultant phasor is opposite but equal in size to the V_B phasor. The resultant loop voltage is zero.

   If however the loop was incorrectly closed and say the blue phase voltage was reversed, then it would be in the same direction as the resultant of V_R and V_Y and thus the resultant voltage in the loop would not be zero. It would be twice V_B. Thus in an incorrectly closed delta system the circulating voltage is twice the line voltage.

22.8 Power dissipation in a three-phase load

For a balanced load the line voltages are equal to each other and the phase voltages are also equal. The current in each line is the same and the phase currents are equal to each other. The balanced load also means that the power factor in each phase is the same. Thus

   power per phase $= I_P V_P \times$ power factor.

Hence

   total power $= 3I_P V_P \times$ power factor.   [22.6]

I_P and V_P are the r.m.s. values of the current and voltage in each phase.

For a star-connected system

$$V_L\sqrt{3}V_P \quad \text{and} \quad I_L = I_P.$$

Thus

$$\text{total power} = 3I_L \frac{V_L}{\sqrt{3}} \times \text{power factor}$$

$$= \sqrt{3}I_L V_L \times \text{power factor}.$$

For a delta-connected system

$$V_L = V_P \quad \text{and} \quad I_L = \sqrt{3}I_P.$$

Thus

$$\text{total power} = 3\frac{I_L}{\sqrt{3}}V_L \times \text{power factor}$$

$$= \sqrt{3}I_L V_L \times \text{power factor}.$$

The expression for the power is the same for the star-connected and the delta-connected systems when the load is balanced.

$$\text{Total power} = \sqrt{3}I_L V_L \times \text{power factor}. \qquad [22.7]$$

Example 22.4 *A three-phase delta connected motor operation off a 400-V system develops 25 kW at an efficiency of 90 per cent and a power factor of 0.8. Calculate the line and the phase current in each winding.*

$$\text{Efficiency} = \frac{\text{output power}}{\text{input power}} \times 100\%.$$

Thus

$$\text{input power} = \frac{90 \times 25 \times 10^3}{100}$$

$$= 22\,500\,\text{W}.$$

But, using equation [22.7],

$$\text{input power} = \sqrt{3}I_L V_L \times \text{power factor}.$$

Since $V_L = 400\,\text{V}$ and power factor $= 0.8$, then

$$22\,500 = \sqrt{3} \times I_L \times 400 \times 0.8$$

$$I_L = 40.6\,\text{A}.$$

For a delta-connected system, equation [22.5] gives

$$I_L = \sqrt{3}I_P$$

and so

$$I_P = \frac{40.6}{\sqrt{3}}$$

$$= 23.4 \text{ A}.$$

Example 22.5 *Three identical 50-Ω resistors are star connected across a 400-V, three-phase supply. Calculate the line current and the total power consumed.*

Using equation [22.1]

$$V_P = \frac{400}{\sqrt{3}}$$

$$= 230.9 \text{ V}.$$

Since $I_P = V_P/R$, then

$$I_P = \frac{230.9}{50}$$

$$= 4.62 \text{ A}.$$

But, according to equation [22.2],

$$I_L = I_P.$$

Thus

$$I_L = 4.62 \text{ A}.$$

Using equation [22.7],

$$\text{total power} = \sqrt{3} I_L V_L \times \text{power factor}$$

and since the power factor for a resistor is 1, then

$$\text{total power} = \sqrt{3} \times 4.62 \times 400$$

$$= 3.2 \text{ kW}.$$

Example 22.6 *Three identical coils, each of resistance 10 Ω and inductance 20 mH are delta connected across a 400-V, 50 Hz, three-phase supply. Calculate the line current and the total power consumed.*

For a coil,

reactance $X = 2\pi f L$ (see section 11.4)

$$= 2\pi \times 50 \times 0.020$$

$$= 6.3 \, \Omega.$$

Impedance of a coil is thus given by (see equation [11.6]),

$$Z = \sqrt{(R^2 + X^2)}$$
$$= \sqrt{(10^2 + 6.3^2)}$$
$$= 11.8\,\Omega.$$

For a delta-connected system, according to equation [22.4],

$$V_L = V_P.$$

Thus $V_P = 400$ V. Hence the phase current is given by

$$I_P = \frac{V_P}{Z} = \frac{400}{11.8}$$
$$= 33.9\,\text{A}.$$

But for a delta-connected system, according to equation [22.5],

$$I_L = \sqrt{3}I_P$$
$$= \sqrt{3} \times 33.9$$
$$= 58.7\,\text{A}.$$

For a coil having both resistance and inductance the phase angle ϕ of the coil is given by (see equation [11.9]),

$$\cos \phi = \frac{R}{Z}$$

$$= \frac{10}{11.8}$$

$$= 0.85.$$

This is the power factor, hence the total power is given by (equation [22.7])

$$\text{total power} = \sqrt{3} \times 58.7 \times 400 \times 0.85$$
$$= 34.6\,\text{kW}.$$

22.9 The measurement of power

The number of wattmeters used to obtain the power in a three-phase system depends on whether the load is balanced or not and whether the neutral point, if there is one, is accessible.

1. Measurement with a star-connected balanced load and neutral point accessible.

    A single wattmeter can be used, see fig. 22.14, with the current coil

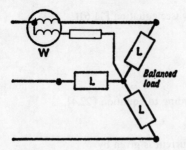

Fig. 22.14

being connected into one line and the voltage circuit between between that line and the neutral point. The wattmeter reading gives the power per phase and so the total power is three times the wattmeter reading.

2. Measurement with a star-connected unbalanced load and the neutral point accessible.

    Three wattmeters can be used, with each wattmeter being connected to a separate phase in the same manner as the one used in fig. 22.14. The total power is then the sum of the three wattmeter readings.

3. Measurement with a star or delta connected, balanced or unbalanced load and regardless of neutral point accessibility.

    In this method two wattmeters are used. The current coils are connected in any two lines and the voltage coils connected between these lines and the third lines (fig. 22.15).

    For the star-connected loads,

    instantaneous current through the current coil of $W_R = i_R$

    instantaneous voltage across voltage coil of $W_R = v_{RN} - v_{YN}$

Thus instantaneous power measured by $W_R = i_R(v_{RN} - v_{YN})$. Similarly for wattmeter W_B,

    instantaneous power measured by $W_B = i_B(v_{BN} - v_{YN})$.

Thus the sum of the instantaneous powers measured by the two wattmeters is

    total instantaneous power

$$= i_R(v_{RN} - v_{YN}) + i_B(v_{BN} - v_{YN})$$
$$= i_R v_{RN} + i_B v_{BN} - (i_R + i_B)v_{YN}.$$

From Kirchhoff's first law we must have

$$i_R + i_Y + i_B = 0.$$

Thus

$$i_R + i_B = -i_Y.$$

Hence

    total instantaneous power $= i_R v_{RN} + i_B v_{BN} + i_Y v_{YN}.$

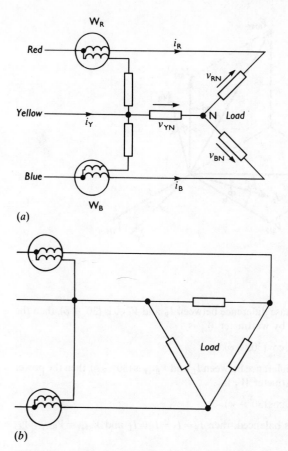

Fig. 22.15 (a) Star connections; (b) Delta connection

But $i_R v_{RN}$ is the instantaneous power for the red load, $i_B v_{BN}$ that for the blue load and $i_Y v_{YN}$ for the yellow load. Thus the sum of the instantaneous powers measured by the two watteters is the total instananeous power. The same result can be deduced for the delta connected loads.

Suppose we have a balanced load, star connected system with sinusoidal voltages and currents. If V_{RN}, V_{YN} and V_{BN} are the r.m.s. values of the phase voltages, and I_R, I_Y and I_B the r.m.s. values of the currents, then fig. 22.16 is the phasor diagram when the currents lag the corresponding phase voltages by ϕ. The potential difference across the voltage circuit of W_1 is the phasor difference of V_{RN} and V_{YN}, represented in the phasor diagram by V_{RNY}. The potential difference across the voltage circuit of W_2 is the phasor difference of V_{BN} and V_{YN}, represented in the phasor diagram by V_{BNY}. The current through W_1 is I_R and that through W_2 is I_B.

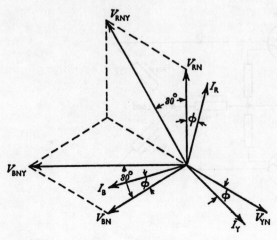

Fig. 22.16

Since the phase difference between I_R and V_{RNY} is $(30° + \phi)$, then the power indicated by wattmeter W_1 is

$$P_1 = I_R V_{RNY} \cos(30° + \phi).$$

Since the phase difference between I_B and V_{BNY} is $(30° - \phi)$, then the power indicated by wattmeter W_2 is

$$P_2 = I_B V_{BNY} \cos(30° - \phi).$$

Since the load is balanced, then $I_R = I_Y = I_B = I_L$ and $V_{RNY} = V_{BNY} = V_L$. Thus

$$P_1 = I_L V_L \cos(30° + \phi) \qquad [22.8]$$

$$P_2 = I_L V_L \cos(30° - \phi). \qquad [22.9]$$

Adding equations [22.8] and [22.9] gives

$$P_1 + P_2 = I_L V_L [\cos(30° + \phi) + \cos(30° - \phi)]$$

$$= I_L V_L (\cos 30° \, . \cos \phi - \sin 30° \, . \sin \phi$$
$$\qquad + \cos 30° \, . \cos \phi + \sin 30° \, . \sin \phi)$$

$$= I_L V_L \times 2 \cos 30° \, . \cos \phi.$$

But $\cos 30° = \frac{1}{2}\sqrt{3}$, hence

$$P_1 + P_2 = \sqrt{3} I_L V_L \cos \phi. \qquad [22.10]$$

This is the expression for the total power in a balanced system, equation [22.7], thus

$$P_1 + P_2 = \text{total power}.$$

This confirms the expression earlier arrived at for instantaneous power, though this expression involves the assumption of a balanced load and sinusoidal voltages and currents.

Subtracting equations [22.8] and [22.9] gives

$$P_2 - P_1 = I_L V_L [\cos(30° + \phi) - \cos(30° - \phi)].$$

Expanding the bracketed terms as before gives

$$P_2 - P_1 = I_L V_L \times 2 \sin 30° . \sin \phi.$$

Since $\sin 30° = \frac{1}{2}$, then

$$P_2 - P_1 = I_L V_L \sin \phi. \qquad [22.11]$$

Thus dividing equations [22.10] and [22.11] gives

$$\tan \phi = \frac{\sin \phi}{\cos \phi} = \frac{\sqrt{3}(P_2 - P_1)}{P_2 + P_1}. \qquad [22.12]$$

While the expression can be used in this form to obtain ϕ it is more useful to have it in the form of $\cos \phi$.

$$\sin^2 \phi + \cos^2 \phi = 1.$$

Dividing throughout by $\cos^2 \phi$, gives

$$\frac{\sin^2 \phi}{\cos^2 \phi} + 1 = \frac{1}{\cos^2 \phi}$$

$$\cos^2 \phi = \frac{1}{(1 + \tan^2 \phi)}.$$

Hence

$$\cos \phi = \frac{1}{\left[1 + 3 \left(\dfrac{P_2 - P_1}{P_2 + P_1} \right)^2 \right]^{1/2}} \qquad [22.13]$$

Example 22.7 *The input power to a three-phase motor was measured by the two-wattmeter method and the readings obtained were 6.5 kW and −2.1 kW. Calculate the total power and the power factor.*

Total power $= 6.5 - 2.1 = 4.4$ kW.

Using equation [22.13],

$$\text{power factor} = \cos \phi = \frac{1}{\left[1 + 3 \left(\dfrac{8.6}{4.4} \right)^2 \right]^{1/2}}$$

$$= 0.28.$$

22.10 The three-phase induction motor

The commonest form of a.c. motor is the induction motor. Its action depends on the fact that when a magnetic field moves past a conductor, the conductor is set into motion and endeavours to follow the field. The motion can be explained in terms of eddy currents. The relative motion between the magnetic field and the conductor induces an e.m.f. in the conductor which then results in eddy currents flowing in the conductor. The direction of these currents is such as to produce an effect which opposes the motion producing them. This means that the currents set up magnetic fields which interact with the magnetic field that started the currents. The outcome is that the conductor moves in order to reduce the relative velocity between the magnetic field and the conductor. All this can be considered to be a consequence of Lenz's law (see section 6.4).

Moving magnetic fields can be produced in a number of ways. One method involves the use of a three-phase supply, this type of induction motor being the one most used in industry. A rotating magnetic field is produced by using three symmetrically placed stationary windings on an iron magnetic circuit, this arrangement being termed the *stator* (fig. 22.17) because it is stationary. A balanced three-phase supply is connected to

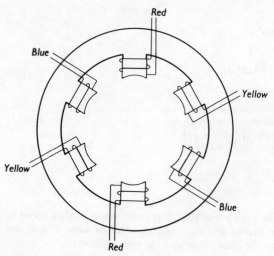

Fig. 22.17   The stator

these windings. Hence as each of the phase currents peaks so the magnetic field can be considered to be moving round the coils and a rotating magnetic field is produced (fig. 22.18).

The conductor, or conductors as more than one is used, are generally in the form of a number of aluminium or copper rods in the form of a cylindrical cage, fig. 22.19. This is known as the *rotor* as it is the part that

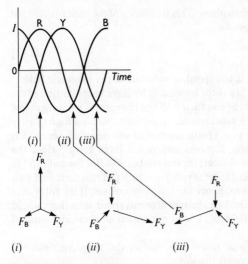

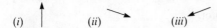

The m.m.f due to each phase

(i)   (ii)   (iii)

The resultant m.m.f. at these times, it is rotating clockwise

Fig. 22.18

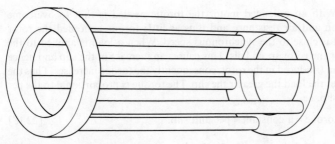

Fig. 22.19 A squirrel-cage rotor

rotates. The rotor is situated inside the stator and as the magnetic field rotates so the rotor rotates.

For the stator shown in fig. 22.17 there are two poles, or sets of stator windings per phase. As is indicated by fig. 22.18 the m.m.f. for such an arrangement goes through one complete revolution when the current in a phase passes through one cycle. The frequency of rotation of the m.m.f. is the same as the frequency of the current. If, however, there are p pairs of poles per phase then the frequency of the m.m.f. is f/p, where f is the frequency of the current in a phase. This frequency of the m.m.f. is referred to as the *synchronous speed* n_s.

$$n_s = \frac{f}{p}.$$ [22.14]

The synchronous speed is the speed of rotation of the magnetic field.

An e.m.f. is induced in a rotor conductor by there being motion of the magnetic field relative to the conductor. When the rotor is rotating at low speeds compared with the synchronous speed then there is a high relative velocity of magnetic field past a rotor conducted and hence a large e.m.f. is induced. This means large currents and hence large torque since the torque is proportional to the current (remember $F = BIl$, formula [5.1]). However, as the rotor speeds up the relative velocity of magnetic field past a rotor conductor decreases, hence the torque decreases. If the rotor ever attained the speed of the field, i.e. the synchronous speed, then there would be no relative velocity of field past a rotor conductor and hence no induced e.m.f. and no torque.

The speed of the rotor relative to that of the magnetic field, the synchronous speed, is termed the *slip*.

$$\text{Slip} = n_s - n_r$$ [22.15]

where n_r is the speed of the rotor.

$$\text{Per unit slip} = \frac{n_s - n_r}{n_s}$$ [22.16]

$$\text{Percentage slip} = \frac{n_s - n_r}{n_s} \times 100.$$ [22.17]

With an unloaded motor the torque needed is that to overcome friction and air resistance and thus as little torque is needed the rotor speed is close to the synchronous speed. When a load is applied to the motor the rotor speed falls. This causes an increase in the slip and hence the e.m.f. induced in the rotor, consequently the torque increases. Thus the greater the load the greater the slip. The torque is proportional to the per unit slip, i.e.

$$\text{torque} = \text{a constant} \times \text{per unit slip.}$$ [22.18]

Example 22.8 *A two-pole, 50-Hz, three-phase induction motor has a rotor speed of 2900 rev/min. What is the percentage slip?*

The number of pairs of poles = 1. Hence

$$\text{synchronous speed} = \frac{f}{p} = \frac{50}{1} = 50 \text{ rev/s.}$$

Rotor speed $n_r = 2900 \text{ rev/min} = 2900/60 \text{ rev/s}$. Hence

$$\text{percentage slip} = \frac{50 - (2900/60)}{50} \times 100$$

$$= 3.3\%.$$

Example 22.9 *A two-pole, 50-Hz, three-phase induction motor has a no-load rotor speed of 2920 rev/min when a torque of 2.0 N m is produced. What will be the rotor speed when there is an external load and a torque of 3.0 N m has to be produced?*

The number of pairs of poles = 1. Hence

$$\text{synchronous speed } n_s = \frac{f}{p} = \frac{50}{1} = 50 \text{ rev/s.}$$

Rotor speed at no load $n_r = 2920$ rev/min = 2920/60 rev/s. Hence, at no load,

$$\text{per unit slip} = \frac{50 - (2920/60)}{50}$$

$$= 0.0267.$$

Using equation [22.18], at no load

$$\text{torque} = 2.0 = \text{a constant} \times 0.0267.$$

At load

$$\text{torque} = 3.0 = \text{a constant} \times \text{per unit slip.}$$

Hence

$$\text{per unit slip at load} = \frac{3.0 \times 0.0267}{2.0}$$

$$= 0.0401.$$

Thus

$$0.0401 = \frac{50 - n_r}{50}$$

and

$$\text{the rotor speed at load} = 48.0 \text{ rev/s}$$

$$= 2880 \text{ rev/min.}$$

22.11 Summary of important formulae

Three-wire star system:

$$V_L = \sqrt{3} V_P \qquad\qquad\qquad [22.1]$$

$$I_L = I_P. \qquad\qquad\qquad [22.2]$$

444   *Principles of Electricity*

Four-wire star system:

$$I_N = \text{phasor sum of } I_R, I_B \text{ and } I_Y. \qquad [22.3]$$

Delta connection:

$$V_L = V_P \qquad [22.4]$$

$$I_L = \sqrt{3} I_P. \qquad [22.5]$$

Total power dissipated in a three-phase load

$$P_l = \sqrt{3} I_L V_L \times \text{power factor.} \qquad [22.7]$$

The two wattmeter measurement of power

$$\text{Total power} = P_1 + P_2 = \sqrt{3} I_L V_L \cos\phi \qquad [22.10]$$

$$\cos\phi = \frac{1}{\left[1 + 3\left(\dfrac{P_2 - P_1}{P_2 + P_1}\right)^2\right]^{1/2}}. \qquad [22.13]$$

The induction motor

$$n_s = \frac{f}{p} \qquad [22.14]$$

$$\text{Slip} = n_s - n_r \qquad [22.15]$$

$$\text{Per unit slip} = \frac{n_s - n_r}{n_s} \qquad [22.16]$$

$$\text{Percentage slip} = \frac{n_s - n_r}{n_s} \times 100 \qquad [22.17]$$

$$\text{Torque} = \text{a constant} \times \text{per unit slip.} \qquad [22.18]$$

22.12 Examples

1. What is the line voltage for a three-phase, balanced, star-connected system having a phase voltage of 3.0 kV?
2. What is the phase voltage for a three-phase, balanced, star-connected system having a line voltage of 415 V?
3. Determine the relationship between the line and phase voltages of a three-phase star-connected alternator.

    If the phase voltage of a three-phase star-connected alternator is 120 V, what is the line voltage?
4. What is the magnitude of I_N and its phase angle with respect to I_R in a three-phase, four-wire star-connected system when $I_R = 20$ A, $I_Y = 30$ A lagging I_R by 120 and $I_B = 32$ A leading I_R by 120°?

5. What is the magnitude of I_N and its phase angle with respect to I_R in a three-phase, four-wire star connected system when $I_R = 25$ A, $I_Y = 15$ A lagging I_R by $150°$ and $I_B = 20$ A leading I_R by $150°$?

6. What is the current in each phase when the line current drawn by a balanced load from a delta connected generator is 20 A?

7. A balanced load takes a current of 30 A at a line voltage of 3.0 kV from a three-phase supply. What are the phase currents and voltages if the load is (a) star connected and (b) delta connected?

8. Calculate the total power supplied by a three-phase generator with a line voltage of 400 V to a balanced delta connected load consisting of 120Ω per phase.

9. Three identical 100-Ω resistors are star connected across a three-phase supply having a line voltage of 400 V. Calculate the line current and the total power consumed.

10. Three identical coils, each having a resistance of 30Ω and an inductance of 50 mH are star-connected across a 400-V, 50-Hz, three-phase supply. Calculate the line current and the total power consumed.

11. Three coils, each of resistance 10Ω are star-connected to a three-phase supply of line voltage 450 V. Calculate (a) the line current, (b) the power factor, (c) the total power.

12. Use phasors to show that (a) the resultant current in a balanced three-phase star-connected load is zero, (b) the resultant voltage in an incorrectly closed delta connected three-phase system is twice the phase voltage.

13. Explain how two wattmeters can be used to determine the total power in a three-phase, three-wire system.

14. When two wattmeters were used to measure the power taken by a balanced three-phase load the readings obtained were 200 W and 80 W. What is the total power and the power factor?

15. The readings given by two wattmeters connected to a balanced three-phase circuit are 2.5 kW and -0.8 kW. What is the total power and the power factor?

16. Explain how a single wattmeter can be used to measure the power in a star-connected balanced three-phase load.

17. Describe the form of construction and principle of operation of the three-phase induction motor.

    A two-pole, 50 Hz, three-phase induction motor has a rotor speed of 2910 rev/min. What is the percentage slip? How will this slip change if the motor loading is increased? Explain the reasons for the change.

18. A four-pole, 50 Hz, three-phase induction motor has a no-load rotor speed of 1470 rev/min when the torque is 1.5 N m. What will be the rotor speed when there is an external load which produces a torque of 2.5 N?

CHAPTER 23

Semiconductors

23.1 Resistivities of materials

The following table shows the resistivities of a range of materials (see section 4.5 for explanation of resistivity). As will be apparent from an examination of the table, the materials fall into three categories when classified by resistivity. The category with resistivities of the order of $10^{-8}\,\Omega\,\text{m}$ is known as *conductors*, the category with resistivities of the order of 10^{10} and upwards is known as *insulators* and the materials with resistivities falling intermediate between those two groups, of the order of 1 to $10^3\,\Omega\,\text{m}$, are known as *semiconductors*.

Material		Resistivities [Ω m]	
Silver		1.5×10^{-8}	
Copper	Metals	1.6×10^{-8}	
Aluminium		2.5×10^{-8}	
Manganin		42×10^{-8}	Conductors
Constantan	Alloys	49×10^{-8}	
Nichrome		108×10^{-8}	
Germanium		0.9	Semiconductors
Silicon		2000	
Glass		10^{10} to 10^{14}	
Mica		10^{11} to 10^{15}	Insulators
PVC		10^{12} to 10^{13}	

Metals have resistivities that increase when the temperature increases, for some alloys the resistivity increases with an increase in temperature while for others it decreases. For semiconductors and insulators the resistivity decreases with an increase in temperature. The changes in resistivity of semiconductors and insulators with temperature are much greater than the change for conductors.

The electrical properties of materials can be expressed in terms of the bonding between electrons and atoms in a solid. In a conductor the atoms have electrons which are so easily detached that we can consider there to be a cloud of free electrons in a good conductor. When a potential difference is applied to a piece of conductor these electrons are able to move through the material and so give a current. On average there is about one free electron for each atom in the material. When the temperature is increased the movement of these free electrons is more impeded by oscillations of the atoms and so they are not able to move so fast through the material and hence the current is reduced. Thus an increase in temperature leads to an increase in resistivity.

In the case of an insulator there are virtually no free electrons available, all being tightly bound to their atoms. Because of this there can be little current, hence a very high resistivity. Increasing the temperature can however shake a few more electrons free from the atoms and so increase the current, hence a decrease in resistivity.

Semiconductors have, at room temperature, some free electrons, generally about one per million atoms. Because of this they have resistivities intermediate between those of insulators and conductors. An increase in temperature releases more electrons and so results in a decrease in resistivity.

23.2 Intrinsic and extrinsic semiconductors

When an electron breaks free from an atom and thus becomes available for a current it leaves the atom with a net positive charge and a vacancy for an electron. This vacancy is called a hole. It represents a location into which an electron from another atom can easily move. However, such a movement would mean that there was a hole in another atom. An electron movement has led to a hole movement. In pure germanium or silicon, semiconductors, there are as many holes as there are free electrons. Thus when a potential difference is applied to such a semiconductor there is a current due to the movement of the free electrons and a current due to the movement of the holes. There are equal numbers of holes and free electrons, each hole being created by an electron becoming free. Such semiconductors are called *intrinsic* semiconductors.

In an intrinsic semiconductor the current is due in equal parts to electron and hole movement.

The introduction of impurities into silicon or germanium can have a great effect on their resistivities, the deliberate introduction of impurities being known as *doping*. When small amounts of phosphorus, arsenic or antimony are added to silicon or germanium the material ends up with more electrons than holes. This is because these materials have easily detached electrons. Such impurities are known as *donors*, because they donate electrons. The materials are known as n-*type*, the n being because the majority carriers of current are electrons which are *n*egatively charged particles. When small amounts of aluminium, gallium or iridium are added to germanium or silicon the material ends up with more holes than free electrons. This is because these atoms supply sites which can be occupied by electrons from the germanium or silicon. Such impurities are known as *acceptors*, because they accept electrons. The materials are known as p-*type*, the p being because the majority carriers of current are holes which behave like *p*ositively charged particles (under a potential difference the holes move in the opposite direction to electrons).

The material produced by doping is known as an *extrinsic* semi-conductor because the impurity introduces charge carriers extra to the intrinsic ones. In such materials there is a majority charge carrier and holes and electrons do not contribute equally to a current.

23.3 The junction diode

The junction diode is a crystal of either germanium or silicon in which a junction between p-type and n-type material has been produced by doping different parts of the crystal with different impurities. When a junction between the two materials is produced a situation occurs where on one side of the junction there are free electrons and on the other side free holes. The result is that a drift occurs across the junction of electrons and holes. The effect of electrons leaving the n-type material is for it to become positively charged due to the loss of these negatively charged electrons. The effect of electrons moving into holes in the p-type material (holes

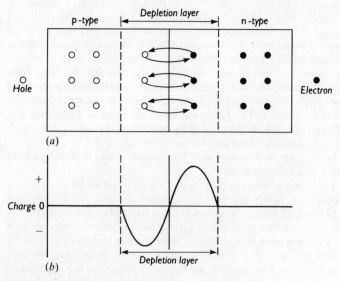

Fig. 23.1 The p–n junction. (*a*) Movement of the majority charge carriers; (*b*) The resulting charge distribution

leaving the p-type material) is that it becomes negatively charged (fig. 23.1).

The layer either side of the junction from which the electrons and holes have moved is called the *depletion layer*, in that the movement has depleted the layer of free electrons and holes. Because the n-side of the junction is positively charged and the p-side negatively charged there is a potential difference across the junction, it being known as the *barrier potential*.

Now consider the effect of applying an external potential difference across the junction, as in fig. 23.2. A significant current can flow through

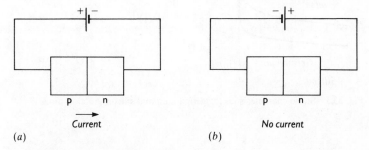

Fig. 23.2   (*a*) Forward bias; (*b*) Reverse bias

the junction when the external potential difference makes the p-side of the junction positive with respect to the n-side. When the external potential difference makes the p-side negative with respect to the n-side then no significant current flows. When the external potential makes the n-side positive the junction is said to be *forward biased*, when it makes it negative then *reverse biased*. Forward bias gives a significant current, reverse bias does not.

With forward bias the external potential difference, when a current flows, is sufficiently large to move charge carriers through the depletion layer, cancelling out the effect of the barrier potential. With reverse bias the external potential difference is in such a direction as just to enhance the barrier potential.

Figure 23.3 shows the static characteristics for germanium and silicon p–n junctions, such components being referred to as junction diodes. Note that the scales used for the reverse and forward bias parts of the graph are not the same. A small current occurs with reverse bias because of the minority charge carriers. Thus in the p-type material we have only referred to the movement of holes, while these are the majority carriers there are some free electrons, minority carriers. In the n-type material the majority carriers are electrons, there are however some holes, minority carriers. The reverse bias condition is one which opposes movement of the majority carriers but does not oppose the movement of the minority carriers, hence the small current.

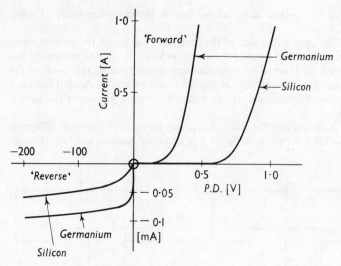

Fig. 23.3 Static characteristics for germanium and silicon junction diodes

23.4 Rectification

The junction diode is a rectifier, this term implying the device has a characteristic which gives a low resistance path in one direction but a high resistance path in the reverse direction. Thus the forward bias junction diode has a relatively low resistance while the reverse bias junction has a high resistance.

Other devices giving similar characteristics, and also used as rectifiers, are vacuum diode valves and metal rectifiers. The vacuum diode valve is a glass envelope containing a filament and an electrode in a vacuum. The heated filament emits electrons which will only move across to the electrode when it is positive with respect to the filament, hence a current and low resistance path in this direction. When the electrode is negative with respect to the filament then no current flows and the valve has a high resistance. The metal rectifier consists of a junction between a metal and a semiconductor. Examples of this are the copper–copper oxide junction and the selenium–selenium alloy junction. The behaviour of such junctions can be explained in a similar way to the p–n junction.

When an alternating potential difference is applied to a rectifier a direct current is produced, i.e. a current which flows in only one direction and does not reverse. The rectifier only gives a current when it is forward biased, hence the current only occurs for half the cycle (fig. 23.4). Such an arrangement is known as a *half-wave rectifier*.

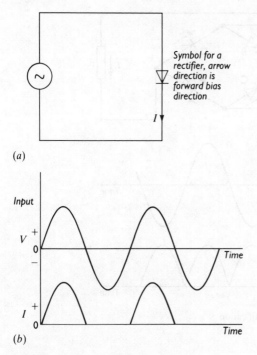

(a)

(b)

Fig. 23.4 Half-wave rectification

The term *full-wave rectifier* is used for those circuits which use both halves of the cycle. Figure 23.5 shows a circuit, known as the *bridge circuit*, which gives full-wave rectification. The rectifiers are so arranged that the alternating e.m.f., when in the half of the cycle represented in the figure by the full arrow, sends a current through B and E but when in the reverse half of the cycle, represented by the dotted arrow, sends a current through C and D. It will be evident from the figure that the current through the load resistor R is unidirectional for both halves of the cycle.

23.5 The bipolar transistor

Bipolar transistors consist of two p–n junctions which are very close together in a single piece of semiconducting material. Such transistors can be n–p–n or p–n–p, as in fig. 23.6. The arrow in the symbol for a transistor indicates the direction of conventional current flow. Of the two forms of transistor the n–p–n one is the most commonly used, with the terms emitter and collector for such a transistor indicating what happens to electrons.

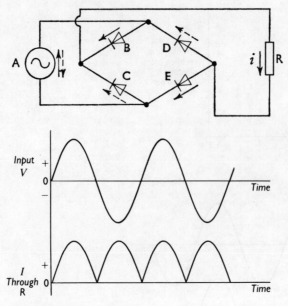

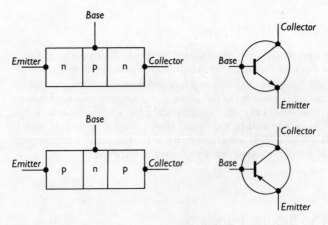

Fig. 23.5   Full-wave rectification

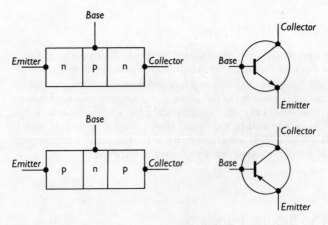

Fig. 23.6   Bipolar transitor forms and symbols

To obtain transistor action the emitter-base junction has to be forward biased and the collector-base junction reverse biased. Figure 23.7 shows the biasing of the p–n junctions for both a n–p–n and a p–n–p transistor, it also shows the movement of the majority charge carriers.

With the n–p–n transistor the majority charge carrier in the n-material is the electron and because the emitter-base junction is forward biased the electrons can flow through the junction and into the base. The

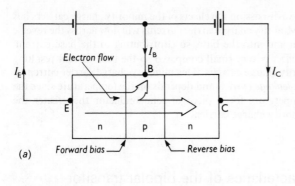

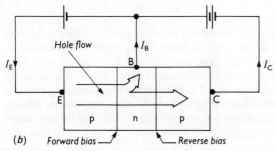

Fig. 23.7 (a) n–p–n transistors; (b) p–n–p transistor

base-collector junction is however reverse biased and so we might not expect a current to flow through that junction. However the base is very thin and most of the electrons entering the base are attracted straight through the reverse biased junction and into the collector. Only a few of the electrons combine with holes in the base and give rise to a base current I_B. Thus a current is produced through a reverse biased junction, a high resistance junction. The effect has therefore been to take a current through a low resistance junction and transfer it to become a current through a large resistance junction. The origin of the term transistor is 'transfer resistor'.

With the p–n–p transistor the majority charge carrier in the p-material is the hole and because the emitter-base junction is forward biased the holes can flow through the junction and into the base. Most of them then flow through the reverse biased collector-base junction because the base is so very thin. Only a few of the holes combine with electrons in the base and give rise to a base current I_B.

Applying Kirchhoff's first law, then

$$I_E = I_C + I_B \qquad [23.1]$$

Approximately 98 per cent of the charge carriers emanating from the emitter make it through to the collector. Thus we might have for the currents $I_E = 1$ mA and $I_C = 0.98$ mA. Thus $I_B = 0.02$ mA.

In the above discussion the effects of the minority charge carriers has been ignored. Minority charge carrier currents will flow across the reverse biased junction and into the base, so contributing to the base current. However, the effect is very small compared to the base current resulting from the majority charge carriers. This minority charge carrier current is referred to as a *leakage current* and depends on the temperature since the higher the temperature for a piece of semiconductor the greater the number of minority charge carriers.

23.6 Characteristics of the bipolar transitor

There are three possible connection arrangements for a transistor, these being referred to as *common base*, *common emitter* and *common collector*. The common-base connection involves the base connection being common to both the emitter and the collector circuits, this being the arrangement indicated in fig. 23.7. With the common emitter circuit the emitter connection is common to both the base and the collector circuits, as in fig. 23.8. With the common collector circuit the collector connection

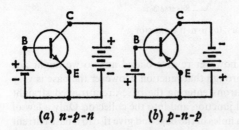

(a) n-p-n   **(b) p-n-p**

Fig. 23.8   Common emitter circuits

is common to both the base and the emitter circuits.

Figure 23.9 shows an arrangement that can be used for the determination of the static characteristics of an n–p–n transistor when in the common-base form of connection. The procedure is to maintain the

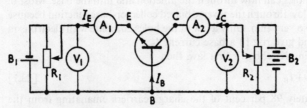

Fig. 23.9   Determination of the characteristics for a common-base n–p–n transitor circuit

value of the emitter current, indicated by A_1, at a constant value by means of R_1 and note the readings on A_2 for different values of the collector-base voltage V_2. This procedure is repeated for a number of values of the emitter current. Figure 23.10 shows the type of results that are obtained.

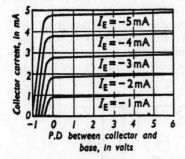

Fig. 23.10 Static characteristic for a common-base n–p–n transistor circuit

Because it cannot be shown on the same scale as the rest of the graph the collector current when the the emitter current is zero has not been shown. This current is generally a few microamps and is the reverse bias leakage current due to the minority charge carriers.

For positive values of the collector-base voltage the collector current remains almost constant. This is because nearly all the electrons entering the base are attracted to the collector and thus increasing the voltage has little effect. For a given collector-base voltage the collector current is practically proportional to the emitter current, this is illustrated by fig. 23.11 where the values of the currents for a collector base voltage V_{CR} of

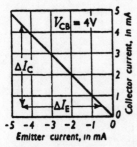

Fig. 23.11 Relationship between collector and emitter currents for a given collector-base voltage

4 V have been plotted. Figure 23.9 is referred to as the *transfer characteristic* and for a fixed value of V_{CB} the quantity I_C/I_E is called the static value of the forward current transfer ratio, symbol h_{FB}.

$$h_{FB} = \frac{I_C}{I_E}.$$ [23.2]

The small signal forward current transfer ratio h_{fb} is defined by

$$h_{fb} = \frac{\Delta I_C}{\Delta I_E}. \qquad [23.3]$$

The term current amplification factor α was previously used for these ratios. The subscript B, or b, used with the ratio symbol indicates that it refers to the common base form of connection, the use of the capital letter indicating that it is the ratio of actual currents while the lower case letter indicating that it refers to the slope of the graph, i.e. current changes.

Figure 23.12 shows an arrangement that can be used for the

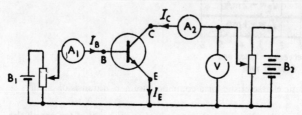

Fig. 23.12　Determination of static characteristics for a common-emitter transistor circuit

determination of the static characteristics of an n–p–n transistor when in the common-emitter form of connection. The procedure is to maintain the base current, indicated by A_1, constant and to note the readings on A_2 for different values of the collector-emitter voltage V. This procedure is repeated for different values of the base current. Figure 23.13 shows the

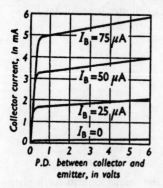

Fig. 23.13　Static characteristic for a common-emitter n–p–n transistor circuit

type of results that are obtained, fig. 23.14 showing the transfer characteristic. For a fixed value of V_{CE} the quantity I_C/I_B is called the static value of the forward current transfer ratio, symbol h_{FE}.

$$h_{FE} = \frac{I_C}{I_B}. \qquad [23.4]$$

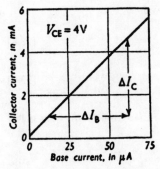

Fig. 23.14 Relationship between collector and base currents for a given collector–emitter voltage

The small signal forward current transfer ratio h_{fe} is defined by

$$h_{fe} = \frac{\Delta I_C}{\Delta I_E}.$$ [23.5]

The term current amplification factor β was previously used for these ratios.

h_{FB} and h_{FE} are related, as also are h_{fb} and h_{fe}. If we consider equation [23.1], then

$$I_E = I_C + I_B$$

and using equations [23.2] and [23.4],

$$\frac{I_E}{I_C} = \frac{I_C}{I_C} + \frac{I_B}{I_C}$$

$$\frac{1}{h_{FB}} = 1 + \frac{1}{h_{FE}}.$$

Hence

$$\frac{1}{h_{FB}} = \frac{h_{FE} + 1}{h_{FE}}$$

$$h_{FB} = \frac{h_{FE}}{h_{FE} + 1}.$$ [23.6]

Similarly the following equation can be derived

$$h_{FE} = \frac{h_{FB}}{1 - h_{FB}}.$$ [23.7]

By considering the equation [23.1] in the form

$$\Delta I_E = \Delta I_C + \Delta I_B$$

the following equations can also be derived

$$h_{\text{fb}} = \frac{h_{\text{fe}}}{h_{\text{fe}} + 1}$$ [23.8]

$$h_{\text{fe}} = \frac{h_{\text{fb}}}{1 - h_{\text{fb}}}.$$ [23.9]

Example 23.1   *For a bipolar transistor, if h_{fe} is 49 what is the value of h_{fb}?*
Using equation [23.8],

$$h_{\text{fb}} = \frac{49}{49 + 1}$$

$$= 0.98.$$

23.7 The thyristor

The term thyristor is used to describe a family of semiconductor devices.
The commonest member is the *reverse blocking triode thyristor* or, as it is
more commonly known, the *silicon controlled rectifier*, abbreviated to
SCR. Other members of the family are the bidirectional trigger diode
(DIAC) and the *bidirectional triode thyristor* (TRIAC).

The SCR is a four-layer semiconductor device, with three terminals
called the anode, the cathode and the gate (fig. 23.15). As the circuit symbol

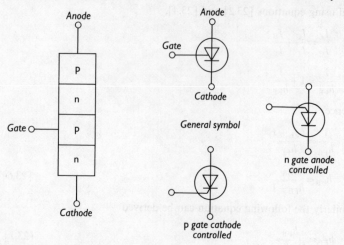

Fig. 23.15   The thyristor (SCR) and its symbols

implies the SCR is a diode with a third electrode, the gate. While the
general symbol can be used to represent the thyristor the alternative

symbols indicate whether the gate contact is fixed to an n-region or a p-region. The gate contact can be made to either of the central n or p regions.

The action of the SCR can be explained in terms of a two transistor model (fig. 23.16), one being a p–n–p transistor and the other an n–p–n transistor.

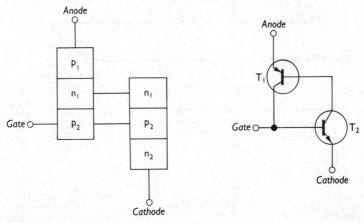

Fig. 23.16 The two transistor analogy

Consider the operation when the gate is disconnected. When a negative potential is applied to the anode the p–n junction $p_1 n_1$ is reverse biased. There is thus no current through the thyristor. The thyristor is then said to be reverse blocking the current. When a positive potential is applied to the anode the p–n junction $p_1 n_1$ is forward biased but the $n_1 p_2$ junction is reverse biased. There can be no current through the thyristor, the thyristor now being said to be forward blocking the current.

Now if while the anode is positive a positive potential is applied to the gate then transistor T_1 is turned on and begins to conduct. Since the collector current of T_2 is the base current of T_1 this has the effect of making T_1 also conduct. Thus making the gate positive when the anode is positive has the effect of making both transistors conduct. But when T_1 conducts its collector current flows into the base of T_2 and would then make it conduct. Once the thyristor starts to conduct each transistor maintains the other in a conducting state and it is then possible to remove the gate signal. It is only needed to start the conduction process and is not necessary to maintain it. It is only positive signals applied to the gate that can trigger this conduction state, and only when the anode is positive.

Figure 23.17 shows the characteristic of a thyristor. Reverse blocking is when the anode is made negative. If however the voltage is made sufficiently negative then breakdown can occur. If the anode is given a positive potential, say V_{A1}, then with the gate disconnected the only current is the forward leakage current and the thyristor is forward blocking. The anode current has to reach the holding current value before conduction occurs. This can be done by a suitable value of gate current I_g

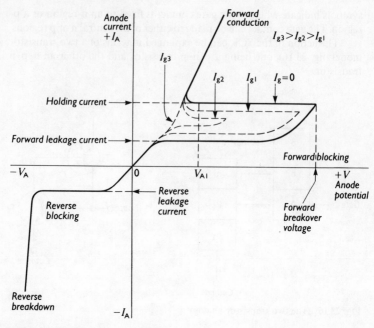

Fig. 23.17 Thyristor characteristic

(an alternative is to increase the anode potential to the forward breakover voltage).

Once the thyristor has been turned on it remains in the conducting state until the anode current falls below the holding current. A thyristor may have a forward current of perhaps 50 A and a holding current of possibly 10 mA.

23.8 Phase control with SCRs

The term phase control is used for the process whereby an a.c. supply is connected across a load for a controlled fraction of each cycle. This enables the power supplied to the load to be controlled. Such control can be exercised by a circuit employing an SCR.

Figure 23.18 shows a half-wave circuit with a single SCR. In the absence of a gate signal there is no conduction, the thyristor blocking in both reverse and forward directions. However, the thyristor can be made to conduct during the positive half cycles by a suitable gate signal. As the figure indicates, a periodic positive gate signal is applied at a suitable time during the cycle and triggers the thyristor into its forward conduction state. In the example shown the gate signal is supplied at a phase angle of

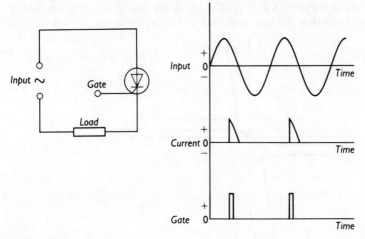

Input

Current

Gate

Fig. 23.18

120° during the positive half of the cycle. The thyristor then conducts between 120° and 180°, i.e. the remaining part of the positive half of the cycle. The control of the current passed by the thyristor can be exercised by determining the phase angle at which the gate signal is supplied.

23.9 Triacs

The triac is a three-terminal device (fig. 23.19) equivalent to two SCRs

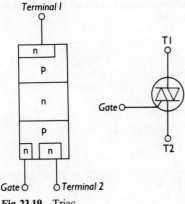

Fig 23.19 Triac

connected in the way implied in the circuit symbol. The characteristic of the triac has the form indicated in fig. 23.20. Conduction between the

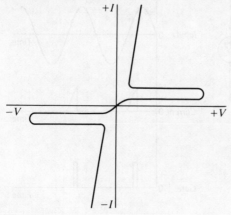

Fig. 23.20 The triac characteristic

terminals T_1 and T_2 can be initiated by a positive or negative pulse applied to the gate or by raising the applied voltage above the breakover value. Unlike the SCR the triac can be triggered with either positive or negative gate pulses. As with the SCR, once the triax has become conducting the gate pulse need no longer be applied.

If a triac were used in the circuit shown in fig. 23.18 in place of the SCR then the triac could be made to fire in both the positive and negative half cycles and so give a current in both parts of the cycle. Unlike the SCR the triac does not rectify.

23.10 Diacs

The diac, or bidirectional diode as it is sometimes referred to, is a two-terminal device (fig. 23.21). The diac switches to a conducting state when

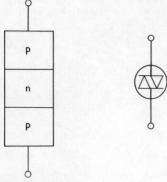

Fig. 23.21 Diac

the applied voltage exceeds the breakover voltage, operating with both positive and negative voltages (fig. 23.22).

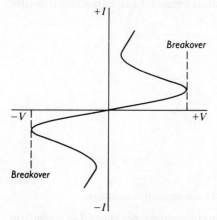

Fig. 23.22   The diac characteristic

23.11 The unijunction transistor

The unijunction transistor (UJT), or double-base diode as it is often referred to, is a diode with three connections. Figure 23.23 shows the basic

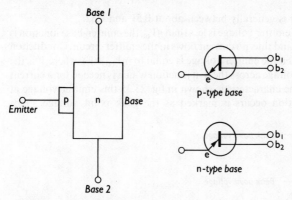

Fig. 23.23   Unijunction transistor

form of the component and its symbols. Two p–n junctions exist, one between the emitter and the base 1 connection and one between the emitter and the base 2 connection.

When connected in a circuit the base 2 connection is made positive with respect to the base 1 connection. The base material can be considered to be a resistor between the b_1 and b_2 connections with the emitter connection being a tapping point along this resistor, as illustrated by fig. 23.24 which shows the equivalent circuit for the unijunction transistor.

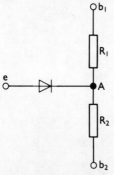

Fig. 23.24   The equivalent circuit for a unijunction transistor

Point A is a point in the base directly opposite the emitter. The potential of point A with respect to b_1 is thus

$$\left(\frac{R_1}{R_1 + R_2}\right) \times V_{bb}$$

where V_{bb} is the potential difference between b_1 and b_2. The ratio of the resistances is known as the *intrinsic stand-off ratio* η.

$$\eta = \frac{R_1}{R_1 + R_2}.$$

The value of η is generally between about 0.51 and 0.82.

When the emitter voltage is less than ηV_{bb} the emitter-base junction is reverse biased and thus no current flows in the emitter circuit. Conduction takes place when the emitter voltage is equal to $(\eta V_{bb} + V)$, where V is the forward bias voltage across the p–n junction which is needed for a current to flow. For the characteristic shown in fig. 23.25 this emitter voltage at which conduction occurs is marked as the peak point voltage. Once

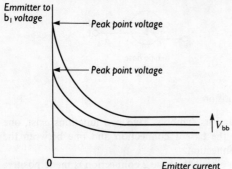

Fig. 23.25   Unijunction transistor characteristic

conduction occurs the resistance between the emitter and b_1 falls to a low value. The value of this peak point voltage depends for a particular unijunction transistor, on just V_{bb}.

23.12 The relaxation oscillator circuit

Figure 23.26 shows the relaxation oscillator circuit that can be obtained using a unijunction transistor. When the circuit is switched on, the

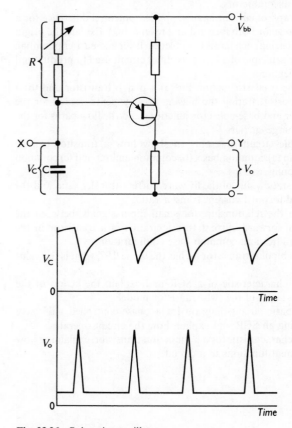

Fig. 23.26   Relaxation oscillator

potential difference across the capacitor V_C rises as the capacitor becomes charged. The time constant CR determines the rate at which V_C increases. When V_C reaches the peak point voltage the resistance of the transistor falls from its initially very high value to a low value. This causes the capacitor to discharge very rapidly. The results of this are that a sawtooth voltage is produced at X and at Y a series of pulses. The output at Y is often used as a thyristor trigger voltage.

23.13 Examples

1. State the order of the resistivities of conductors, semiconductors and insulators and describe how the resistivities vary with temperature.
2. Describe the basis of electrical conduction in (a) indrinsic and (b) extrinsic semiconductors.
3. Sketch a graph of current against applied potential difference for a junction diode for both forward and reverse bias. Explain the origin of the graph using such terms as depletion layer and barrier potential.
4. Explain the principle of a bridge rectifier circuit used to produce full wave rectification.
5. Describe the transistor action for (a) a p–n–p transistor and (b) a n–p–n transistor, stating the biasing arrangements needed for the base-emitter and base-collector junctions and the flow paths for the majority charge carriers.
6. Draw simple circuit diagrams to show how a transitor can be connected in (a) common base, (b) common emitter and (c) common collector configurations.
7. Define the static value of the forward current transfer ratio and the small signal forward current transfer ratio.
   Derive the relationship, for small input signals, between the small signal forward current transfer ratios of a transistor in the common emitter and common base configurations.
   If for a bipolar transistor h_{fb} has the value 0.97, what is the value of h_{fe}?
8. Sketch the characteristic of a SCR and explain the action of the thyristor in terms of the two transistor model.
9. Draw the basic circuit diagram for a phase-controlled half wave rectifier using an SCR and explain how the circuit operates.
10. Sketch the characteristic for a unijunction transistor, explaining how the component functions in a circuit.

Answers To Examples

Chapter 1, Page 13

1. 33.15 K. 46.85°C
2. 98.1 N.
3. 408 g.
4. 50 N.
5. 0.6 m/s².
6. 120 J.
7. 314 kJ.
8. 15.3 kg, 150 W.
9. 125 W.
10. 3.6 MJ, 1 kW h.
11. 19.62 kJ.
12. 48 kW h, 172.8 MJ.
13. 153 N m.
14. 637 rev/min.
15. 754 kW, 2262 kW h.
16. 840 N m, 21.2 kW.
17. 147.2 N m, 6.17 kW.
18. 224 kW, £201.60.
19. 62.85 MJ.
20. 500 J/kg K.
21. 679 kJ.
22. 4.58 min, 0.153 kW h, 550 kJ.
23. 11.06 kW, 7.37 p.

Chapter 2, Page 24

1. 0.05 A high.
2. 2.78 A.
3. 0.0222 mm.
4. 21.2 min.
5. 3.22 g, 2880 C, 0.8 A h.
6. 3.02 h, 87 000 C, 24.16 A h.
7. 0.095 mm.
8. 0.33 mg/C.
9. 0.012 25 mm.
10. 2.64 g.
11. 3 A.
12. 0.0932 mg/c, 0.681 mg/c, 0.3675 mg/c.

Chapter 3, Page 38

1. 0.435 A, 529 Ω.
2. 1.232 kW.
3. 9.5 Ω, 1.52 W.
4. 53.8 Ω, 19 per cent.
5. 15.33 mA, 7.05 W.
6. 6.39 μA, 735 μW.
7. 36 000 000 C.
8. 0.24 A, 1.44 W, 1728 J.
9. 4.08 V, 19.6 mW.
10. 20 A, 19.2 kW h, 5.65 kW.
11. 113.6 A, 50 kW, £4.50.
12. 9.17 A, 550 C, 211.2 kW h.
13. 111 kW.
14. 2.15 kg.
15. 9.4 min, 1.44 kW, 0.316 p.
16. 40.5 A, 5.92 Ω, 9.7 kW.
17. 4.83 A, 49.6 Ω, 1.16 kW, 0.386 kW h.

Chapter 4, Page 66

1. 32 Ω, 7.875 Ω.
2. 0.847 A, 116.5 V, 103.5 V.
3. 20.83 Ω, 27.78 Ω, 90 W.
4. 44 per cent, 16 Ω, 115.2 Ω.
5. 44 Ω, 4 Ω, 1 A, 11 A.
6. 4.444 A, 3.333 A, 2.222 A.

7. 8 A, 10 A.

8. 9 Ω, 90 W, 21.6 W.

9. 1.8 A, 1.2 A, 81 V, 243 W.

10. 0.793 A, 1.24 A, 2.033 A.

11. 5 A; 2 A, 1.8 A, 1.2 A, 2.5 A, 2.5 A; 22.5 W; 36 V.

12. 0.016 25 Ω.

13. 0.833 Ω.

14. 0.25 V, 1.2 V.

15. 9.6 A, 0.014 58 Ω, 19.776 W, 18.432 W, 1.344 W.

16. 0.4 A, 0.75 Ω, 4.8 W.

17. 0.5 A, 5 V, 300 J.

18. 0.3 A, 0.2 A, 0.278 A.

19. 2.945 A, 1.718 A, 1.227 A, 34.35 V.

20. 30 C.

21. 0.54 A, 0.225 A, 0.315 A, 0.852 V, 0.722 V.

22. 0.182 A.

23. 5 Ω.

24. 9600 J.

25. 5.333 Ω, 1.059 A, 5.65 V.

26. 0.99 Ω.

27. 356 mm^2.

28. 0.439 $\mu\Omega$ m.

29. 100 Ω.

30. 83.6 Ω.

31. 0.016 96 $\mu\Omega$ m.

32. 36.8 m, 12 Ω.

33. 109.5 Ω.

34. 50.2°C.

35. 0.251 Ω.

36. 65.3°C.

37. 162 Ω.

38. I_A, 1.625 A, discharge; I_B, 0.75 A, charge; 0.875 A, 8.75 V.

39. 19.7 A, 8.5 A, charge; 224.25 V.

40. I_A, 1.226 A, charge; I_B, 3.548 A, discharge; I_C, 2.322 A, charge.

41. 4.91 A, 614 A, 11.05 A.

42. 455.4 V, 36.43 kW, 0.768 kW.

43. 13.2 V.

44. 132 mm^2, 231 V.

45. 236.53 V, 235.61 V, 976 W.

46. 230.3 V, 226.53 V.

47. 118.7 A, 38.7 A, 81.3 A; 234.3 V, 232.68 V; 1.334 kW.

48. 150 A, 30 A, 120 A; 241 V, 238.48 V; 2.808 kW.

49. 0.50.

Chapter 5, Page 95

8. 1.15 T

9. 160 μWb.

10. 300 N/m.

11. 0.444 T.

12. 22.6 mN.

13. 90 N, horizontally.

14. 74×10^{-6} N m.

15. 8.43 mA.

16. 48.6×10^{-6} N m.

17. 173 N m, 12.7 kW.

Chapter 6, Page 107

1. 24 V, 224 V.

2. 37.5 V.

3. 90 V.

4. 0.32 V.

5. 5.87 V.

6. 1305 V.

7. 0.267 V.

8. 2.33 mV.

9. 0.333 T, 0.267 N.

10. 12 N, 1.2 V, 120 W.

11. 0.2 V.

13. 4.24 μV.

14. 5.6 V.

15. 1400 rev/min.

Chapter 7, Page 130

1. 318 A/m, 0.004 T.

2. 6000 A.

3. 6370 A/m, 8 mT; 7580 A/m, 9.52 mT; 0, 0; 3030 A/m, 3.81 mT.

4. 382 000 A/Wb, 306 A, 255 000 A/m.

5. 3.53 A.

6. 557 000 A/Wb, 334 A.

7. 500 A/m, 955; 1500 A/m, 637.

8. 750 A, 1875 A/m, 0.001 06 Wb, 707 000 A/Wb.

9. 0.731 T, 1828

10. 0.81 T, 1215

11. 10.05 divisions.
13. 6000 A; 500 000 A/Wb, 2650.
15. 12.05 A.
17. 2.2 A.
19. 990 A.
21. 41.5 A.
23. 3090 A.

12. 216 A, 3540
14. 3 A.
16. 0.86 A, 1.92 A, 2 880 000 A/Wb.
18. 3.65 A.
20. 4.9 A.
22. 1520 A.

Chapter 8, Page 152

1. 0.375 H.
3. −160 A/s.
5. 1.25 mH, 0.1 V.
7. 0.0833 H, 0.041 65 s, 1.667 V.
10. 3 A/s, 12 A, 120 V.
13. 12.5 A/s, 0.984 A, 0.183 s, 12.5 J.
17. 0.4 H, 0.1 H.
19. 12 V.
21. −10 V.
23. 1.35 mH.
25. 0.848.

2. 0.15 H.
4. −3.5 V.
6. 0.129 mH, 0.0258 V.
9. 0.15 H, 30 V.
11. 2.85 A.
16. 500 μJ, 5.5 V.
18. 9.375 J.
20. 0.0625 H.
22. 0.0344 mH.
24. 100.5 μH, 40.2 mV.
26. 80 μH.

Chapter 9, Page 183

1. 0.0885 μF.

3. 1327 pF, 0.531 μC, 8.85 μC/m^2, 200 kV/m.
5. (a) 300 ε_r; (b) 600 ε_r, where ε_r = relative permittivity of insulating material.
7. 0.415 μF.

9. 2.81; 30 kV/m, 0.7425 μC/m^2; 0.4455 μJ.
11. 1.416 mm.
13. 0.3 μF, 0.0667 μF.
15. 6.06 μF.

17. 260.9 V, 87 V, 52.1 V.
19. 300 V; 4 μJ, 16 μJ.

21. 357 V; 1.25 J, 0.893 J.
24. 5.9 μA, 236 V.
26. 0.5 μC, 3.12 × 10^{12} electrons.
28. 11.2 mA.
30. 18.72 × 10^{15} electrons/second; 6.74 × 10^{20} electron-volts, 108 J.

2. 4.71 m/strip, neglecting outer surface of outer layer.
4. 1.33 mm.

6. 2.

8. 8.8 × 10^{-22} F/m; 100 kV/m, 0.88 μC/m^2.
10. 177 pF, 1062 pF, 66.7 V, 0.0708 μC, 14.16 μJ, 2.36 μJ.
12. 2124 pF.
14. 1200 μC, 120 V, 80 V, 6 μF.
16. 500 μC; 250 V, 166.7 V, 83.3 V, 0.0208 J.
18. 3.75 μF, 1500 μC.
20. 48 V; 960 μC, 240 μC; 0.036 J, 0.0288 J.
22. 0.006 C, 1.8 J.
25. 50 μA, 100 V/s, 0.0025 J.
27. 3.744 × 10^{21} electrons/second.
29. 115.2 J, 7.2 × 10^{20} electron-volts.

Chapter 10, Page 198

1. 900 rev/min.
3. 60 Hz.
5. 2400 rev/min, 0.0478 T, 32.4 V, 50.9 V.
7. 188.5 V.
10. 20 ms, 11.44 A, 1.395 ms.
12. 25 Hz, 141.4 V, 347 W.

2. 16 poles.
4. 10 Hz, 0.1 s; 0.1332 V, 0.1884 V.
6. 0.48 V.

8. 12.74 A, 14.14 A.
11. 1.414 A.
13. 3.535 A, 225 000 J.

14. 27.3 A, 66.7 Hz.

16. 2.93 A, 50 Hz.
18. 2.61 kV, 3 kV.
20. 7.77 mA, 11 mA.

15. 3.05 A, 3.27 A, assuming points joined by straight lines.
17. 6.45 A, 5.75 A.
19. 5 A, 5.77 A, 1.154, 1.73.

Chapter 11, Page 223

1. 14.8 A.
4. 31 A, 18.8°.
6. 41.2 A.
8. 160.5 sin $(\omega t + 0.0213)$ V,
   125.8 sin $(\omega t + 1.185)$ V.
10. 147.5, 23.75° lead.

12. 2.65 A, 0.159 A.
14. 63.6 Ω, 62.8 Ω, 1.57 A, 81°.
16. 4 A, 5.3 A, 6.64 A, 52.95°.

3. 132.3 V, 93.5 V.
5. 86.6 V, 50 V.
7. 100 Hz, 9.51 A.
9. 22.14 sin $(314t + 0.11)$, 15.65,
   50 Hz.
11. 127.4 Hz, 1.59 sin $(800t − 1.01)$ A,
    1.125 A.
13. 0.0552 H.
15. 0.1718 H, 69.6°.
17. 11.58 A, 20.7 Ω, 39.5° lag.

Chapter 12, Page 233

1. $v = 339.4 \sin 314t$ volts,
   $i = 14.14 \sin (314t − \pi/6)$ amperes,
      12.25 A,   7.07 A,
      12.25 A.
3. 1 440 000 J.
5. 10.75 A, 57.5°, 1156 W.
7. 50 Ω, 0.5 W, 100 Ω, 0.6, 60 Ω.
9. 0.0729 H, 200 W.
11. 0.085 H, 0.599 lag; 4.13 A, 5.52 A;
    1270 VAr.
13. 0.9 A, 81 W.
15. 7.09 A, 750 W.

17. 69.5 per cent.

2. 500 W, 10 watt seconds.

4. 20 Ω, 34.64 Ω.
6. 199 V, 0.243 H, 0.5 lag.
8. 0.857 lag, 0.8 lag, 0.824 lag.
10. 16.67 Ω.
12. 4 A, 3.54 A, 5.34 A, 41.5°,
    1.6 kW, 0.749 lag.
14. 25.21 kW, 27.9 kVA, 0.903 lag.
16. 18.4 A, 14.72 A, 11.04 A,
    4.42 kVAr.

Chapter 13, Page 257

1. 106.2 Ω, 1.036 A.

3. 50 Ω, 79.6 μF, 691 W, 0.6 lead.
5. 1.54 A, 1.884 A, 2.435 A, 50.7°,
   308 W, 0.633 lead
7. 2.96 A, 29.6 V, 93 V, 188.5 V,
   72.8°, 0.296 lead, 87.6 W.
9. 159.2 V; 16 Ω, 0.375 H, 0.1346,
   0.24 lag.
11. 1590 Hz, 0.05 W, 0.05 W, 49.9.

13. 159 μF.
15. 69.4 Hz, 4.36; 2.24 A, 2.18 A;
    0.5 A; 1.0.
17. 324 μF.

2. 5.32 A, 106.4 V, 169.2 V, 57.9°,
   556 W, 0.532 lead.
4. 12.75 μF, 0.5 lead.
6. 19.9 μF, 0.625 lead.

8. 408 V, 0.875 lead.

10. 50.7 μF, 6 A, 378.6 V, 377 V, 1.0,
    12.56
12. 10 A, 10.61 A, 4.71 A, 11.6 A,
    0.862 lag, 1 kW.
14. 145 μF, 11.
16. 1.265 A, 503 Hz.

18. 11.6 A, 0.905 lag.

Chapter 14, Page 274

1. 240 A, 498 V; 720 A, 166 V;
   119.4 kW.

2. 120 V, 301 V, 36.12 kW; 240 A,
   150.5 V, 36.12 kW.

3. 432 V.
5. 946 rev/min.

4. 0.02 Wb.

Chapter 15, Page 283

1. 8, 0.0321 Wb.
3. 91 V.
5. 502 V, 85 Ω.

2. 467 V, 373 V, 0.240 Ω.
4. 260 V, 25 Ω.

Chapter 16, Page 300

1. 1140 rev/min.
3. 208 V, 41 A.
5. 10.48 mWb, 104 N m.
7. 469 rev/min, 268 N/m.
9. 44.4 Ω, 842.5 rev/min.
11. 785 rev/min, 30.8 A.
13. 4.34 Ω.
15. 4.35 per cent increase.
17. 595 rev/min.
19. 2.767 Ω, 76.7 V.

2. 640 rev/min.
4. 140 A.
6. 1005 rev/min, 438 N m.
8. 529 rev/min.
10. 25 A, 979 rev/min.
12. 779 rev/min.
14. 940 rev/min.
16. 325 N m.
18. 1.56 Ω, 77 V.
20. 76.7 A, 1.15 A; 76.6 A, 0.096 A.

Chapter 17, Page 318

1. 80 V, 15 A.
3. (a) 72, (b) 6.0 A.
5. (a) 0.10 V per turn, (b) 20 V,
    (c) 3000 ampere turns.
7. (a) 50 A, 200 A, (b) 0.0113 Wb.
9. (a) 60 W, (b) 0.208, (c) 1.17 A.
13. 97.9 per cent.
15. 96.9 per cent.
17. 0.17.

2. 400 V, 5 A.
4. (a) 62.5 V, (b) 40 A.
6. (a) 64, (b) 1.41 T.

8. 275, 20, 0.054 Wb.
11. 240 W.
14. 96.6 per cent.
16. 0.16.
18. 11.0.

Chapter 18, Page 360

1. 0.006 003 Ω, 19 988 Ω.
3. 5005 μΩ, 995 Ω.
5. 45.6×10^{-6} N m.
7. 25 turns.
9. 33 800 Ω.
11. 62.4 Ω, 23.2 mA in P and Q,
    14.8 mA in R and X.
13. 0.7 V, B positive; 9.6 Ω.
15. 2.98 per cent low.
17. 1.443 V.
19. 68.6 Ω, 26.4×10^{-3} H.

2. 9.995 mm.
4. 25.6 mA.
6. 10 turns
8. 3.6 per cent, 65.8 per cent.
10. 2.5 Ω, 2.6 Ω.
12. 8.58 Ω, 51.75 cm.

14. 3.08 mA from D to B.
16. 1.24 per cent high.
18. 2.17 Ω.
20. 4.0 μF.

Chapter 19, page 380

1. 0.182 A.
3. 0.016 25 Ω.
5. 110 cells, 125 cells.
7. 32.5 A, 23.75 A.
9. 1.0625 Ω, 0.6875 Ω.
11. 83.3 per cent, 72.7 per cent.
13. 21 A h, 42 A h.

2. 33 600 mm³.
4. 0.0133 Ω.
6. 141 cells.
8. 26.4 Ω.
10. 1.35 Ω, 1.54 A.
12. 79.4 per cent, 64.4 per cent.

Chapter 20, page 403

1. 0.11 A.
3. 0.18 A.
5. 1.25 A.
7. 0.33 A.
9. 1.0 A.
11. 7.2 W.
13. 0.44 Ω, 2.65 W.

2. 0.94 A.
4. 0.11 A.
6. (a) 0.96 A, (b) 4.32 A and 3.36 A.
8. 0.22 A.
10. 0.65 A.
12. 2.4 Ω.
14. 1.8 Ω, 16.5 W.

Chapter 21, Page 421

1. (a) 16 s, (b) 12 μA, (c) 0.75 V s^{-1},
    (d) 11.4 V, 0.59 μA, (e) 1.4 V,
    10.6 μA.
3. 4.55 μF.

5. 4.72 V, 7.28 V.

7. (a) 0.0393 A, (b) 0.0632 A,
    (c) 0.0777 A.
9. 10 A s^{-1}, 0.4 A, 0.0277 s.
11. 76.9 ms.

2. (a) 3.2 s, (b) 0.12 mA, (c)
    7.5 V s^{-1}, (d) 0.016 mA, 20.8 V,
    (e) 0.034 mA, 17.1 V.
4. (a) 72.8 mA, (b) 44.1 mA,
    (c) 26.8 mA.
6. (a) 0.017 s, (b) 6 A s^{-1}, (c) 0.10 A,
    (d) 0.063 A, (e) 0.095 A, (f) 0.63 V.
8. 0.84 ms.
10. (a) 48 Ω, (b) 10 A s^{-1}, (c) 0.576 s.

Chapter 22, Page 444.

1. 5.2 kV.
3. 207.8 V.
5. 2.8 A leading by 25.2°.
7. (a) 30 A, 1.7 kV, (b) 17.3 A, 3.0
    kV.
9. 2.3 A, 1.6 kW.
11. (a) 11.8 A, (b) 0.45, (c) 4.1 kW.
15. 1.7 kW, 0.29.
18. 450 rev/min.

2. 239.6 V.
4. 11.1 A lagging by 8.9°.
6. 11.5 A.
8. 2.3 kW.

10. 5.1 A, 2.3 kW.
14. 280 W, 0.80.
17. 0.03 per cent, increase.

Chapter 23, Page 466

7. 32.3.

Index